Revolutionizing Construction and Environmental Engineering With Materials Innovation

Roberto Alonso González-Lezcano
https://orcid.org/0000-0002-6185-4929
Universidad CEU San Pablo, Spain

Mehmet Serkan Kirgiz
Academia of ACMCEN, USA

IGI Global
Scientific Publishing
Publishing Tomorrow's Research Today

Published in the United States of America by
IGI Global Scientific Publishing
701 East Chocolate Avenue
Hershey, PA, 17033, USA
Tel: 717-533-8845 | Fax: 717-533-7115
Website: https://www.igi-global.com E-mail: cust@igi-global.com

Library of Congress Cataloging-in-Publication Data

Names: González-Lezcano, Roberto Alonso, 1971- editor | Kirgiz, Mehmet Serkan editor
Title: Revolutionizing construction and environmental engineering with materials innovation / edited by: Roberto Alonso González-Lezcano, Mehmet Kirgiz.
Description: Hershey, PA : IGI Global Scientific Publishing, [2026] | Includes bibliographical references. | Summary: "This book seeks to present groundbreaking research in material innovations that support sustainable engineering solutions, particularly in the civil engineering and environmental sectors"-- Provided by publisher.
Identifiers: LCCN 2025024359 (print) | LCCN 2025024360 (ebook) | ISBN 9798337324142 hardcover | ISBN 9798337324159 paperback | ISBN 9798337324166 ebook
Subjects: LCSH: Sustainable engineering--Technological innovations | Materials--Environmental aspects
Classification: LCC TA163 .R48 2026 (print) | LCC TA163 (ebook)
LC record available at https://lccn.loc.gov/2025024359
LC ebook record available at https://lccn.loc.gov/2025024360

British Cataloguing in Publication Data
A Cataloguing in Publication record for this book is available from the British Library.

Table of Contents

Detailed Table of Contents

The long-term durability of Engineered Cementitious Composites (ECC) under diverse exposure conditions has garnered significant attention in both research and practical applications. ECC, a high-performance, fibre-reinforced material, exhibits superior strain-hardening behaviour, self-healing capacity, and multiple micro-cracking under tensile loading, distinguishing it from conventional concrete. This review comprehensively analyses the durability performance of ECC, emphasizing properties such as permeability resistance, shrinkage resistance, sulphate and acid erosion resistance, thermal and impact resistance, electrical resistivity, self-healing capacity, surface drainage, and skid resistance. The influence of key parameters including water-to-cement ratio, fibre type and dosage, sand-to-binder ratio, and Supplementary Cementitious Material content on ECC durability is critically examined. Recent advancements such as the incorporation of biochar, super absorbent polymers, graphene nanoparticles, ceramic-based microfibres, and carbon-based fillers have demonstrated significant potential in enhancing ECC resistance to chemical, thermal, and mechanical degradation. Additionally, innovations like bacteria-induced self-healing and the use of 3D-printable ECC formulations are paving the way for multifunctional and smart construction materials. This review not only consolidates existing knowledge but also bridges research gaps in the integration of sustainable and smart materials for future infrastructure.

This study investigates the effects of organic zirconia nanoparticle reinforcement (0.5–5 wt%) on the mechanical and thermal, properties of polybutylene terephthalate (PBT) composites. Mechanical tests revealed enhanced strength, toughness, and hardness, with optimal performance at 2% zirconia loading, balancing mechanical reinforcement and thermal stability. The elongation at break decreased with increasing zirconia content, indicating reduced ductility, while Young's modulus remained stable, preserving stiffness. Impact strength peaked at 2% zirconia, enhancing toughness, and Shore D hardness increased up to 2% filler loading before declining. Thermogravimetric analysis showed improved thermal stability, as indicated by a shift in degradation temperatures. These results demonstrate that zirconia nanoparticles significantly enhance the mechanical and thermal properties of PBT, making them suitable for advanced applications in automotive components, electronic housings, and aerospace parts requiring high strength, thermal resistance, and durability.

The construction industry plays a pivotal role in global energy consumption and environmental impact, making the adoption of energy-efficient materials essential for sustainable development. This chapter explores the latest advancements in energy-efficient construction materials, examining their properties, applications, and contributions to reducing energy use and carbon emissions. Key materials discussed include advanced insulation solutions such as aerogels and spray foam, high-performance glazing like Low-E and smart windows, and sustainable concrete alternatives such as geopolymer and recycled aggregate concrete. Natural and renewable resources, including bamboo, hempcrete, and straw bale, are also highlighted for their ecological benefits and performance. The chapter delves into innovative technologies, such as phase-change materials, nanotechnology, and 3D-printed components, that enhance energy performance in modern buildings. It further reviews energy efficiency standards and certifications like LEED, BREEAM, and Passive House, offering guidance on best practices for material selection. Additionally, it addresses challenges in material adoption, including cost, supply

extended to industry stakeholders. The purpose of these suggestions is to hasten transition. The ultimate goal is to move to more ecological construction practices.

Nanotechnology has revolutionized various industrial sectors, and construction is no exception. The incorporation of nanoparticles and nanomaterials in building materials significantly enhances their mechanical, thermal, and durability properties while promoting sustainability in the sector. This chapter examines the impact of nanotechnology in construction, focusing on advancements in materials such as carbon nanotube-reinforced concrete, self-cleaning coatings with titanium dioxide nanoparticles, and nanoscale additives to improve the energy efficiency of buildings. Furthermore, the benefits of nanomaterials are explored in terms of structural strength, weight reduction, and decreased porosity, leading to lower water absorption and increased infrastructure lifespan. The role of nanotechnology in developing smart materials, such as self-healing materials, is also discussed, highlighting their potential to reduce maintenance costs and improve structural safety. The chapter also addresses the challenges associated with the application of nanotechnology in construction, including production costs, scalability, and environmental considerations. Finally, case studies and practical examples of current applications are presented, along with a future perspective on how nanotechnology can transform the construction industry, making it more efficient and sustainable.

This research aims to address the housing deficit in Honduras, with a focus on the Central District. This is a complex issue exacerbated by the effects of climate change and the shortcomings of traditional construction methods. The research centres on analysing the vernacular architecture of the Lenca indigenous people, as it is considered a source of profound knowledge regarding bioclimatic and sustainable solutions. This study employs a structured research methodology comprising several stages: an ethnographic review of building traditions; an analysis of local climatic conditions; and the execution of detailed dynamic energy simulations using Design Builder software. The experimental stage has validated the significantly higher thermal performance and energy efficiency of Lenca vernacular-inspired housing typologies compared to contemporary conventional alternatives. Based on these findings, the chapter presents an innovative model of progressive social housing that systematically integrates traditional construction principles with the functional and programme-related needs of today's society. Finally, the research concludes with an argument for crucial policy reforms, an update to existing building regulations and the active promotion of local materials and craftsmanship. It suggests that these measures are vital for developing housing solutions that are culturally appropriate, resilient to climate variations, and energy efficient. This would offer a tangible way to address the nation's housing deficit and substantially improve the quality of life of its most vulnerable populations.

This chapter presents the development of binderless thermal-acoustic insulation panels using rice straw, water hyacinth, and wastepaper as locally sourced waste materials. Utilizing Persea kurzii powder as a natural adhesion enhancer, the panels were produced without synthetic binders or energy-intensive curing, aligning with sustainable manufacturing. Thermal conductivity values ranged from 0.040 to 0.067 W/m·K, and Noise Reduction Coefficients (NRC) between 0.48 and 0.51, comparable to conventional materials like expanded polystyrene (EPS) and glass wool. Economic analysis showed production costs 25–40% lower than imported synthetic options. Additional environmental advantages include biodegradability, low embodied carbon, and compatibility with circular economy models. The chapter discusses performance data, material economics, and long-term sustainability implications, concluding with recommendations for life cycle assessment and policy integration to enable broader market adoption.

The advancement of sustainable construction relies on next-generation eco-friendly and high-performance materials that minimize environmental impact while enhancing structural efficiency. Innovations in biodegradable composites, recycled materials, self-healing concrete, and energy-efficient insulation contribute to reducing carbon

Preface

INTRODUCTION

We are pleased to introduce this edited volume, *Revolutionizing Construction and Environmental Engineering With Materials Innovation*, a collection that captures the extraordinary momentum driving today's advancements in materials science. In an era defined by climate pressures, rapid urbanization, and increasing demands on infrastructure, the construction and environmental engineering sectors are undergoing profound transformation. At the heart of this shift lies a powerful evolution in materials research—one that prioritizes durability, biocompatibility, energy efficiency, and environmental sustainability alongside traditional structural performance.

Recent decades have marked a significant departure from the sole reliance on conventional materials. Where steel, concrete, and stone once dominated engineering practice, the contemporary landscape calls for multifunctional solutions capable of addressing complex environmental and operational challenges. From adaptive composites and protective surface treatments to eco-friendly coatings and high-performance hybrids, materials innovation has become an essential driver of sustainable progress. These advances offer new pathways to reduce environmental impact, improve longevity, and enhance the overall resilience of built environments.

This book brings together pioneering research aimed at developing and analyzing the next generation of construction and environmental materials. Our goal is to highlight not only technical breakthroughs in material synthesis and modification but also the practical implications of these innovations for real-world engineering applications. By exploring surface engineering methods, hybrid composites, and advanced coatings, the chapters collectively reveal how modern materials can better withstand harsh environmental conditions, support eco-conscious design, and adapt to the evolving needs of infrastructure and environmental management systems.

The work presented here reflects the collaborative spirit of a multidisciplinary field—one where engineers, material scientists, environmental researchers, and public health professionals converge to address global challenges. It is our hope that this volume serves as both a resource and an inspiration to its diverse readership: researchers pushing the boundaries of materials science, academics shaping the next generation of engineers, professionals in construction and environmental sectors seeking innovative solutions, and graduate students contributing new perspectives to a rapidly advancing discipline.

As editors, we are honored to curate this body of work and to support the scientific and practical advancements that will shape the future of sustainable engineering. We invite readers to engage deeply with the ideas and innovations presented here, recognizing that material science holds not only the promise of improved structures, but the potential to strengthen the relationship between human development and the natural environment.

ORGANIZATION OF THE BOOK

In this volume, each chapter contributes a vital perspective on the evolving relationship between materials innovation, sustainability, and the future of construction and environmental engineering. Together, they illustrate a field in rapid transformation—one where performance, resilience, and ecological responsibility are deeply intertwined.

Chapter 1 offers a comprehensive review of the durability of Engineered Cementitious Composites (ECC), a material heralded for its strain-hardening behavior, micro-cracking control, and self-healing capacity. The chapter evaluates ECC's performance under a wide range of environmental and mechanical stressors—from permeability and shrinkage to thermal, chemical, and impact resistance. By analyzing how variables such as fibre type, supplementary cementitious materials, and emerging additives like graphene nanoparticles and biochar influence performance, the authors illuminate promising avenues for next-generation ECC, including bacteria-induced self-healing and 3D-printable formulations. This review underscores ECC's potential as a cornerstone of resilient and sustainable infrastructure.

Chapter 2 examines the mechanical and thermal enhancements achieved by reinforcing polybutylene terephthalate (PBT) with organic zirconia nanoparticles. Through robust testing across multiple nanoparticle loadings, the authors reveal significant improvements in strength, toughness, hardness, and thermal stability, particularly at an optimal 2% reinforcement level. While ductility diminishes at higher concentrations, the overall performance gains position these composites as high-value materials for automotive, electronics, and aerospace applications. This work demonstrates the promise of nanoparticle reinforcement for engineering durable, multifunctional polymers.

Chapter 3 presents a sweeping review of innovative energy-efficient materials critical to sustainable construction. From aerogels and high-performance glazing to geopolymers, bamboo, and hempcrete, the chapter highlights materials that reduce energy consumption and carbon footprints. Emerging technologies—including phase-change materials, nanotechnology applications, and 3D printing—are examined for their transformative potential. The authors also contextualize these innovations within global certification frameworks such as LEED and Passive House, providing readers with a clear roadmap for sustainable material selection despite challenges in cost, regulation, and supply chains.

Chapter 4 continues the sustainability narrative by examining circular economy strategies and eco-friendly material alternatives. This chapter emphasizes the environmental shortcomings of traditional construction materials and champions biodegradable composites, recycled resources, and functional coatings as viable solutions. The role of digital innovation—particularly 3D printing—is highlighted for its ability to reduce waste and enhance efficiency. Addressing regulatory and economic challenges, the chapter concludes with practical recommendations for accelerating sustainable adoption within the industry.

Chapter 5 explores the transformative power of nanotechnology in construction. Through a detailed review of nanoscale enhancements—from carbon nanotube-reinforced concrete to titanium dioxide self-cleaning coatings—the chapter illustrates how nanomaterials dramatically improve strength, durability, and energy performance. Particularly promising are smart materials with self-healing capabilities, which offer long-term reductions in maintenance and increased structural resilience. This chapter points toward a future where nanotechnology plays a foundational role in intelligent, sustainable infrastructure systems.

Chapter 6 brings a socio-environmental perspective by examining vernacular Lenca architecture from Honduras as a model for bioclimatic, culturally rooted housing solutions. Using climate analysis, ethnographic insights, and dynamic energy simulations, the chapter demonstrates how traditional design principles outperform modern construction in thermal comfort and energy efficiency. It proposes a hybrid model for social housing that blends ancestral wisdom with contemporary needs, advocating for policy reforms that support local materials, craftsmanship, and climate resilience—a compelling pathway toward equitable and sustainable housing.

Chapter 7 proposes an innovative approach to thermal-acoustic insulation through panels made from agricultural and paper waste. By utilizing rice straw, water hyacinth, and wastepaper without synthetic binders and leveraging Persea kurzii as a natural adhesive, the authors present a low-cost, low-carbon alternative to conventional insulation. With thermal and acoustic performance comparable to EPS and glass wool—and significantly lower production costs—these panels exemplify circular economy principles. The chapter highlights their environmental benefits, economic feasibility, and the need for life cycle assessments and supportive policy frameworks for broader adoption.

Chapter 8 concludes the collection with a forward-facing examination of next-generation eco-friendly and high-performance construction materials. Highlighting biodegradable composites, recycled resources, self-healing concrete, and adaptive smart materials, the chapter presents a vision of construction rooted in ecological responsibility and engineering excellence. The authors explore challenges to large-scale adoption—ranging from cost to material availability—while emphasizing the transformative potential of technological innovation for achieving a more sustainable, energy-efficient built environment.

Collectively, these chapters underscore the central message of this volume: materials innovation is not merely an engineering challenge but a profound opportunity to reshape the future of construction and environmental stewardship. Through scientific rigor and visionary thinking, each contribution brings us closer to a sustainable, resilient, and technologically empowered world.

CONCLUSION

As we conclude this volume, *Revolutionizing Construction and Environmental Engineering With Materials Innovation*, we reflect on the remarkable breadth and depth of research presented across its chapters. Collectively, these contributions affirm a central truth: the future of construction and environmental engineering will be shaped not only by structural ingenuity, but by our capacity to innovate responsibly, sustainably, and with a clear vision for the long-term resilience of our built environment.

The chapters in this book illuminate an interdisciplinary landscape where material science, environmental stewardship, digital technologies, and engineering practice converge. From the advanced durability of engineered cementitious composites, to the promise of nanotechnology, bio-based insulation, and next-generation eco-friendly composites, the research demonstrates how modern materials transcend traditional functions. They enable structures that are not only stronger and more durable, but more efficient, adaptive, and attuned to ecological constraints.

Equally important are the chapters that remind us of the social and cultural dimensions of sustainable engineering—whether through the revival of vernacular architecture, the integration of bioclimatic principles, or the adoption of circular economy strategies. These contributions underscore that innovation is most effective when it respects context, heritage, and human well-being.

Across the volume, a recurring theme emerges: sustainability must be approached holistically. Innovations in construction materials cannot be isolated from environmental conditions, socio-economic realities, or ethical imperatives. Whether employing AI for structural monitoring, evaluating the thermal behavior of mineral-added concretes, or leveraging ultrasonic testing for composite integrity, each study emphasizes thoughtful adoption, rigorous testing, and the need for scalable solutions grounded in research and practice.

As editors, we are encouraged by the momentum reflected in this book. The diversity of perspectives—from polymer engineering to atmospheric science, from waste-derived materials to digital fabrication—demonstrates a thriving global effort to transform construction into a more sustainable, efficient, and innovative sector. We hope that this collection inspires researchers, practitioners, policymakers, and students to continue pushing the boundaries of materials science with purpose and responsibility.

In closing, we extend our deepest appreciation to the authors for their scholarly rigor and forward-thinking insights, and to our readers for joining us in this exploration. The challenges facing our world—climate change, resource scarcity, rapid urbanization—demand solutions rooted in creativity and scientific excellence. It is our hope that the knowledge presented in this book serves as both catalyst and guide, empowering the next generation of sustainable engineering innovation.

Chapter 1

Advances in Durability of Engineered Cementitious Composites (ECC): A Review Across Different Conditions

N. Shanmugasundaram
https://orcid.org/0000-0002-2312-5442
Sri Krishna College of Engineering and Technology, India

Praveen Kumar S.
https://orcid.org/0000-0002-3931-6104
PSG College of Technology, India

S. Divya
PSG College of Technology, India

ABSTRACT

The long-term durability of Engineered Cementitious Composites (ECC) under diverse exposure conditions has garnered significant attention in both research and practical applications. ECC, a high-performance, fibre-reinforced material, exhibits superior strain-hardening behaviour, self-healing capacity, and multiple micro-cracking under tensile loading, distinguishing it from conventional concrete. This review comprehensively analyses the durability performance of ECC, emphasizing properties such as permeability resistance, shrinkage resistance, sulphate and acid erosion resistance, thermal and impact resistance, electrical resistivity, self-healing capacity, surface drainage, and skid resistance. The influence of key parameters including water-to-cement ratio, fibre type and dosage, sand-to-binder ratio, and

DOI: 10.4018/979-8-3373-2414-2.ch001

Supplementary Cementitious Material content on ECC durability is critically examined. Recent advancements such as the incorporation of biochar, super absorbent polymers, graphene nanoparticles, ceramic-based microfibres, and carbon-based fillers have demonstrated significant potential in enhancing ECC resistance to chemical, thermal, and mechanical degradation. Additionally, innovations like bacteria-induced self-healing and the use of 3D-printable ECC formulations are paving the way for multifunctional and smart construction materials. This review not only consolidates existing knowledge but also bridges research gaps in the integration of sustainable and smart materials for future infrastructure.

1. INTRODUCTION

To enrich the mechanical and durability properties of conventional concrete, many investigators successfully used different types of fibres and Supplementary Cementitious Materials (SCM) in various civil engineering applications (Muralidharan *et al.*, 2021; Ashraf *et al.*, 2022; de Siqueira and Cordeiro, 2022). Correspondingly, fibre and SCM-based Engineered Cementitious Composite (ECC) development increased worldwide in recent years (Muralidharan *et al.*, 2021; Shanmugasundaram and Praveenkumar, 2021; N.Shanmugasundaram et.al, 2025). Many investigators used ECC in different applications due to its high tensile and strain-hardening behaviour where the Conventional Concrete (CC) fails. Due to the incorporation of the high volume of different types of fibres (1.5% to 2%), such as polyvinyl alcohol fibre (PVA), polypropylene fibre (PP), polyethylene fibre (PE), steel fibre (SF), etc., in ECC is the reason for strain hardening behaviour compared to Ultra-High Performance Concrete (UHPC) and conventional concrete (Yoo and Banthia, 2022), which is shown in Figure 1. Similarly, Figure 2 shows the stress-strain relation between CC, Fibre Reinforced Cementitious Composites (FRCC) and ECC under loading condition. It was observed from Figure 2 that multiple micro-cracks were developed in ECC related to conventional concrete and FRCC under strain hardening behaviour as a result of high fibre-bridging ability. Nevertheless, the role of fibres, SCM, binding materials, and mix ratio are the important factor for attaining improved performance of ECC (Kewalramani, Mohamed and Syed, 2017; Arulanandam *et al.*, 2020). In connection to these, the materials such as fly ash, PVA fibre, silica sand, Ordinary Portland Cement (OPC), High Range Water Reducer (HRWR) and water were added with proper mix proportions to attain better properties of ECC by many researchers. The standard mix proportions and essential mechanical properties of ECC are enlisted in Table 1. In general, the shrinkage properties of ECC are high due to the higher amount of cement (1254kg/m^3) used in ECC. Similarly, the modulus of elasticity is very low because of neglecting coarse aggregate in ECC. Therefore, PVA fibres with high physical properties and a high percentage of fly ash are used in ECC to enrich the modulus of elasticity and reduce the cement content, respectively (Singh, Saini and Chalak, 2019; Zhang, Yu, *et al.*, 2020). Various researchers attained the significant properties of ECC, such as compressive strength (20MPa to 150MPa), ultimate tensile strength (4MPa to 16MPa), first cracking strength (3MPa to 10MPa) and flexural strength (10MPa to 50MPa). Similarly, the ECC tensile strain attained about 3% to 10%, which is much higher than CC, UHPC and FRCC (Shanmugasundaram *et al.*, 2025).

On the other hand, durability properties are essential to operating Infrastructures effectively and safely in the long term. Moreover, to achieve real-world engineering applications, it is crucial to accurately assess and comprehend ECC endurance under different environmental circumstances (Shanmugasundaram et.al, 2025). Nevertheless, the durability properties of ECC are enhanced at later age with the result of combination of the high volume of SCM such as fly ash, slag, stone slurry powder, glass powder and locally available materials (Yu *et al.*, 2018; Isyaka.A et.al., 2020; A.Adesina, 2021). For reasons mentioned above, the durability properties of ECC differed depends on the blending of materials and environmental conditions. Based on the high properties, ECC is used in different environmental conditions and different applications, such as Repair and rehabilitation works, precast construction, different pavements, marine construction, Acid or chemical resistance industries, Earthquake resistant structures, Self-compacting construction, High rise structures, high durability required structures, etc (N.Shanmugasundaram et.al, 2025).

Figure 1. Mechanical behaviour of ECC compared to UHPC and CC

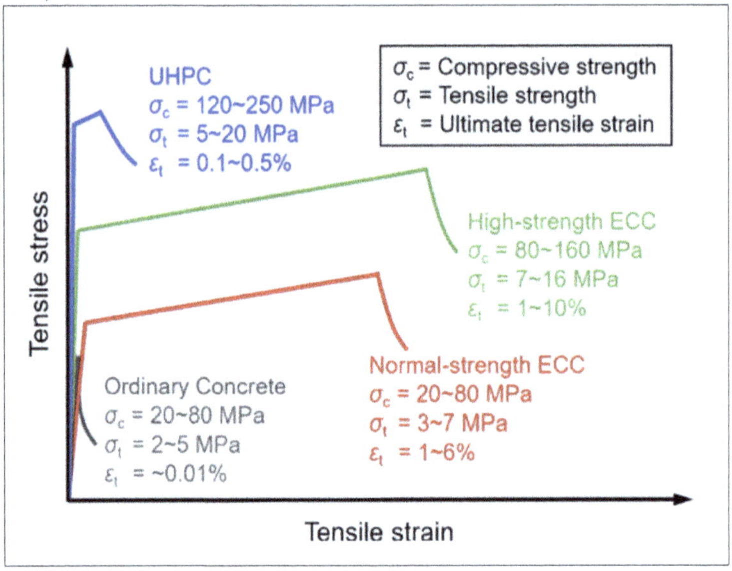

Source: Shanmugasundaram and Praveenkumar (2023)

Figure 2. Stress–strain curves of CC, FRCC and ECC

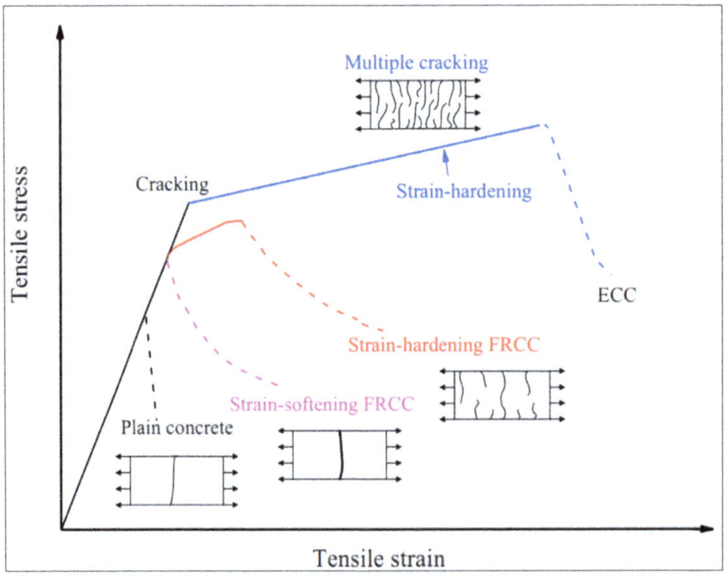

Source: Ma, Yi, and Wu (2021)

Table 1. Standard mix proportions and essential mechanical properties of ECC (Yu et al., 2018)

Standard mix proportions					**Mechanical Properties**	
Designation of mixture	**M45**	**M46**	**M47**	**M48**	**Compressive strength**	**20-160 MPa**
Cement	1	1	1	1	First cracking strength	3-10 MPa
Fly ash	1.2	1.2	1.2	1.2	Ultimate tensile strength	4-16 MPa
Fine Aggregate	0.8	1.2	1.4	1.6	Ultimate tensile Strain	3-10%
Water	0.56	0.58	0.59	0.6	Flexural strength	10-50 MPa
HRWR	0.012	0.012	0.012	0.012	Density	0.95-2.3 g/ml
Fibre percentage	0.02	0.02	0.02	0.02	Young's modulus	18-40 GPa

In recent years, the application of ECC has expanded beyond traditional cast-in-place construction techniques into advanced manufacturing approaches such as 3D printing. The integration of ECC with additive manufacturing technologies offers unprecedented control over material placement, enabling the creation of intricate and customized structural forms while simultaneously reducing material waste and construction time. More importantly, 3D-printed ECC has demonstrated improved crack resistance and homogeneity due to precise extrusion control and optimized mix design, thereby enhancing structural performance and long-term durability (Asghari *et al.*, 2024). Parallel to advancements in fabrication methods, the inclusion of nanomaterials such as carbon nanotubes (CNTs) and nano-titanium dioxide (nano-TiO_2) in ECC matrices has gained significant attention for improving environmental durability. These nanomaterials refine the pore structure, reduce microcrack widths, and enhance resistance to ultraviolet (UV) radiation and carbonation. The addition of CNTs and nano-TiO_2 improves both the mechanical strength and resistance to environmental degradation, which is crucial for infrastructure exposed to aggressive climatic conditions (Ye *et al.*, 2015). Collectively, these innovations highlight a shift towards multifunctional ECC systems that are not only mechanically robust and durable but also compatible with sustainable and smart construction practices.

For conventional concrete, durability is measured in terms of crack width, which defines the permeability of concrete. Therefore, various researchers increase the water tightness in concrete, thus increasing the density and improving the durability of conventional concrete (Yu *et al.*, 2018; Isyaka.A et.al., 2020; Andrade Neto, De la Torre and Kirchheim, 2021). In contrast, the high durability of ECC is attained due to its self-healing ability and micro-cracking behaviour under different environmental conditions (Adesina, 2021; Ma, Yi and Wu, 2021). Furthermore, ECC has superior durability performance than conventional concrete in terms of permeability, electrical resistivity, carbonization, freeze-thaw and high-temperature conditions because of the bridging effect of different high-volume fibres (Li *et al.*, 2020; Zheng *et al.*,

2021). In these connections, the present article reviews the durability properties of ECC with the combination of various blending materials using chloride ion penetration, water absorption, water permeability, shrinkage resistance, sulphate and acid erosion resistance, temperature resistance, impact resistance, electrical resistivity, self-healing ability, skid resistance and water surface drainage resistance. In addition, observations on the durability of ECC were also discussed based on various literature surveys.

1.2 Fibre-matrix bonding and crack width control of ECC

The strength of the ECC matrix, types of fibres, properties of fibres, percentage of fibre and the interfacial bonding properties influence the significant impact on the width and number of cracks in ECC. Additionally, the bonding of the fibre-matrix interface is significant for the longevity of ECC (Cai *et al.*, 2021). The fibre-matrix interface, which is directly related to the performance of the ECC endurance, may progressively alter during prolonged exposure to external conditions is the reason for fibre-matrix bonding. Likewise, the crack propagation and strain-hardening properties of ECC materials, as well as the relevant design parameters of ECC, are beneficial for study and various applications. ECC steady-state crack development and strain-hardening properties are attained when the stress parameters and energy are fulfilled (Shanmugasundaram and Praveenkumar, 2024). Correspondingly, the relationship between crack opening and crack bridging stress plays a vital role in the durability properties, as shown in Figure 3. The maximum crack opening should be kept at less than 100 lm to ensure excellent durability (Ma, Yi and Wu, 2021). Although, the crack width and multiple cracking mechanisms significantly affect the durability properties of ECC such as fatigue, permeability, freeze-thaw cycle, etc.

Figure 3. Relationship between crack opening and crack bridging stress

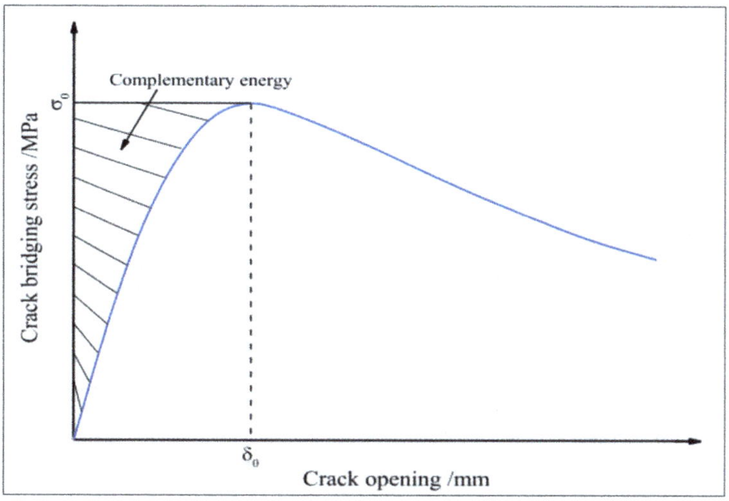

Source: Ma, Yi, and Wu (2021)

The fibre-matrix interface in ECC plays a pivotal role in regulating crack propagation, controlling crack width, and ensuring long-term durability. Fibre-matrix bonding is influenced by the type of fibre used (e.g., PVA, PE, PP, steel), the surface texture and chemical affinity of the fibre, fibre volume fraction, and the surrounding cementitious matrix properties, as listed in Table 2. The interaction at this interface determines the stress transfer efficiency from the matrix to the fibre during tensile loading. In well-designed ECC, a controlled de-bonding followed by frictional sliding mechanism allows fibres to bridge cracks effectively, promoting multiple micro-cracking rather than single wide crack failure. This behavior leads to strain-hardening and tight crack widths (typically <100 μm), which are critical for preventing ingress of aggressive agents such as chlorides and sulfates. Over time, environmental exposure may weaken this interface; however, using coated or hybrid fibres and incorporating mineral admixtures can enhance interfacial bonding. Thus, controlling fibre-matrix bonding is essential not only for mechanical performance but also for enhancing ECC resistance to permeability, fatigue, and freeze-thaw cycles (Mensah *et al.*, 2021; Yang *et al.*, 2021; Amir Ali et.al., 2025).

Table 2. Effect of fibre type and volume on crack width control

Fibre Type	Volume Fraction (%)	Avg. Crack Width (μm)	Observed Behavior
PVA	2.0	60	Tight microcracks
PE	1.5	70	Multiple cracking
Steel	1.0	>150	Fewer, wider cracks

2. SIGNIFICANCE AND FLOWCHART FOR PRESENT REVIEW

Many investigators have successfully reviewed the mechanical properties of ECC with the result of fibre volume, different materials, w/c ratio, curing conditions, etc. (Booya et al, 2020; G. Ajith et.al, 2021; Shanmugasundaram and Praveenkumar, 2024). In contrast, very few researchers have studied the durability properties of ECC in terms of self-healing ability, permeability, fatigue, abrasive, frost and salt control resistance. This review article bridges the gap between the effects of durability properties by incorporating various materials in ECC at different exposure conditions. Also, the review about water surface drainage, skid resistance and electrical conductivity of ECC is entirely new. Figure 4 shows the flowchart of this review article for better understanding. This review addresses a critical gap in ECC research by focusing extensively on its durability characteristics, an area less explored compared to its mechanical behavior. While many studies emphasize the tensile and compressive properties of ECC, fewer explore its performance under aggressive environmental conditions such as exposure to chlorides, acids, sulfates, extreme temperatures, and mechanical wear. This paper synthesizes the latest advancements in material innovations such as biochar, SAPs, graphene, bacteria-based self-healing agents, and carbon fillers and connects them to specific durability functions. Additionally, novel aspects such as surface water drainage and skid resistance are highlighted, which are scarcely covered in prior ECC reviews. The flowchart provided outlines the structure of the review, from material selection and mix proportions to different durability properties and performance outcomes under varied exposures. This framework aids readers in understanding the holistic performance of ECC across different durability dimensions.

Figure 4. Flowchart for review article

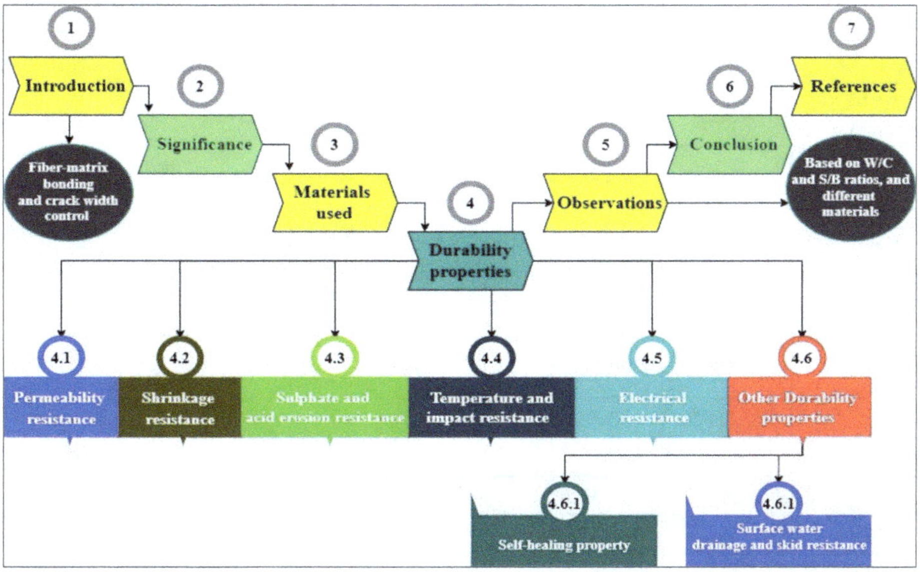

3. MATERIALS USED IN ECC

ECC with the incorporation of diverse types of materials, such as SCM, fine aggregate and fibres, w/c ratio, length, and percentage of fibres used by many researchers are listed in Table 3. Also, the standard codal provisions used by various researchers for finding durability properties are listed in Table 4. It was observed from Table 3 that various researchers used similar types of materials in their research, such as fly ash, silica sand and PVA fibre. Similarly, 2% of PVA with 12mm length was used by many investigators to attain high tensile properties of ECC. The minimum and maximum ranges of w/c ratios, fibres' percentages and fibres' length from 0.2 to 0.535, 0.25% to 2.65% and 8mm to 20mm are used to prepare ECC. On the other hand, the different types of SCM, fibres and fine aggregate, namely, GGBS, rice husk ash, silica fume, metakaolin, bagasse ash, limestone powder, stone slurry powder, recycled glass powder, nano silica, polypropylene fibre, polyethylene fibre, steel fibre, tire wire fibre, silica and micro silica sand, corundum, river sand, recycled asphalt, cenosphere, stone sand, crumb rubber, recycled concrete, modified polyvinyl chloride aggregate, quartz sand, zeolite, crushed rounded siliceous gravel, mortar sand, dolomitic limestone sand, lake sand, etc., were used to prepare the ECC.

Table 3. Materials used in production of ECC

SCM	Fine aggregate	Fibre	Length of fibre (mm)	Percentage of fibre	W/C ratio	Investigators
FA,SL	LS	PVA	8	2	0.25	(Booya et al, 2020)
GGBS,FA,SF, MK,RHA,BA	Msand	PVA	12	2	0.35	(G. Ajith et.al, 2021)
FA	SS	PVA	8	2	0.27	(Sherir, Hossain and Lachemi, 2018)
SS,RS	SS,RS	PVA	12	2	0.27	(Soe, Zhang and Zhang, 2013)
FA, MK,LS	SS	PVA	8	2	0.25	(Zhang, Jaworska, *et al.*, 2020)
FA	RS	PP,ST	16	1	0.4	(Ammasi and Ragul, 2018)
GGBS	MSS,SSP,RS	PVA	12	2	0.27	(M.Sing et.al., 2020)
SF,GGBS,FAC	RFA	PE	18	2	0.527	(Zhou *et al.*, 2021)
FA,LS	FSS	PVA	12	2	0.36	(Guan *et al.*, 2020)
FA,RGP	SS	PVA	12	2	0.27	(H.Said et.al., 2018)
FA,MK,LS	-	PP	10	2	0.3	(Zhu *et al.*, 2020)
FA	RS,CBFS	PVA	16	2	0.3	(Yogesh Biyani, L. G. Patil, 2019)
FA,NS	RS	PVA	12	2	0.145	(Mohammed, Achara and Liew, 2018)
FA	MSS,Cor	PVA	12	2	0.265	(Bawono, Lechner and Yang, 2019)
FA,LS	SS	PVA	12	2	0.269	(Siad *et al.*, 2015)
FA	SS,RAS	PVA	8	2	0.25	(A. Adesina, 2021)
FA,GS	RS	-	-	-	0.5	(Saloni *et al.*, 2019)
FS,CS	STS	-	12,13,14	-	0.35	(Danish and Mosaberpanah, 2021)
GGBS	RS	PVA	12	2	0.535	(Younsi *et al.*, 2021)
FA	RS,CR	PVA	12	2.414	0.267	(Alaloul *et al.*, 2020)
FA	SS,RS,RC	PVA	8	2	0.25	(Adesina and Das, 2021)
FA	RS	PVA	12	2	0.215	(Sabapathy *et al.*, 2020)
FA	QS,M-PVC	PVA,PP	12	2	0.247	(Hu *et al.*, 2021)
SF,SL,CS	SS	UHMWPE	12	2	0.13-0.24	(Ye *et al.*, 2019)
FA,GGBS,SF	QS,Z	PVA	12	2	0.2	(Wang, Zhang and Ho, 2020)
FA	RS	PVA,TWF	12,20	0.25-2.65	0.145	(Bashar Khed et, 2018)
FA	SS,CDL, RSG	PVA	8	2	0.27	(Sahmaran *et al.*, 2009)

Note: SSP- Stone slurry powder, SCM-Supplementary cementitious materials, FA-Fly ash, RGP- Recycled glass powder, SS-Silica sand, RHA-Rice husk ash, UHMWPE-Ultra-high molecular weight polyethylene, BA-

Bagasse ash, RS-River sand, MK-Metakaolin, LS- Limestone, PVA- Polyvinyl alcohol fibre, RSG-Rounded siliceous gravel, PP- Polypropylene fibre, PE- Polyethylene fibre, ST-Steel fibre, GGBS-Ground granulated blast furnace slag, MSS-Micro silica sand, FAC- Fly ash cenosphere, SF-Silica fume, FSS-Fine silica sand, CBFS-Crushed blast furnace slag, NS-Nano silica, Cor-Corundum, RAS-Recycled asphalt, STS-Stone sand, CS-Cenosphere, CR-Crumb rubber, RC-Recycled concrete, M-PVC- modified Polyvinyl chloride aggregate, SL-Slag, QS- Quartz sand, Z-Zeolite, TWF-Tire wire fibre, CDL-Crushed dolomitic limestone sand, LS-Lake sand, M.S-Mortar sand.

Table 4. Standard codal provision for various durability tests of ECC

Details of test	Standard code
Permeability resistance	(ASTM C1585-13, 2013), (ASTM C1403-15, 2016)
Shrinkage resistance	(ASTM C1012/C1012M-15, 2015), (ASTM C596-18, 2018), (ASTM C 157/C 157M–17, 2017), (ASTM C490-07, 2008)
Absorption and Voids	(ASTM C642-21, 2021)
Temperature and fire resistance	(ASTM E119, 1997)
Electrical resistance	(ASTM C1202, 2012)

4. DURABILITY PROPERTIES OF ECC

4.1 Permeability resistance

The crack width, depth and the pore structure are related to permeability resistance in concrete. According to various researchers, the crack width and depth influence the permeability of concrete at all curing ages.(Soe, Zhang and Zhang, 2013) compared the permeability resistance of ECC with silica sand and river sand (Zone II - Indian sand), from which ECC with both sand attained a similar intensity of chloride ion penetration about 925 and 930 coulombs, respectively, due to high packing density with cementitious materials. (Zhang, Jaworska, *et al.*, 2020) examined the pore structure of ECC with limestone calcined clay cement (LC3). The study found that ECC with LC3 attained a coarser pore structure than ECC with conventional cement. The lower clinker content and higher w/c ratio is the reason for achieving coarser porous in ECC with LC3. Similarly, (Siad *et al.*, 2015) studied the water absorption and chloride ion penetration of ECC with cement replacement by high volume fly ash (HVFA). The study revealed that ECC with HVFA exhibited lesser water absorption and chloride ion penetration at 28 and 90 days. Also, the permeability of ECC is decreased with increasing the fly ash content and ages of curing. The dense matrix developed at a more extended period because of the high pozzolanic action of fly ash, which improve the permeability of ECC. Also, (M.Sing et.al., 2020) revealed that replacing river sand and MSS with stone slurry powder in ECC decreases the water absorption due to the high filling ability of stone slurry powder. Additionally, ECC with river sand exhibits higher water absorption than micro silica sand. The additional volume of micro pores developed in hardened ECC with river sand due to less packing density is the reason for higher water absorption.

(Zhou *et al.*, 2021) studied the effect on porosity of ECC with different sizes of recycled fine aggregate (RFA), such as 0 to 0.42mm, 0 to 0.72mm and 0 to1mm. The porosity of the cementitious matrix decreased with decreasing the size of RFA. In contrast, the water absorption increases by reducing the size of RFA. The irregular surface area and presence of capillary pores in RFA could impact the water absorption and porosity of the cementitious matrix. (H.Said et.al., 2018) advanced the high volume fly ash ECC with recycled glass powder (RGP). The replacement of fly ash by RGP exhibited a lower permeability resistance compared to ECC with 1.2% and 2.2% of fly ash. Also, ECC with 2.2% fly ash showed higher permeability resistance than 1.2% owing to higher pozzolanic activity of unreacted fly ash at a later age. (Şahmaran and Li, 2010) found that the water absorption of ECC decreased after 72 hours compared to conventional mortar and concrete, which was primarily due to the higher bonding of fibre and cementitious matrix.

(A. Adesina, 2021) attained higher water absorption and chloride ion penetration by replacing silica sand with more than 50% of recycled asphalt in ECC. The high volume of porosity developed in ECC due to more voids in recycled asphalt and improves the durability properties. Similarly, (Saloni *et al.*, 2019) reported that the chloride ion penetration decreased when increasing the ground river sand and fly ash percentage in ECC at 28 and 90 days of testing as a result of higher filling and packing effect of ground river sand's. (Danish and Mosaberpanah, 2021) studied the impact of cenosphere and fly ash on durability properties of ECC at different curing regimes such as water, dry and temperature curing (20°C to 80°C). It was observed that compared to cenosphere 500 grade, the water absorption of cenosphere 300 grade with fly ash replacement in cement was higher because of the porosity and lesser size of SCM. Also, compared to water and dry curing, the number of capillary pores was increased in more elevated temperature curing, which increased the water absorption. Furthermore, (Adesina and Das, 2021) revealed that owing to the high porosity of recycled concrete, the chloride ion penetration was slightly increased from 6% to 19% when the silica sand was replaced by 75% and 100% by recycled concrete in ECC. A similar effect was also observed in a water penetration test, and the penetration value was attained from 8.5% to 57.9% when the replacement of recycled concrete increased from 25% to 100%. The maximum penetration depth of 1.43mm was attained at 100% replacement, which is the acceptable value for various mechanical and durability applications. Moreover, the percentage of permeable voids and absorption of ECC increased when increasing the replacement of silica sand by

25% to 100% of recycled concrete due to the greater porous of recycled concrete, absence of coarse aggregate and high volume of fly ash. (Booya et al, 2020) studied the impact of ECC with fly ash and slag on the durability properties of ECC. Due to the high pozzolanicty of slag, the rate of water absorption and chloride ion penetration decreased at 28 and 90days. Additionally, 100% replacement of fly ash by slag in ECC achieved less water absorption and chloride ion penetration than partial replacement. A similar statement was also stated by (A. Adesina, 2021). However, the sorptivity index of ECC with mortar sand exhibited higher water absorption capacity due to its high volume.

In recent advancements, the incorporation of biochar as a partial replacement for cement in ECC has been increasingly explored as an effective means of enhancing durability. It has been observed that the addition of biochar significantly reduces permeability and improves resistance to chloride ion penetration. This improvement is attributed to the pore-refining ability of biochar, which enables a denser and more discontinuous pore structure within the cementitious matrix. When biochar particles are integrated, the internal microstructure is altered such that the connectivity of capillary pores is disrupted, thereby limiting the transport of aggressive agents through the matrix. ECC specimens modified with biochar demonstrated lower water absorption and superior chloride resistance compared to conventional fly ash-based ECC (Amir Ali et.al., 2025). This finding indicates that biochar, a renewable and carbon-sequestering material, not only contribute to sustainability, but also enhances the impermeability of ECC under severe exposure conditions. As such, biochar is being recognized as a promising supplementary cementitious material (SCM) in the design of high-performance, durable ECC mixtures.

4.2 Shrinkage resistance

Shrinkage is an essential factor that causes the cracks in concrete on different curing days. Several researchers reported that higher shrinkage occurred in ECC due to complex material proportions than conventional concrete. (A. Adesina, 2021) stated that incorporating recycled asphalt in ECC causes a higher shrinkage due to the absorption of higher water content. Also, the lower w/c ratio and higher fly ash content in ECC developed the hardened cementitious blend, which could be the reason for higher shrinkage in ECC. Likewise, (Danish and Mosaberpanah, 2021) revealed that cement mortar with cenospheres exhibited lower drying shrinkage than cement mortar with fly ash due to higher lime content in cenospheres. On the other hand, (Younsi *et al.*, 2021) studied the early drying behaviour of ECC blended with Ordinary Portland Cement (OPC), GGBS and siliceous sand. It was observed that premature drying causes less hydration, lower density and more pores in the mortar. Due to early drying, the water porosity and portlandite decrease near the surface by +8% and -41%, respectively. Also, the shape of the profile is affected due to early drying. Nevertheless, the risk of carbonation is higher due to increasing carbonation potential in the surface layer of ECC because the risk of carbonation and carbonation potential both are directly propositional.

(Alaloul *et al.*, 2020) studied the effect of replacing fine aggregate with crumb rubber in ECC. It was revealed that the shrinkage increased due to the porosity of crumb rubber. The highest shrinkage value of 0.25% was attained in ECC with 30% crumb rubber and 0% PVA fibre, respectively. Similarly, (Adesina and Das, 2021) reported that the drying shrinkage was positive when replacing silica sand with recycled concrete (RC) at 25%, 50%, 75% and 100% in ECC. The ECC with RC exhibited lower drying shrinkage of 6.5%, 21.7%, 30.4% and 36.9% than the ECC with silica sand at 28 days. This decrease in the drying shrinkage is due to fly ash's high pozzolanic reaction and pore-filling effect. (Ye *et al.*, 2019) compared the impact of sand to cement (s/c) and w/c ratio on different shrinkage resistance of ECC such as chemical, autogenous and drying shrinkage. The chemical and autogenous shrinkage resistance decreases with the increases in w/c and s/c ratios. In contrast, DS decreases with the rise of s/b ratio and decrease of w/c ratio. The low w/c ratio and high s/c ratio develop the porous and cracks in the specimen, increasing the water losses and decreasing shrinkage resistance. Figure 5 shows the classification of shrinkage with sealed and external drying conditions concerning time.

(Wang, Zhang and Ho, 2020) advanced the high strength ECC with the inclusion of zeolite to improve the shrinkage resistance property. It was observed that the shrinkage value decreases from 1560μ to 1142μ and 2283μ to 1848μ in sealed and drying conditions, respectively. The high pore volume in zeolite allows the additional space for free water after the calcination period, and thereby attaining higher shrinkage performance than ECC with quartz powder. (Bashar Khed et, 2018) examined the shrinkage performance of ECC with tire wire fibre (TWF) and PVA fibres. Shrinkage resistance of ECC decreases with increases in fibre volume, and also ECC with TWF fibres attained better shrinkage resistance than ECC with PVA fibre. It is due to the high fibre matrix improving the interfacial interaction and increasing shrinkage resistance during the hydration process.

In recent developments, the use of super absorbent polymers (SAPs) in Engineered Cementitious Composites (ECC) has been investigated as a viable strategy for mitigating shrinkage-related deterioration. It has been demonstrated that both autogenous and drying shrinkage can be significantly reduced when SAPs are incorporated into ECC mixtures. These polymers function as internal curing agents, whereby absorbed water is gradually released during the hydration process, maintaining internal humidity and thus minimizing self-desiccation. ECC specimens modified with SAPs exhibited lower shrinkage strains compared to control mixes, particularly during the early stages of curing (Mensah *et al.*, 2021). The internal curing effect provided by SAPs contributed to the reduction of capillary tension and pore shrinkage, which are commonly responsible for microcrack formation in cementitious materials. Furthermore, improved matrix integrity and reduced crack width were achieved, indicating that SAPs play a vital role in enhancing the dimensional stability and long-term durability of ECC. Accordingly, the incorporation of SAPs has been recognized as a promising approach for controlling shrinkage without compromising other mechanical properties.

Figure 5. Classification of shrinkage

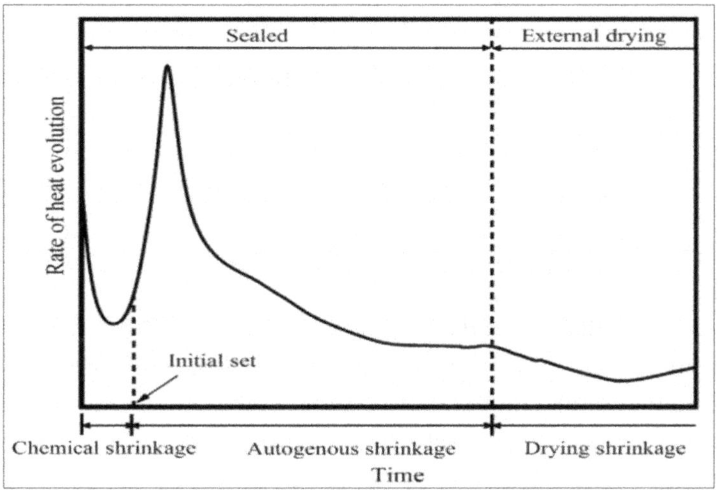

Source: Zhu et al. (2020)

On the other side, (Sahmaran *et al.*, 2009) attained a better drying shrinkage in ECC with high fly ash by cement ratio (2.2) and crushed sand (2.38mm) at 90 days. It was observed that drying shrinkage decreased with an increase in the size of fine aggregate and fly ash by cement ratio. The restraining action of the larger aggregate size in the cementitious matrix and secondary pozzolanic reaction of high volume fly ash at later ages is the reason for the lower shrinkage value of ECC. (Booya et al, 2020) reported that ECC with fly ash by cement (F/C) ratio of 2.2 attained less shrinkage value than the F/C ratio of 1.2. Also, fly ash replacement by 30%, 70%, and 100% of slag achieved a higher shrinkage value at 56 days. The high volume of fly ash improves the dense particle packing and thereby increases the pore distribution which is the major reason for less shrinkage value.

4.3 Sulphate and acid erosion resistance

(Danish and Mosaberpanah, 2021) reported that weight of cementitious blends decreased at 20°C and 80°C curing conditions under 5% of H_2SO_4. The soluble salts of calcium developed during the reaction of calcium hydroxide in cement, which increases the disintegration and spalling of the cementitious matrix and thereby decreases the strength of ECC. Correspondingly, (Sabapathy *et al.*, 2020) developed the rubberized ECC with crumb rubber (CR) and graphene oxide (GO) to examine the sulphate and acid attack. It was observed that the weight of cementitious blends decreased from 3.7% to 29.8% with the inclusion of 0% to 10% of CR in ECC under the 5% of sulfuric acid curing (H_2SO_4). Several micro-cracks were developed in ECC under acid curing due to the CR particles, allowing more acid penetration in ECC. Therefore, the weight losses occurred in ECC. However, adding GO in ECC improves durability due to the higher cementitious matrix. Investigators also observed a similar effect in the sulphate attack test regarding the change in length of the specimens immersed in Na_2SO_4 solution at 28 days.

In recent studies, the inclusion of graphene-based nanomaterials in ECC has been investigated to enhance resistance against aggressive chemical environments. It has been reported that the acid resistance of ECC is significantly improved when graphene nanoplatelets are incorporated into the cementitious matrix. In particular, exposure to sulphuric and nitric acid solutions resulted in markedly reduced mass loss and fewer visible surface cracks in graphene-enhanced ECC, as compared to conventional ECC mixes. This enhancement has been attributed to the formation of a denser and more refined microstructure, which was facilitated by the high surface area and impermeability of graphene particles. Moreover, the pozzolanic reaction and filler effect provided by graphene contributed to the reduction of calcium hydroxide leaching and the mitigation of acid-induced degradation. In the study conducted by (Anwar *et al.*, 2020), it was observed that the presence of graphene limited the penetration of acidic ions, thereby preserving the mechanical integrity and dimensional stability of the ECC specimens. As a result, graphene has been recognized as an effective nano-additive for increasing the acid and sulphate resistance of ECC in chemically aggressive environments.

ECC resistance to chemical attacks, particularly sulphate and acid exposure, is significantly improved through the use of mineral and nano-additives. Sulphate attack causes expansion and cracking due to ettringite formation, while acid erosion leads to the leaching of calcium compounds. ECC mixes incorporating supplementary cementitious materials like fly ash, silica fume, and more recently, graphene nanoplatelets, have shown enhanced resistance due to reduced calcium hydroxide content and a denser matrix. Graphene, due to its impermeability and high surface area, effectively blocks the ingress of acidic ions and limits mass loss. Studies demonstrate that ECC with graphene retains structural integrity and exhibits fewer surface cracks under prolonged exposure to H_2SO_4 and HNO_3 solutions. The addition of rubber particles, though beneficial for ductility, may reduce resistance if not properly proportioned. Therefore, an optimized mix using nano-enhanced materials offers a promising pathway for achieving chemical resilience in ECC (Bheel *et al.*, 2023, 2025; Göçügenci and Keskin, 2023).

4.4 Temperature and impact resistance

The mechanical properties of conventional concrete were affected under elevated temperature, i.e. the spalling of concrete may occur while raising the temperature from 200°C to 350°C. (Mohammed, Achara and Liew, 2018) observed the result of elevated temperature with the incorporation of different percentages of PVA fibre and nano-silica in ECC. ECC with 2% and 3% of nano-silica and 1.5% and 2% of PVA eliminate the spalling behaviour of ECC at a higher temperature. Nano silica in ECC develops fewer internal cracks than the control mix, thereby decreasing the temperature effect of ECC. Also, the compressive strength of ECC with PVA fibres (0.5%) is decreased due to the melting of fibre at low temperature, which reduces the binding of the cementitious matrix. The presence of high percentages of PVA fibres and nano-silica in ECC controlled the internal cracks up to 200°C is the reason for the less porosity of the cementitious matrix. Similarly, (Danish and Mosaberpanah, 2021) found that ECC with fly ash and cenosphere achieved higher compressive strength at 200°C and 400°C owing to the development of higher C-S-H gel. In contrast, it decreased at 600°C because of the development of extreme vapour pressure, which causes several cracks in ECC and decreases the strength. Figure 6 shows the temperature effect on ECC relative strength and stiffness.

(Hu *et al.*, 2021) examine the impact resistance of ECC with fly ash, quartz sand, PVC aggregate, PP and PVA fibre. The impact strength of ECC with PVA fibres produced higher impact strength than ECC with PP fibre and PVC aggregates. The impact failure energy of PVA-ECC was achieved as 15000 J, and it was 276 times higher than ECC without fibres. Moreover, the impact energy of ECC with both PVC aggregates and fibres was also increased compared to ECC without fibres, especially M-PVC aggregate-based ECC was produced higher impact resistance due to the increased bonding and fewer porosity properties. Enhanced fire resistance in ECC has been achieved through the incorporation of ceramic-based microfibres. It has been observed that these fibres significantly reduce spalling and maintain structural integrity at temperatures up to 800°C. The thermal stability and low thermal conductivity of ceramic fibres have been found to contribute to improved heat resistance and reduced internal cracking. ECC specimens reinforced with ceramic fibres retained higher compressive strength after exposure to elevated temperatures compared to those without such reinforcement (Xu *et al.*, 2021).

Figure 6. Temperature effect on relative strength and stiffness of ECC

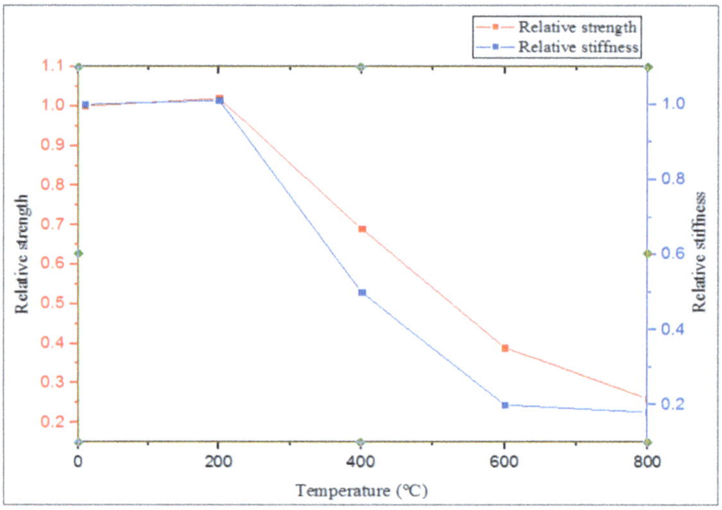

Source: Ma, Yi, and Wu (2021)

ECC demonstrates superior resistance to elevated temperatures and dynamic loads when engineered with the appropriate fibre-matrix composition. Exposure to high temperatures typically results in spalling and degradation in ordinary concrete. However, ECC with thermally stable fibres such as ceramic microfibres or PVA shows significant improvement in residual strength and thermal stability. Nano-silica further contributes by refining the microstructure and minimizing internal cracking under heat. Experimental data reveal that ECC can retain substantial compressive strength even after exposure to temperatures up to 800°C. In terms of impact resistance, ECC ability to undergo multiple microcracking without catastrophic failure ensures superior energy absorption (Liu *et al.*, 2023). Fibres like PVA and polyethylene enhance crack-bridging, while recycled PVC aggregates and hybrid fibre systems improve the overall toughness (Deng and Liu, 2025). These properties make ECC ideal for applications in fire-exposed structures and high-impact zones such as pavements, tunnel linings, and military infrastructure (Rafiei *et al.*, 2021).

4.5 Electrical resistance (ER)

According to (M.Sing et.al., 2020), ER measures ions' movement and diffusion inside the cement-based matrix system. ER outcomes of cementitious material are a good benchmark or indicator for chloride ions penetration, corrosion risk and durability of concrete. Furthermore, this study found that replacing river sand and MSS with 25% and 50% of stone slurry powder in ECC enhanced the ER by about 12.1% and 15.9%, respectively. The dense hardened matrix developed due to the filling effect of stone slurry powder, which increases the flow of current ions in the ECC matrix, which improve the higher ER enhancement. Similarly, (H.Said et.al., 2018) found that replacing fly ash by 30% and 45% of RGP in ECC increases the ER at all curing ages due to the high filling effect of RGP. Improved electrical conductivity and self-sensing capability in ECC have been achieved through the incorporation of carbon black and carbon fibre. It has been demonstrated that these conductive fillers form a percolated network within the cementitious matrix, allowing electrical resistance changes to be measured in response to mechanical strain. Also, the enhanced sensitivity and durability were observed, indicating the potential of such composites for structural health monitoring applications under dynamic loading and service conditions (Han *et al.*, 2024).

Electrical resistance in ECC serves as a critical indicator of its durability and self-monitoring capabilities. High ER values typically suggest a dense microstructure and low ion permeability, correlating with improved durability. The incorporation of fine fillers such as stone slurry powder and recycled glass powder increases ER by reducing the connectivity of pore channels. Moreover, functional ECC embedded with carbon black and carbon fibres have been developed to enable self-sensing properties. These conductive networks allow real-time monitoring of strain and crack development through resistivity changes. Such ECC not only offer structural performance but also act as smart materials for condition assessment in critical infrastructure. The synergy between mechanical integrity and electrical performance is especially valuable in environments prone to corrosion or dynamic loading (Dhandapani *et al.*, 2020).

4.6. Other properties

4.6.1 Self-healing property

(Guan *et al.*, 2020) examined the self-healing ability of ECC by adding limestone powder (LP) in different exposure conditions. The study revealed that ECC with LP enhanced the performance of self-healing ability more than ECC without LP. The presence of high calcium carbonate in LP enhances the pozzolanic activity of the cementitious matrix, thereby increasing the self-healing performance. Additionally, better self-healing performance of ECC was attained by $Ca(OH)_2$ exposure condition than Na_2SO_4 and water. Similarly, (H.Said et.al., 2018) achieved a better self-healing performance by replacing fly ash with 15% to 30% RGP in ECC. The existence of high silica, sodium and alumina in RGP developed the high amount of C-S-H gel in the cement matrix, thereby enhancing the self-healing performance of ECC. Alternatively, (Zhu *et al.*, 2020) developed the low carbon and low-cost ECC incorporating LC3 and PP fibre. ECC with LC3 and PP fibre attained a better self-healing ability than the control mix due to larger strain capacity and tight crack width. (Siad *et al.*, 2015) examined the self-healing ability of ECC with the use of microstructural analysis, chloride ion penetration and electrical resistivity test. The self-healing ability of ECC with LP exhibits better results than that without LP. The higher pozzolanic activity and presence of monocarboaluminate in the mix is the reason for healing cracks at all ages of curing.

Enhanced self-healing performance in ECC has been achieved through the incorporation of bacteria-based agents. It has been observed that biogenic healing of microcracks occurs as bacterial spores become activated in the presence of moisture, leading to the precipitation of calcium carbonate. Improved crack sealing, tensile strength recovery, and reduced permeability were reported by (Dwi, Huang and Hsu, 2024), confirming the potential of bacterial additives for long-term durability enhancement in ECC. The self-healing capability of ECC is one of its most innovative and valuable durability features. ECC tight crack widths (<100 μm) allow for autogenous healing through continued hydration and precipitation of calcium carbonate in the presence of water. This process can be further enhanced by incorporating limestone powder, recycled glass powder, or LC3 binders, which supply additional reactive silica and calcium ions. More recently, bacteria-based self-healing ECC has emerged, where microbial agents embedded within the matrix are activated upon moisture ingress, producing calcite that seals cracks (Qiu, Aw-Yong and Yang, 2018). This biogenic approach has demonstrated long-term sealing of cracks, recovery of tensile strength, and reduction in permeability. Thus, self-healing ECC reduces maintenance needs, extends service life, and supports sustainable infrastructure.

4.6.2 Surface water drainage and skid resistance

(Bawono, Lechner and Yang, 2019) studied the surface water drainage performance (SWD) and skid resistance (SR) of ECC with the inclusion of PVA fibre, MSS, fine aggregate (2.38mm) and corundum. The ECC with different percentages of corundum exhibited better SWD performance. Similarly, the skid resistance of ECC increased while increasing the rate of corundum. It is due to the impact of the surface exposure process on road surfaces and the shape, size and gradation of corundum. Improved surface drainage and skid resistance in ECC pavements have been achieved through the use of porous recycled ceramic aggregates. It has been shown that the angular shape and high porosity of these aggregates enhance surface texture and water flow. Similarly, ECC incorporating ceramic aggregates exhibited higher skid resistance values and more efficient water drainage compared to mixes with conventional aggregates (Yücel, Dutkiewicz and Yıldızhan, 2022). Surface water drainage and skid resistance are critical performance parameters for ECC used in pavements and transportation infrastructure. These properties are largely influenced by the surface texture, porosity, and aggregate type in the ECC mix. The use of coarse, angular, and porous aggregates such as recycled ceramic or corundum significantly enhances drainage by facilitating surface water runoff and improving texture depth. Simultaneously, these aggregates increase frictional resistance, resulting in improved skid resistance under wet conditions. Research has shown that ECC mixes incorporating such aggregates exhibit higher skid values and better water dispersal without compromising structural strength. These attributes make ECC a sustainable and high-performing alternative for urban roads, airport runways, and bridge decks.

5. OBSERVATIONS ON DURABILITY PROPERTIES

Based on the interpretations from various investigators, the higher and lower durability properties of ECC with the effect of W/C ratio, S/B ratio, curing conditions and different materials such as fibres, fine aggregates and SCM are enlisted in Table 5. Comprehensive analysis of ECC durability studies reveals that performance is intricately linked to multiple material and environmental parameters. High SCM content enhances sulphate and acid resistance due to refined microstructure and secondary C-S-H formation. Fine aggregates with optimized gradation improve packing density and permeability resistance. Fibres contribute significantly by limiting crack width and enabling multiple microcracks that promote self-healing. However, improper fibre selection or inadequate volume can lead to brittle failure and poor impact resistance. The W/C and S/B ratios must be finely balanced to achieve dense microstructures while preventing shrinkage or excessive porosity. Curing conditions also play a decisive role; prolonged moist curing enhances hydration and durability, while improper exposure can induce microcracks and salt leaching. These observations suggest that ECC design should adopt an integrated mix design approach tailored to specific environmental exposures and performance objectives. Table 6 provides the application suitability matrix for ECC based on durability criteria to enhance the clarity and visual impact of this review.

Table 5. Observation on durability properties of ECC

Effects/ Materials	Reason for high durability properties	Reason for low durability properties
SCM	High formation of secondary C-S-H gel due to high volume of SCM in later ages Existence of high silica and CaO in SCM Development of dense matrix due to pozzolanic activity of SCM High surface area and less porous developed during mix	High percentage of cement replacement Less specific surface area Large number of internal crack development Fewer filling and dense matrix Less chemical compositions Lower volume of SCM
Fine aggregate	Better filling and packing density of fine aggregates Less capillary pores and similar surface area Larger size of fine aggregate High impact and fire resistance	High porous and irregular surface area Shape and size of aggregate Irregular gradation and several internal crack formation Improper selection
Fibres	High bonding ability of fibre with cementitious matrix Higher tensile strength capacity Larger strain capacity and tight crack width	Less fibre matrix interaction Less percentage of fibres Fewer elongation of fibre Development of more micro cracks
W/C and S/B ratio	Optimum W/C and S/B ratio	Insufficient or higher W/C or S/B ratio
Curing conditions	Development high bonding under exposure conditions Less macro cracks development during curing	Several micro and macro crack formation under exposure conditions Development of high soluble salts

Table 6. Application suitability matrix for ECC based on durability criteria

Additive Material	Durability Improved	Mechanism of Action	Applications
Fly ash + Graphene	Chloride ingress	Pore refinement	Coastal infrastructure
Graphene + Dense matrix additives	Acid exposure	Matrix densification	Industrial flooring
Ceramic aggregates + PVA fibres	Impact + skid resistance	Matrix densification	Airport runways
Nano-silica + Ceramic microfibres	Thermal degradation	Thermal stability	Fire-resilient structures
Carbon black + Conductive fibre	Self-monitoring	Conductive network formation	Smart structures

5. CONCLUSION

This article reviewed the durability properties of ECC, such as chloride ion penetration, water absorption, pore structure analysis, hardened shrinkage resistance, sulphate resistance, acid resistance, temperature resistance, impact resistance, skid resistance, surface water drainage and self-healing properties. The following conclusions are drawn based on the literature review and discussion.

- The durability properties of ECC depend on several factors such as materials used in ECC, physical and chemical properties of materials, exposure condition, w/c ratio, sand to binder ratio, SCM to cement ratio, etc.
- In long term curing conditions, the amount of SCM plays a significant role to improve the chloride ion penetration and water absorption due to the development of the secondary pozzolanic activity.
- Partial replacement of SCM or fine aggregate with finer particles, namely, stone slurry powder, slag, concrete dust, fine recycled aggregate, recycled glass powder, limestone powder and crumb rubber, exhibited better durability under various exposure conditions.
- Incorporating 1.5% to 2% of PE, PP PVA and hybrid fibres in ECC attained better temperature resistance up to 400°C. Also, shrinkage resistance of cementitious matrix decreases with an increase in the volume of fibres. ECC with tire wire fibre attained better shrinkage resistance than PVA fibre due to its high bonding matrix.
- The sulphate and acid erosion resistance improved notably with the incorporation of graphene oxide in cementitious blends. Similarly, the addition of finer particles improves the electrical resistivity due to its filling effect.
- Various investigators proved the self-healing ability of ECC with the effective use of limestone powder, LC3, glass powder, etc. Also, the surface drainage properties improved remarkably with the incorporation a larger size of fine aggregates in ECC.

6. EMERGING TRENDS AND FUTURE DIRECTIONS IN ECC DURABILITY

In recent years, several innovative strategies have been adopted to enhance the durability of ECC under diverse environmental conditions. Advanced materials have been incorporated to improve ECC resistance to mechanical, chemical, and thermal degradation. Notably, permeability has been reduced and chloride ion resistance improved through the partial replacement of cement with biochar, owing to its pore-refining capability and environmental benefits (Guan *et al.*, 2020). Shrinkage-induced cracking has been minimized by incorporating super absorbent polymers (SAPs), which provide internal curing by gradually releasing absorbed water during hydration (Zhu *et al.*, 2020). The resistance of ECC to acid and sulphate attack has been significantly improved with the addition of graphene nanoplatelets, which densify the microstructure and limit aggressive ion penetration (Bheel *et al.*, 2023; S. Divya et.al, 2024). Similarly, the inclusion of ceramic-based microfibres has been shown to enhance ECC fire resistance by reducing internal spalling and improving thermal stability at temperatures exceeding 800°C (Bahraq, Maslehuddin and Al-Dulaijan, 2020). Furthermore, multifunctional ECC capable of structural health monitoring have been developed by integrating carbon black and carbon fibres, which enable self-sensing behaviour through changes in electrical resistance under strain (Shi *et al.*, 2018). Enhanced self-healing performance has also been achieved through the incorporation of bacteria-based agents, which precipitate calcium carbonate within cracks upon moisture activation, promoting long-term sealing and strength recovery (Victor Li et al, 2017). Additionally, the development of 3D-printable ECC formulations has been pursued to enable complex structural applications and improved quality control during placement (Asghari *et al.*, 2024). This evolving research landscape reflects a shift toward the development of sustainable, intelligent, and highly durable ECC systems for modern infrastructure.

CREDIT AUTHORSHIP CONTRIBUTION STATEMENT

N. Shanmugasundaram: Conceptualization, Methodology, Investigation and Writing original draft. **S. Praveenkumar:** Writing - review & editing. **Divya S:** Writing - review & editing.

DECLARATION OF COMPETING INTEREST

The authors declare that they have no known competing financial interests or personal relationships that could have appeared to influence the work reported in this paper.

DATA AVAILABILITY

The authors declare that the data supporting the findings of this study are available within the article

ACKNOWLEDGEMENT

The authors wish to thank Dr.K.Prakasan, Principal, PSG College of Technology, Coimbatore for the facilities and support provided for carrying out this research work at Advanced Concrete Research Laboratory

REFERENCES

Adesina, A. (2021). Performance of cementitious composites reinforced with chopped basalt fibres—An overview. *Construction & Building Materials*, *266*, 120970. Advance online publication. DOI: 10.1016/j.conbuildmat.2020.120970

Adesina, A. (2021). Overview of the influence of waste materials on the thermal conductivity of cementitious composites. *Cleaner Engineering and Technology*, *2*, 100046. Advance online publication. DOI: 10.1016/j.clet.2021.100046

Adesina, A., & Das, S. (2021). Sustainable utilization of recycled asphalt as aggregates in engineered cementitious composites. *Construction & Building Materials*, *283*, 122727. Advance online publication. DOI: 10.1016/j.conbuildmat.2021.122727

Adesina, A., & Das, S. (2021). Evaluation of the durability properties of engineered cementitious composites incorporating recycled concrete as aggregate. *Journal of Materials in Civil Engineering*, *33*(2), 04020439. Advance online publication. DOI: 10.1061/(ASCE)MT.1943-5533.0003563

Ajith, G., Kumar, S., & Eligehausen, R. (2021). Effect of mineral admixtures and manufactured sand on compressive strength of engineered cementitious composite. *Journal of Building Pathology and Rehabilitation*, *6*(1), 37. Advance online publication. DOI: 10.1007/s41024-021-00137-y

Alaloul, W. S., Musarat, M. A., Haruna, S., Law, K., Tayeh, B. A., Rafiq, W., & Ayub, S. (2020). Mechanical and deformation properties of rubberized engineered cementitious composite (ECC). *Case Studies in Construction Materials*, *13*, e00385. Advance online publication. DOI: 10.1016/j.cscm.2020.e00385

Amir Ali, A., Kumar, R., Khan, M. A., & Chauhan, D. P. S. (2025). Biochar-amended high-strength engineered cementitious composites. *Cement and Concrete Composites*, *164*, 106219. Advance online publication. DOI: 10.1016/j.cemconcomp.2025.106219

Ammasi, A. K., & Ragul, G. (2018). Strength and durability of high volume fly ash in engineered cementitious composites. *Materials Today: Proceedings*, *5*(11), 24050–24058. DOI: 10.1016/j.matpr.2018.10.198

Andrade Neto, J., De la Torre, A. G., & Kirchheim, A. P. (2021). Effects of sulfates on the hydration of Portland cement—A review. *Construction & Building Materials, 279*, 122428. Advance online publication. DOI: 10.1016/j.conbuildmat.2021.122428

Anwar, A., Nasir, M., Hussain, I., & Shah, S. A. R. (2020). Enhanced properties of cementitious composite tailored with graphene oxide nanomaterial—A review. *Developments in the Built Environment, 1*, 100002. Advance online publication. DOI: 10.1016/j.dibe.2019.100002

Arulanandam, P. M., Arunachalam, S., & Gobinath, R. (2020). Behavior of engineered cementitious composite structural elements—A review. *Indian Concrete Journal, 94*(6), 5–28.

Asghari, Y., Alaghebandian, A., & Zhang, Y. (2024). 3D printing and implementation of engineered cementitious composites—A review. *Case Studies in Construction Materials, 21*, e03462. Advance online publication. DOI: 10.1016/j.cscm.2024.e03462

Ashraf, M., Khan, A. N., Ali, M., & Khan, M. A. (2022). Developing a sustainable concrete incorporating bentonite clay and silica fume: Mechanical and durability performance. *Journal of Cleaner Production, 337*, 130315. Advance online publication. DOI: 10.1016/j.jclepro.2021.130315

. ASTM International. (1997). ASTM E119: Standard test methods for fire tests of building construction and materials.

. ASTM International. (2007). ASTM C490-07: Standard practice for use of apparatus for the determination of length change of hardened cement paste, mortar, and concrete.

. ASTM International. (2012). ASTM C1202-12: Standard test method for electrical indication of concrete's ability to resist chloride ion penetration. https://doi.org/ DOI: 10.1520/C1202-12

. ASTM International. (2013). ASTM C1585-13: Standard test method for measurement of rate of absorption of water by hydraulic cement concretes.

ASTM International. (2015). *ASTM C1012/C1012M-15: Standard test method for length change of hydraulic-cement mortars exposed to a sulfate solution*. ASTM International.

. ASTM International. (2015). ASTM C1403-15: Standard test method for rate of water absorption of masonry mortars.

. ASTM International. (2017). ASTM C157/C157M-17: Standard test method for length change of hardened hydraulic-cement mortar and concrete. *Annual Book of ASTM Standards* (Vol. 04.01).

. ASTM International. (2018). ASTM C596-18: Standard test method for drying shrinkage of mortar containing hydraulic cement.

. ASTM International. (2021). ASTM C642-21: Standard test method for density, absorption, and voids in hardened concrete. https://doi.org/DOI: 10.1520/C0642-21

Bahraq, A. A., Maslehuddin, M., & Al-Dulaijan, S. U. (2020). Macro- and micro-properties of engineered cementitious composites (ECCs) incorporating industrial waste materials: A review. *Arabian Journal for Science and Engineering*, *45*(10), 7869–7895. DOI: 10.1007/s13369-020-04729-7

Bawono, A. A., Lechner, B., & Yang, E. H. (2019). Skid resistance and surface water drainage performance of engineered cementitious composites for pavement applications. *Cement and Concrete Composites*, *104*, 103387. Advance online publication. DOI: 10.1016/j.cemconcomp.2019.103387

Bheel, N., Mahro, S. K., & Adesina, A. (2023). Effects of graphene oxide on the properties of engineered cementitious composites: Multi-objective optimization technique using RSM. *Buildings (Basel, Switzerland)*, *13*(8), 2018. Advance online publication. DOI: 10.3390/buildings13082018

Bheel, N., Mahro, S. K., & Mohammed, B. S. (2025). Optimization of durability characteristics of engineered cementitious composites combined with titanium dioxide as a nanomaterial applying RSM modelling. *Scientific Reports*, *15*(1), 94382. Advance online publication. DOI: 10.1038/s41598-025-94382-7

Biyani, Y., Patil, L. G., & Chandrasekhar, C. N. K. (2019). Engineered cementitious composites (ECC). In *Engineered cementitious composites (ECC)* (pp. 206–214). Springer. https://doi.org/DOI: 10.1007/978-3-662-58438-5

Booya, E., Gorospe, K., Dashti, A., & Ghaednia, H. (2020). The influence of utilizing slag in lieu of fly ash on the performance of engineered cementitious composites. *Construction & Building Materials*, *256*, 119412. Advance online publication. DOI: 10.1016/j.conbuildmat.2020.119412

Cai, Z., Liu, X., Chen, H., & Yu, J. (2021). Development of ultra-high ductility engineered cementitious composites as a novel and resilient fireproof coating. *Construction & Building Materials*, *288*, 123090. Advance online publication. DOI: 10.1016/j.conbuildmat.2021.123090

Danish, A., & Mosaberpanah, M. A. (2021). Influence of cenospheres and fly ash on the mechanical and durability properties of high-performance cement mortar under different curing regimes. *Construction & Building Materials*, *279*, 122458. Advance online publication. DOI: 10.1016/j.conbuildmat.2021.122458

de Siqueira, A. A., & Cordeiro, G. C. (2022). Sustainable cements containing sugarcane bagasse ash and limestone: Effects on compressive strength and acid attack of mortar. *Sustainability (Basel)*, *14*(9), 5683. Advance online publication. DOI: 10.3390/su14095683

Deng, Z., & Liu, D. (2025). Impact resistance of engineered cementitious composites reinforced with hybrid fibers. *Journal of Advanced Concrete Technology*, *23*(7), 296–309. DOI: 10.3151/jact.23.296

Dhandapani, Y., Sakthivel, T., Santhanam, M., & Gettu, R. (2020). Perspectives on blended cementitious systems with calcined clay-limestone combination for sustainable low carbon cement transition. *Indian Concrete Journal*, *94*(2), 31–45.

Divya, S., Thomas, B. S., & Anand, K. B. (2024). Pozzolanic attributes of hydraulic cement paste hybridized with agricultural by-product and nano-carbon. *Construction & Building Materials*, *411*, 134119. Advance online publication. DOI: 10.1016/j.conbuildmat.2023.134119

Dwi, M., Huang, W., & Hsu, H. (2024). Improving properties of engineered cementitious composite (ECC) using Bacillus subtilis immobilized in silica gel. *Case Studies in Construction Materials*, *20*, e03165. Advance online publication. DOI: 10.1016/j.cscm.2024.e03165

Göçügenci, A., & Keskin, S. B. (2023). The effect of graphene nanoplatelet addition on the mechanical, durability and self-healing properties of engineered cementitious composites. *MATEC Web of Conferences, 378*, Article 02024. https://doi.org/DOI: 10.1051/matecconf/202337802024

Guan, X., Li, M., & Yang, E. H. (2020). Effect of exposure conditions on self-healing behavior of engineered cementitious composite incorporating limestone powder. *Cement and Concrete Composites*, *114*, 103808. Advance online publication. DOI: 10.1016/j.cemconcomp.2020.103808

Han, J., Zhang, J., Kim, D., & Lee, S. (2024). Development of self-sensing engineered cementitious composite sensors for monitoring flexural performance of reinforced concrete beam. *Developments in the Built Environment*, *18*, 100407. Advance online publication. DOI: 10.1016/j.dibe.2024.100407

Hu, S., Zhang, Y., & Liu, X. (2021). Performance test and microstructure of modified PVC aggregate-hybrid fiber reinforced engineering cementitious composite (ECC). *Materials (Basel)*, *14*(8), 1856. Advance online publication. DOI: 10.3390/ma14081856

Isyaka, A., Ibrahim, A., & Bello, A. (2020). A review of the effect of waste tire rubber on the properties of ECC. *International Journal of Advanced and Applied Sciences*, *7*(8), 105–116. DOI: 10.21833/ijaas.2020.08.011

Kewalramani, M. A., Mohamed, O. A., & Syed, Z. I. (2017). Engineered cementitious composites for modern civil engineering structures in hot arid coastal climatic conditions. *Procedia Engineering*, *180*, 767–774. DOI: 10.1016/j.proeng.2017.04.237

Khedher, B. I., & Al-Sulaimi, S. K. (2018). Optimization of hybrid fibres in engineered cementitious composites. *Construction & Building Materials*, *190*, 24–37. DOI: 10.1016/j.conbuildmat.2018.08.188

Li, V. C., Wu, C., Wang, S., Ogawa, A., & Saito, T. (2020). On the emergence of 3D printable engineered, strain hardening cementitious composites (ECC/SHCC). *Cement and Concrete Research*, *132*, 106038. Advance online publication. DOI: 10.1016/j.cemconres.2020.106038

Li, V. C., Yang, E. H., & Li, M. (2017). Self-healing of microcracks in engineered cementitious composites under sulfate and chloride environment. *Construction & Building Materials*, *153*, 948–956. DOI: 10.1016/j.conbuildmat.2017.07.126

Liu, D., Li, X., & Chen, Y. (2023). Mechanical performance of high-strength engineering cementitious composites (ECC) with hybriding PE and steel fibers. *Case Studies in Construction Materials*, *18*, e01961. Advance online publication. DOI: 10.1016/j.cscm.2023.e01961

Ma, H., Yi, C., & Wu, C. (2021). Review and outlook on durability of engineered cementitious composite (ECC). *Construction & Building Materials*, *287*, 122719. Advance online publication. DOI: 10.1016/j.conbuildmat.2021.122719

Mensah, R. A., Xiao, J., Asamoah, R. B., & Amponsah, K. (2021). Biochar-added cementitious materials—A review on mechanical, thermal, and environmental properties. *Sustainability (Basel)*, *13*(16), 9073. Advance online publication. DOI: 10.3390/su13169336

Mohammed, B. S., Achara, B. E., & Liew, M. S. (2018). The influence of high temperature on microstructural damage and residual properties of nano-silica-modified (NS-modified) self-consolidating engineering cementitious composites (SC-ECC) using response surface methodology (RSM). *Construction & Building Materials*, *192*, 450–466. DOI: 10.1016/j.conbuildmat.2018.10.114

Muralidharan, R., Sathyan, D., & Anand, K. B. (2021). Review of the effects of supplementary cementitious materials and chemical additives on the physical, mechanical and durability properties of hydraulic concrete. *Materials (Basel)*, *14*(23), 7270. Advance online publication. DOI: 10.3390/ma14237270

Qiu, J., Aw-Yong, W. L., & Yang, E. H. (2018). Effect of self-healing on fatigue of engineered cementitious composites (ECCs). *Cement and Concrete Composites*, *94*, 145–152. DOI: 10.1016/j.cemconcomp.2018.09.007

Rafiei, P., Nili, M., & Amini, M. (2021). Temperature impact on engineered cementitious composite containing basalt fibers. *Applied Sciences (Basel, Switzerland)*, *11*(15), 6848. Advance online publication. DOI: 10.3390/app11156848

Sabapathy, L., Mahalingam, B., & Ravichandran, P. T. (2020). Acid and sulphate attacks on a rubberized engineered cementitious composite containing graphene oxide. *Materials (Basel)*, *13*(14), 3125. Advance online publication. DOI: 10.3390/ma13143125

Şahmaran, M., Lachemi, M., Hossain, K. M. A., Ranade, R., & Li, V. C. (2009). Influence of aggregate type and size on ductility and mechanical properties of engineered cementitious composites. *ACI Materials Journal*, *106*(3), 308–316. DOI: 10.14359/56556

Şahmaran, M., & Li, V. C. (2010). Engineered cementitious composites. *Transportation Research Record: Journal of the Transportation Research Board*, *2164*(1), 1–8. DOI: 10.3141/2164-01

Said, H. S., Ibrahim, A., & Elrahman, M. A. (2018). Use of recycled glass powder to improve the performance properties of high volume fly ash-engineered cementitious composites. *Construction & Building Materials*, *163*, 53–62. DOI: 10.1016/j.conbuildmat.2017.12.067

Saloni, S., Singh, A., Jatin, , Kumar, J., & Parveen, . (2019). Influence of fly ash, cement and ground river sand on compressive strength and chloride penetration of mortar. *Materials Today: Proceedings*, *33*, 1690–1694. DOI: 10.1016/j.matpr.2020.07.269

Shanmugasundaram, N., & Praveenkumar, S. (2021). Influence of supplementary cementitious materials, curing conditions and mixing ratios on fresh and mechanical properties of engineered cementitious composites—A review. *Construction & Building Materials*, *309*, 125038. Advance online publication. DOI: 10.1016/j.conbuildmat.2021.125038

Shanmugasundaram, N., & Praveenkumar, S. (2023). Influence of manufactured sand gradation and water cement ratios on compressive strength of engineered cementitious composites. *Materials Today: Proceedings*. Advance online publication. DOI: 10.1016/j.matpr.2023.05.024

Shanmugasundaram, N., & Praveenkumar, S. (2024). Prediction on mechanical properties of engineered cementitious composites: An experimental and machine learning approach. *Structural Concrete*. Advance online publication. DOI: 10.1002/suco.202400231

Shanmugasundaram, N., Praveenkumar, S., & Kumar, K. V. (2025). Structural behavior of engineered cementitious composite substrate slab overlays for bridge deck and pavement applications. *Construction & Building Materials*, *459*, 139736. Advance online publication. DOI: 10.1016/j.conbuildmat.2024.139736

Shanmugasundaram, N., Praveenkumar, S., & Kumar, K. V. (2025). Effects of basalt fiber content on the properties of ambient-cured engineered geopolymer composites. *Fibers and Polymers*, *26*(7), 3147–3168. Advance online publication. DOI: 10.1007/s12221-025-01019-7

Shanmugasundaram, N., Praveenkumar, S., & Kumar, K. V. (2025). Development of sustainable engineered cementitious composites using limestone calcined clay cement (LC3): A review. *Civil Engineering (Shiraz)*. Advance online publication. DOI: 10.1007/s40996-025-01966-4

Sherir, M. A. A., Hossain, K. M. A., & Lachemi, M. (2018). Fresh state, mechanical & durability properties of strain hardening cementitious composite produced with locally available aggregates and high volume of fly ash. *Construction & Building Materials*, *189*, 253–264. DOI: 10.1016/j.conbuildmat.2018.08.204

Shi, L., Li, Z., & Wu, Z. (2018). Artificial neural network based mechanical and electrical property prediction of engineered cementitious composites. *Construction & Building Materials*, *174*, 667–674. DOI: 10.1016/j.conbuildmat.2018.04.127

Siad, H., Lachemi, M., Sahmaran, M., & Hossain, K. M. A. (2015). Influence of limestone powder on mechanical, physical and self-healing behavior of engineered cementitious composites. *Construction & Building Materials*, *99*, 1–10. DOI: 10.1016/j.conbuildmat.2015.09.007

Singh, M., Saini, B., & Chalak, H. D. (2019). Performance and composition analysis of engineered cementitious composite (ECC)—A review. *Journal of Building Engineering, 26*, 100851. Advance online publication. DOI: 10.1016/j.jobe.2019.100851

Singh, M., Saini, B., & Chalak, H. D. (2020). Long term evaluation of engineered cementitious composite containing stone slurry powder. *Construction & Building Materials, 264*, 120183. Advance online publication. DOI: 10.1016/j.conbuildmat.2020.120183

Soe, K. T., Zhang, Y. X., & Zhang, L. C. (2013). Material properties of a new hybrid fibre-reinforced engineered cementitious composite. *Construction & Building Materials, 43*, 399–407. DOI: 10.1016/j.conbuildmat.2013.02.021

Wang, Q., Zhang, J., & Ho, J. C. M. (2020). Zeolite to improve strength-shrinkage performance of high-strength engineered cementitious composite. *Construction & Building Materials, 234*, 117335. Advance online publication. DOI: 10.1016/j.conbuildmat.2019.117335

Xu, S., Li, H., & Li, Q. (2021). Thermal stability study of fiber-reinforced cementitious material with high ductility under high-temperature casting. *Construction & Building Materials, 282*, 122700. Advance online publication. DOI: 10.1016/j.conbuildmat.2021.122700

Yang, X., Liu, J., Li, H., & Xu, L. (2021). Behavior of biochar-modified cementitious composites exposed to high temperatures. *Materials (Basel), 14*(18), 5414. Advance online publication. DOI: 10.3390/ma14185414

Ye, B., Zhang, Y., & Yu, J. (2019). Effect of water to binder ratio and sand to binder ratio on shrinkage and mechanical properties of high-strength engineered cementitious composite. *Construction & Building Materials, 226*, 899–909. DOI: 10.1016/j.conbuildmat.2019.07.303

Ye, S., Li, V. C., & Li, H. (2015). Nano-modification to improve the ductility of cementitious composites. *Cement and Concrete Research, 76*, 170–179. DOI: 10.1016/j.cemconres.2015.05.026

Yoo, D.-Y., & Banthia, N. (2022). High-performance strain-hardening cementitious composites with tensile strain capacity exceeding 4%: A review. *Cement and Concrete Composites, 125*, 104325. Advance online publication. DOI: 10.1016/j.cemconcomp.2021.104325

Younsi, A., Turcry, P., Aït-Mokhtar, A., & Staquet, S. (2021). Hydration-drying interactions in a high-volume ground granulated blast-furnace slag mortar. *Construction & Building Materials, 279*, 122427. Advance online publication. DOI: 10.1016/j.conbuildmat.2021.122427

Yu, K., Wang, Y., & Yu, J. (2018). Direct tensile properties of engineered cementitious composites: A review. *Construction & Building Materials, 165*, 346–362. DOI: 10.1016/j.conbuildmat.2017.12.124

Yücel, H. E., Dutkiewicz, M., & Yıldızhan, F. (2022). Application of ECC as a repair/retrofit and pavement/bridge deck material for sustainable structures: A review. *Materials (Basel), 15*(24), 8752. Advance online publication. DOI: 10.3390/ma15248752

Zhang, D., Jaworska, B., Zhu, Y., Yu, J., & Li, V. C. (2020). Engineered cementitious composites (ECC) with limestone calcined clay cement (LC3). *Cement and Concrete Composites, 114*, 103766. Advance online publication. DOI: 10.1016/j.cemconcomp.2020.103766

Zhang, D., Yu, J., Wu, H., Jaworska, B., & Li, V. C. (2020). Discontinuous microfibers as intrinsic reinforcement for ductile engineered cementitious composites (ECC). *Composites. Part B, Engineering, 184*, 107741. Advance online publication. DOI: 10.1016/j.compositesb.2020.107741

Zheng, P. D., Li, Y., & Li, Z. (2021). Compressive behavior of steel stirrups-confined square engineered cementitious composite (ECC) columns. *Advances in Concrete Construction, 11*(3), 193–206. DOI: 10.12989/acc.2021.11.3.193

Zhou, Y., Zhang, J., Li, Y., & Wu, Z. (2021). Feasibility of incorporating recycled fine aggregate in high performance green lightweight engineered cementitious composites. *Journal of Cleaner Production, 280*, 124445. Advance online publication. DOI: 10.1016/j.jclepro.2020.124445

Zhu, H., Zhang, D., Wu, C., & Li, V. C. (2020). Mechanical and self-healing behavior of low carbon engineered cementitious composites reinforced with PP-fibers. *Construction & Building Materials, 259*, 119805. Advance online publication. DOI: 10.1016/j.conbuildmat.2020.119805

Chapter 2

Enhancement in Thermo-Mechanical Properties of Polybutylene Terephthalate with Organic Nano Particle Reinforcement

Jeenendra Kumar Shiv
Birla Institute of Technology, Mesra, India

Chikesh Ranjan
https://orcid.org/0000-0003-1197-601X
National Institute of Technology, Rourkela, India

Kaushik Kumar
https://orcid.org/0000-0002-4237-2836
Birla Institute of Technology, Mesra, India

Sudeepan Jayapalan
Birla Institute of Technology, Mesra, India

ABSTRACT

This study investigates the effects of organic zirconia nanoparticle reinforcement (0.5–5 wt%) on the mechanical and thermal, properties of polybutylene terephthalate (PBT) composites. Mechanical tests revealed enhanced strength, toughness, and hardness, with optimal performance at 2% zirconia loading, balancing mechanical

DOI: 10.4018/979-8-3373-2414-2.ch002

reinforcement and thermal stability. The elongation at break decreased with increasing zirconia content, indicating reduced ductility, while Young's modulus remained stable, preserving stiffness. Impact strength peaked at 2% zirconia, enhancing toughness, and Shore D hardness increased up to 2% filler loading before declining. Thermogravimetric analysis showed improved thermal stability, as indicated by a shift in degradation temperatures. These results demonstrate that zirconia nanoparticles significantly enhance the mechanical and thermal properties of PBT, making them suitable for advanced applications in automotive components, electronic housings, and aerospace parts requiring high strength, thermal resistance, and durability.

1. INTRODUCTION

Polybutylene terephthalate (PBT) is a high-performance, semi-crystalline thermoplastic polymer belonging to the polyester family. It is known for its excellent mechanical strength, thermal stability, chemical resistance, and electrical insulating properties. PBT is widely used in automotive, electrical, and electronic applications due to its durability, dimensional stability, and resistance to moisture and solvents. Its properties can be further enhanced by incorporating reinforcements such as organic nanoparticles, making it suitable for advanced engineering applications (Mishra, et al. 2022).

Organic nanoparticles are nanoscale materials composed of organic compounds, typically polymers, lipids, or carbon-based structures. They are lightweight, biocompatible, and can be tailored for specific interactions with polymer matrices (He, et al. 2024). In material science, they are used to enhance mechanical strength, thermal stability, and other functional properties of polymers like PBT. Their uniform dispersion and strong interfacial bonding with the polymer matrix contribute to improved performance, making them valuable reinforcements in advanced composite materials (Visakh, 2015).

Zirconia nanoparticles (ZrO_2) are advanced ceramic nanoparticles known for their exceptional mechanical strength, thermal stability, and chemical resistance. They exhibit high hardness, wear resistance, and low thermal conductivity, making them suitable for reinforcing polymers and composites (Mulla, et al. 2021). When incorporated into materials like Polybutylene Terephthalate (PBT), zirconia nanoparticles enhance mechanical properties, such as tensile strength and toughness, while also improving thermal stability and resistance to degradation. Their applications extend to biomedical, aerospace, and structural engineering fields due to their superior performance characteristics (Alhotan, et al. 2021).

Polybutylene terephthalate (PBT) is a widely used thermoplastic polymer in various industries due to its excellent mechanical and thermal properties. However, its performance can be further enhanced by incorporating reinforcing materials (Hong, et al. 2015). Organic nanoparticles have shown great potential as reinforcements due to their lightweight nature, compatibility with polymer matrices, and superior mechanical properties. This paper explores the impact of organic nanoparticle reinforcement on the thermo-mechanical behavior of PBT.

Polybutylene Terephthalate (PBT) reinforced with organic nanoparticles is a high-performance composite material with enhanced thermo-mechanical properties (Mendes, et al. 2024). The incorporation of organic nanoparticles improves PBT's tensile strength, thermal stability, and durability by promoting better interfacial interactions within the polymer matrix. These nanoparticles contribute to uniform stress distribution, reduced thermal degradation, and improved crystallization behavior. Such reinforced PBT composites are widely used in automotive, electronics, and high-performance engineering applications due to their superior mechanical and thermal characteristics (Schmidt, et al. 2000).

The reinforcement of polybutylene terephthalate (PBT) with organic nanoparticles has emerged as a promising approach to enhance its thermo-mechanical properties for advanced engineering applications (Paul, et al. 2008). Various studies have explored the effects of incorporating nanofillers such as organoclays, zirconium-based frameworks, and other functionalized nanoparticles into the PBT matrix, leading to improved mechanical strength, thermal stability, and structural integrity (Gupta, et al. 2024). This lecture survey reviews recent advancements in PBT-based nanocomposites, focusing on the role of nanoparticle dispersion, interfacial interactions, and processing techniques in optimizing material performance.

(Kumar, et al. 2021) investigated the effects of ZrO_2 nanoparticle content on the viscoelastic, mechanical, and thermal properties of epoxy resin. Using an ultrasonic dual mixing technique, they prepared ZrO_2–epoxy nanocomposites and analyzed them through DMA, tensile testing, and TGA. The study found that a 4 wt.% ZrO_2–epoxy composite showed a 70% increase in storage modulus and a 34% rise in tensile strength, along with improved thermal stability and uniform nanoparticle dispersion.

(Khan, et al. 2022) reviewed the potential of sepiolite as a cost-effective reinforcing filler in thermoplastic composites, offering advantages over graphene and CNTs. Due to its high surface area and unique channel structure, sepiolite enhances mechanical and thermal properties, including tensile strength, stiffness, and flammability resistance. The study highlights that the performance of sepiolite-reinforced composites depends on filler content, matrix interactions, and particle characteristics, with further research needed for broader applications.

(Ogbonna, et al. 2022) reviewed the advancements in epoxy nanocomposites for mechanical and tribological applications, highlighting their lightweight, high strength, and wear resistance. The study addressed challenges related to mechanical property degradation over time and examined the effects of nano-inorganic and carbon-based fillers on thermal, mechanical, and frictional performance. The review concluded with key advancements, challenges, and recommendations for further improvements.

(Idumah, 2023) discussed the growing need for polymeric nanocomposites with enhanced thermomechanical properties for advanced engineering applications (Utracki, et al. 2007). The study emphasized the importance of controlling thermal expansion by incorporating low-CTE or negative-CTE fillers to reduce material mismatches. Thermal Mechanical Analysis (TMA) was highlighted as a key technique for evaluating thermal expansivity, with insights into PNC applications in adhesives, coatings, and electronic packaging.

(Owen, et al. 2024) investigated the effects of polybenzimidazole (PBI) nanoparticle reinforcement on the mechanical, thermal, flammability, and thermo-mechanical properties of HDPE composites. Using a melt intercalation process with varying PBI concentrations (1, 3, and 5 wt%), the study revealed significant improvements in mechanical strength, thermal stability, and flame resistance. Characterization through SEM, TGA, DSC, and DMA confirmed enhanced toughness, increased Tg, and improved damping properties, making these composites suitable for engineering, aerospace, and defense applications.

(Fukushima, et al. 2012) studied PLA-based nanocomposites reinforced with 5 wt.% of CLO, SEP, and ZrP using melt blending. The study found that ZrP enhanced PLA crystallization, while CLO and SEP improved melt viscosity and dynamic shear moduli by forming an interconnected structure within the matrix. CLO also reduced oxygen permeability by 30%, maintaining PLA's transparency, making these nanocomposites promising for sustainable packaging applications.

(Barletta, et al. 2022) reviewed the development of polybutylene succinate (PBS) as a biodegradable alternative to fossil-based plastics, highlighting its compostability and favorable mechanical and thermal properties. Due to its low Young's modulus and processing challenges, PBS is often blended with polymers like PLA and reinforced with additives to enhance stiffness, thermal resistance, and processability. The study explored recent advances in PBS-based blends, their industrial applications, and future prospects for improving performance while maintaining compostability.

(Verma, et al. 2022) reviewed the potential of agro-waste, particularly natural fibers (NFs) and natural particulates (NPs), for polymer composite development, emphasizing the challenge of hydrophilicity. Various chemical and physical treatments were discussed to enhance adhesion between NFs, NPs, and the polymer matrix, thereby improving mechanical properties. The study explored the physical characteristics of NFs and NPs, different surface treatments, composite processing techniques, and the resulting mechanical performance (Chen et al. 2023).

(Sudhakar, et al. 2021) investigated the enhancement of biodegradability in synthetic and natural polymers by incorporating organically modified clay as a nano-filler. The study examined poly (methyl methacrylate) (PMMA), poly(vinyl chloride) (PVC), poly(vinyl acetate) (PVAc), and cellulose acetate butyrate (CAB), revealing improved intercalation, mechanical strength, and hydrolytic degradation. Characterization techniques like DSC, SEM, XRD, and FTIR confirmed the structural and thermal modifications, making these polymer-clay nanocomposites promising for eco-friendly packaging applications.

(Bas, et al. 2024) examined the effect of two types of organoclays on the mechanical, thermal, and morphological properties of PBT nanocomposites fabricated via melt mixing. The study found that NC 130 clay enhanced tensile strength, hardness, and melt flow behavior, while NC 140 improved impact resistance, thermo-mechanical performance, and density reduction. These findings highlight the potential of tailored PBT-organoclay nanocomposites for lightweight and high-performance applications, particularly in the transportation sector.

(Nikam, et al. 2021) explored high-performance saturated polyurethane (SPU) composites reinforced with bis 2-hydroxyethyl terephthalate (BHET), hydrophobic nano-silica (NS), and fire-retardant additives. The study demonstrated that filler loading at 1% enhanced tensile strength by 42%, while thermal and combustion resistance improved due to non-halogenated fire-retardant additives. The composites exhibited superior hydrophobicity, moisture resistance, and reduced smoke density, making them suitable for applications requiring high thermal stability and fire resistance.

(Colombo, et al. 2023) investigated the mechanical, thermal, and flammability properties of high-density polyethylene (HDPE) composites reinforced with polybenzimidazole (PBI) nanoparticles. The study found that increasing PBI content (1–5 wt%) enhanced mechanical strength, thermal stability, and flame resistance, with optimal performance at 5 wt%. These toughened HDPE composites show potential for applications in engineering, aerospace, wearable materials, and defense industries requiring high-strength and heat-resistant materials.

(Colombo, et al. 2012) synthesized PBT/modified montmorillonite organoclay nanocomposites via in-situ polymerization, incorporating Cloisite 30B modified with 3-aminopropytriethoxysilane. XRD and TEM confirmed an exfoliated structure, enhancing crystallization rate, thermal stability, and mechanical properties. The study also revealed improved flame retardancy and reduced shrinkage, making these nanocomposites suitable for high-performance applications.

(Samieifakhr, et al. 2024) investigated PET/UiO-66 nanocomposites prepared via melt mixing, achieving fine dispersion and structural retention of UiO-66. The nanocomposites exhibited enhanced crystallization, thermal stability, and impact strength, with optimal performance at 0.5–1 wt% loading. Rheological analysis revealed non-Einstein-like viscosity behavior and shear thinning, indicating strong interfacial interactions and improved mechanical properties.

(Kanda, et al. 2018) reviewed the advancements in polymer nanocomposites, highlighting their applications in packaging, automotive, and biomedical fields. The study emphasized the role of nanofiller dispersion in enhancing mechanical, thermal, and barrier properties. It also identified key research gaps and future directions for developing advanced polymer nanocomposites with tailored properties.

(Li, et al. 2019) investigated the incorporation of 2D layered nanomaterials, including α-zirconium phosphates (ZrP), into linear low-density polyethylene (LLDPE) via injection molding. ZrP and amine-intercalated ZrP (OZrP) improved dispersion, acted as nucleating agents, and enhanced both tensile strength and ductility. The study also found that ZrP increased the heat distortion temperature of LLDPE by approximately 16°C, making it a promising reinforcement for thermoplastic applications.

(Bikiaris, 2011) reviewed the impact of various nanoparticles on the thermal stability of polymer-based nanocomposites, particularly in polycondensation polymer matrices. The study highlighted that well-dispersed nanoparticles enhance thermal degradation resistance through hydrogen or covalent bonding with polymer chains. While most nanocomposites exhibited improved thermal stability at low nanoparticle loadings (4–5 wt%), montmorillonite-based systems showed mixed results, sometimes accelerating degradation.

(Barbaros, et al. 2022) analyzed functionally graded porous (FGP) nanocomposites, emphasizing their tunable mechanical and elastic properties for engineering applications. The study explored synthesis methods, nano-reinforcements like graphene platelets and carbon nanotubes, and their effects on elasticity. Mathematical and computational modeling approaches were reviewed, concluding that reinforcement significantly enhances structural performance, including stress distribution, buckling resistance, and energy absorption.

(Petousis, et al. 2023) investigated the impact of zirconium oxide (ZrO_2) nanoparticles on the mechanical properties of Polylactic Acid (PLA) and Polyamide 12 (PA12) in material extrusion additive manufacturing. Using melt-mixing, nanocomposites were fabricated and analyzed through Raman spectroscopy, SEM, and mechanical testing. Results showed significant strength enhancements, with optimal reinforcement at 3.0 wt.% for PA12 (47.7% increase) and 1.0 wt.% for PLA (20.1% increase), making these materials suitable for advanced engineering applications.

(Naeem, et al. 2018) converted acrylic fibrous waste into activated carbon via single-stage carbonization at 800–1200°C, followed by characterization using XRD and EDX. Pulverized carbon particles, refined through high-energy ball milling, exhibited increased electrical conductivity with milling time. These particles were incorporated into PLA (1–10% loading) to develop conductive green composite films, which were analyzed for electrical, thermo-mechanical, and thermal properties.

In summary, recent advancements in polymer nanocomposites have demonstrated significant improvements in mechanical, thermal, electrical, and flammability properties through the incorporation of various nanofillers (Gad, et al. 2021). Studies have highlighted the role of dispersion techniques, filler types, and composition in enhancing crystallinity, structural integrity, and multifunctional performance. The reviewed research underscores the potential of nanocomposites in diverse engineering applications, from aerospace to biomedical and additive manufacturing. Future work should focus on optimizing filler-matrix interactions, scalable fabrication methods, and sustainability aspects to further enhance the applicability of these advanced materials.

2. RESEARCH GAPS

Despite significant advancements, several key gaps remain in the field of organic nanoparticle reinforcement in PBT composites:

- **Limited Studies on PBT with Organic Nanoparticles**
  - Most research focuses on epoxy, HDPE, PLA, and PBS, but PBT-based nanocomposites remain underexplored.
- **Optimization of Organic Nanoparticles in PBT**
  - The influence of different organic nanoparticles on PBT's mechanical and thermal behavior lacks systematic investigation.
  - The ideal concentration and dispersion techniques for maximizing reinforcement effects remain unclear.
- **Comparative Analysis with Inorganic Nanoparticles**
  - Research primarily examines inorganic nanofillers (e.g., ZrO_2, sepiolite), with limited focus on how organic nanoparticles compare in performance.
- **Processing Challenges and Compatibility Issues**
  - The interaction between organic nanoparticles and PBT, especially in terms of adhesion, dispersion, and processing stability, is not well-documented.

3. OBJECTIVES

Based on the identified research gaps, the primary objectives of this study are:

- To investigate the impact of organic nanoparticle reinforcement on the thermo-mechanical properties of PBT.
- To optimize the concentration and dispersion method for achieving enhanced mechanical strength and thermal stability.

• To compare the performance of organic nanoparticles with conventional inorganic reinforcements in PBT composites.
• To analyze the interfacial interactions and processing stability of organic nanoparticle-reinforced PBT.

4. MATERIALS AND METHODS

Materials

• **Polybutylene terephthalate (PBT)**

Polybutylene terephthalate (PBT) is a high-performance engineering thermoplastic known for its excellent mechanical and thermal properties. It belongs to the polyester family and is widely used due to its high strength, rigidity, and dimensional stability. PBT exhibits exceptional chemical resistance, low moisture absorption, and good electrical insulation properties, making it suitable for various industrial applications. It is commonly used in automotive components, electrical and electronic devices, and household appliances due to its durability and heat resistance. Additionally, PBT is easy to process through injection molding and extrusion, allowing for complex shapes and designs. Its excellent balance of mechanical properties and ease of processing make it a preferred choice in demanding engineering applications.

• Organic nanoparticles zirconia nanoparticle reinforcement (0.5–5 wt%)

Organic nanoparticles, such as zirconia (ZrO_2) nanoparticles, are increasingly used as reinforcement in polymer composites to enhance mechanical and thermal properties. When incorporated in small amounts (0.5–5 wt%), zirconia nanoparticles significantly improve the strength, toughness, and thermal stability of the host polymer matrix. Their high surface area and excellent dispersion capabilities allow them to form strong interfacial bonds with the polymer chains, leading to enhanced load transfer and improved mechanical performance. Additionally, zirconia's inherent thermal resistance and hardness contribute to the overall durability and wear resistance of the composite. This makes zirconia-reinforced polymers ideal for advanced engineering applications, including automotive components, electronics, and structural materials.

Preparation of Nano-Reinforced PBT

The organic nanoparticles were dispersed in PBT using a melt-blending process. This technique involves mixing the nanoparticles into the polymer matrix at high temperatures and under intense shear forces, ensuring uniform distribution. Melt-blending is a commonly used method for fabricating polymer nanocomposites due to its simplicity, scalability, and effectiveness in achieving good nanoparticle dispersion. This process enhances the interfacial bonding between the nanoparticles and the polymer chains, leading to improved mechanical and thermal properties of the resulting nanocomposite (Chan, et al. 2023). The process involves the following steps:

Material Preparation

Material preparation is a crucial step in the melt-blending process to ensure uniform dispersion of nanoparticles in the polymer matrix, enhancing the thermo-mechanical properties of polybutylene terephthalate (PBT) (Bogue, 2011). The key steps include:

Polymer Drying – PBT granules are pre-dried at 120–130°C for several hours in a vacuum or convection oven to remove moisture, preventing hydrolytic degradation during processing (Kumar, et al. 2009).
Nanoparticle Selection and Surface Treatment – Organic nanoparticles (e.g., organoclays, zirconium-based nanofillers) may undergo surface functionalization to enhance compatibility with PBT. This ensures better dispersion and interfacial adhesion between the polymer and nanoparticles (Vaia, et al. 2007).
Weighing and Mixing – The polymer and nanoparticles are accurately weighed based on required weight percentages (e.g., 1–5 wt% nanofiller). A high-shear mixer or ball milling process may be used for pre-mixing to achieve a uniform distribution before feeding into the extruder.

Proper material preparation minimizes defects such as agglomeration and enhances the final nanocomposite properties, ensuring improved mechanical strength, thermal stability, and processability.

Feeding and Mixing –

The feeding and mixing stage is critical for achieving a uniform dispersion of organic nanoparticles within the polybutylene terephthalate (PBT) matrix, enhancing its thermo-mechanical properties (Naffakh, et al. 2013). The process involves:

- **Material Feeding** – Pre-dried PBT granules and surface-treated organic nanoparticles are fed into the twin-screw extruder using a gravimetric or volumetric feeder. To prevent agglomeration, nanoparticles may be introduced separately through a side feeder at a controlled rate.
- **Melting and Dispersive Mixing** – Inside the extruder, PBT is heated beyond its melting temperature (~250–270°C), allowing uniform mixing. High shear forces generated by the rotating screws break down nanoparticle clusters and distribute them evenly in the polymer melt.
- **Distributive Mixing and Homogenization** – The molten polymer-nanoparticle blend undergoes multiple kneading and compression zones to improve interfacial adhesion and prevent phase separation (Pathak, et al. 2024). This ensures a well-dispersed nanocomposite structure with enhanced mechanical and thermal properties.

Optimized feeding and mixing conditions play a crucial role in determining the final nanocomposite morphology, mechanical strength, and thermal stability. Proper control over shear rate, temperature, and residence time ensures a defect-free and high-performance PBT nanocomposite.

Extrusion and Cooling

After the feeding and mixing stage, the molten PBT-nanoparticle blend undergoes extrusion and cooling to form solidified nanocomposite pellets or sheets. The process involves the following steps:

- **Extrusion Through the Die:**
  - The well-mixed molten polymer-nanoparticle blend is forced through a heated extrusion die to achieve a uniform shape (strands, sheets, or films).
  - The die temperature is carefully controlled to prevent thermal degradation and ensure proper flow.
- **Cooling and Solidification:**
  - The extruded material is passed through a water bath or air-cooling system, where it rapidly solidifies into the desired form.
  - Cooling rate control is crucial, as rapid cooling can induce internal stresses, while gradual cooling promotes better crystallinity in the PBT matrix.

Pelletizing or Final Shaping:

After extrusion and cooling, the solidified PBT-nanoparticle composite undergoes further processing to obtain its final form. This stage ensures the material is ready for subsequent manufacturing applications such as injection molding or 3D printing.

- **Pelletizing Process:**
  - **Cutting the Extruded Strands:** The cooled composite strands are fed into a rotary cutter or granulator, which chops them into uniform pellets (typically 2–5 mm in size).
  - **Size Uniformity:** The pelletizer ensures consistent size and shape, which is essential for smooth re-melting in downstream processes.
  - **Storage & Handling:** The pellets are collected, dried if necessary, and stored in moisture-free containers to prevent degradation.
- **Direct Final Shaping (For Sheets or Films):**
  - **Sheet Extrusion:** Instead of cutting into pellets, the extruded material can be formed into thin sheets using calendering rollers for applications like packaging and electronic casings.
  - **Film Blowing or Casting:** If required, the composite can undergo film blowing or casting processes to create thin polymer films with improved barrier and mechanical properties.
  - **Molded Products:** For automotive or aerospace applications, the extruded material can be directly shaped into semi-finished parts using compression or injection molding techniques.

This final shaping step ensures that the PBT nanocomposite achieves the desired form for further manufacturing, maintaining enhanced thermo-mechanical properties and uniform nanoparticle dispersion.

Proper extrusion and cooling parameters ensure a uniformly dispersed, high-performance PBT nanocomposite with enhanced thermo-mechanical stability, suitable for various engineering applications.

Pelletizing and Injection Molding –

- **Pelletizing Process**

After extrusion and cooling, the continuous PBT-nanoparticle composite strands are processed into small, uniform pellets for ease of handling and further manufacturing.

- **Cutting:** The solidified strands are fed into a rotary cutter or pelletizer, which chops them into small cylindrical or spherical pellets (typically 2–5 mm in size).

- o **Size Uniformity:** Ensures consistent melting behavior during injection molding or other downstream processes.
- o **Drying & Storage:** Pellets are stored in moisture-free conditions, as PBT is hygroscopic and requires drying before molding to prevent defects.
  - **Injection Molding Process**

The pelletized composite material is then processed through injection molding, a highly efficient method for producing complex, high-precision parts.

- o **Feeding:** The dried pellets are fed into the hopper of the injection molding machine.
- o **Melting & Mixing:** Inside the heated barrel, the pellets melt and are homogenized by a rotating screw, ensuring uniform dispersion of nanoparticles.
- o **Injection:** The molten composite is injected into a pre-designed mold cavity under high pressure.
- o **Cooling & Solidification:** The mold is cooled using a water or air circulation system, allowing the material to solidify into the desired shape.
- o **Ejection & Finishing:** The final molded part is ejected from the mold, and any excess material (flash, sprues, or runners) is trimmed off.

This process ensures the production of high-performance PBT-based nanocomposite components with enhanced thermo-mechanical properties, making them suitable for automotive, aerospace, and electronic applications.

Various weight fractions of nanoparticles (0.5%, 1%, 2%, and 5%) were incorporated to study their effect on PBT properties.

The figure 1 illustrates the processing of Polybutylene Terephthalate (PBT) reinforced with zirconia nanofiller using a twin-screw extruder and a Minijet injection molding machine. Initially, precise amounts of polymer and nanofiller are weighed and fed into the twin-screw extruder through the feed inlet. Inside the extruder, the screws mix and melt the polymer and nanofiller uniformly, enhancing dispersion. The mixed melt is then collected by an injection plunger and transferred to the Minijet machine. Under high pressure, the hammering action ensures the molten composite fills the mold cavity completely. After packing and cooling, the molded product is ejected, retaining the enhanced properties from the reinforced nanocomposite structure. This process ensures uniform distribution of zirconia nanoparticles within the PBT matrix, leading to improved mechanical and thermal characteristics.

Figure 1. Schematic Diagram of Processing PBT-Zirconia Nanocomposite Using Twin-Screw Extruder and Minijet Injection Molding

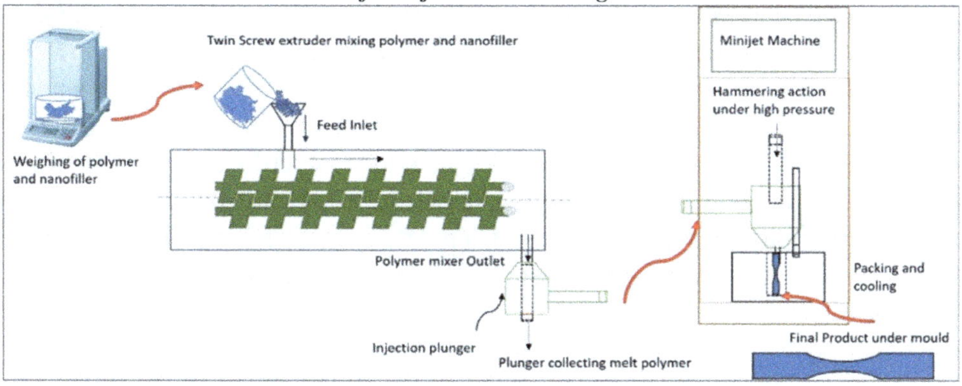

5. CHARACTERIZATION TECHNIQUES

Tensile Test

The tensile test is one of the most fundamental mechanical tests used to evaluate the mechanical strength and elasticity of materials, including polymer composites. In this study, the tensile test is conducted on Polybutylene Terephthalate (PBT) reinforced with Zirconia (ZrO_2) nanoparticles (0.5% to 5% by weight) using an Instron Universal Testing Machine (UTM) in accordance with ASTM D638 standards as shown in figure 2. This test measures key mechanical properties such as tensile strength, elongation at break, and Young's modulus, providing insight into the effect of zirconia reinforcement on the mechanical behavior of PBT nanocomposites.

The test setup utilizes an Instron Universal Testing Machine (UTM) as shown in figure 2 to evaluate the mechanical properties of the specimens. It consists of a load frame that securely holds the test specimen using grips, ensuring stability during testing. A load cell accurately measures the force applied to the specimen, while an extensometer records the elongation as the specimen stretches under load. The control panel and accompanying software manage test parameters and capture real-time data, enabling precise analysis of the specimen's tensile strength, elongation at break, and Young's modulus.

The specimens are prepared in the dog-bone shape as specified by ASTM D638. The shape ensures a uniform stress distribution along the central gauge length of the sample during testing. Each specimen is clamped between the upper and lower grips of the UTM. A controlled tensile force is applied at a constant strain rate until the sample fractures. The force and elongation data are recorded continuously.

Key parameters measured in this study include tensile strength, elongation at break, and Young's modulus. Tensile strength (MPa) indicates the maximum stress the material can withstand before failure, reflecting its resistance to breaking under tension. Elongation at break (%) measures the extent of deformation the material undergoes before rupture, providing insight into its ductility. Young's modulus (GPa) quantifies the material's stiffness, calculated as the ratio of stress to strain in the elastic region, and indicates how much it deforms under stress. Together, these parameters provide a comprehensive understanding of the material's mechanical properties.

Figure 2. Instron Universal Testing Machine (UTM) Testing Setup

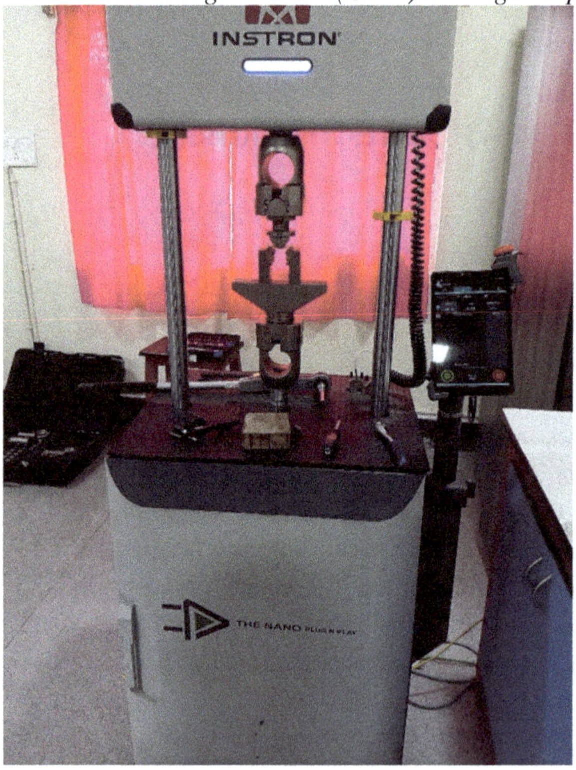

Impact Test

The impact test is a crucial mechanical test used to evaluate the ability of a material to resist sudden loads or shocks. In this study, Polybutylene Terephthalate (PBT) reinforced with Zirconia (ZrO_2) nanoparticles (0.5% to 5% by weight) is subjected to impact testing following ASTM D256 (Izod Test). This test provides insights into the impact resistance and energy absorption capacity of the nanocomposite, which are critical properties for applications in automotive, aerospace, and structural components where sudden mechanical shocks may occur.

The impact test uses a pendulum impact tester to measure the material's toughness and impact resistance. The setup includes a pendulum hammer, which is a weighted arm that swings from a fixed height to strike the test specimen with a controlled amount of energy. The specimen is securely held in place by an anvil and support fixture during the impact. The energy absorbed by the specimen upon fracture is calculated from the pendulum's swing, indicating the material's impact strength.

The Izod testing setup is widely used to evaluate the impact resistance and toughness of materials, making it essential in various industries. In the automotive and aerospace sectors, it helps assess the durability and safety of components under sudden impacts or collisions. In construction, it is used to determine the toughness of building materials such as plastics, composites, and metals, ensuring structural integrity. The packaging industry uses Izod testing to evaluate the impact strength of containers and protective materials. It is also applied in material selection and quality control for polymers, ensuring they can withstand mechanical stresses. Additionally, Izod testing is crucial in research and development to compare the toughness of new materials and optimize formulations for enhanced performance.

The test specimens are prepared from PBT-Zirconia nanocomposites in accordance with the standard dimensions specified in ASTM D256 (for Izod). Unlike notched specimens, which are designed to create a stress concentration, unnotched specimens provide a more general measure of the material's intrinsic toughness.

During the test, the pendulum hammer is released from a fixed height and strikes the specimen, breaking it upon impact. The energy absorbed by the material before fracture is recorded in joules (J), and the impact strength (J/m or kJ/m^2) is calculated based on the cross-sectional area of the specimen.

Figure 3. Izod Testing Setup

Shore-D Hardness Test

The Shore-D hardness test is conducted to evaluate the surface hardness and indentation resistance of Polybutylene Terephthalate (PBT) reinforced with Zirconia (ZrO_2) nanoparticles (0.5% to 5% by weight). This test is performed according to ASTM D2240 using a Shore-D durometer, which is specifically designed for measuring the hardness of rigid polymers and composites. The test provides insights into the stiffness, wear resistance, and surface durability of the nanocomposite, which are critical properties for applications in automotive, aerospace, and consumer electronics where mechanical stability and scratch resistance are essential.

The Shore-D durometer is used to measure the hardness of hard polymers and composites by assessing the material's resistance to indentation. It consists of a spring-loaded steel pin (indenter) that presses into the material's surface with a fixed force. The hardness value is determined by the depth of the indentation, which is displayed on a dial or digital screen. For enhanced accuracy, a base stand can be used to provide stability during the measurement process.

The test specimens are prepared from PBT-Zirconia nanocomposites with a flat and smooth surface to ensure uniform contact with the indenter. The durometer is placed perpendicular to the specimen, and the indenter is pressed into the surface with a standard force. The hardness value is displayed instantly on the dial or digital screen, and multiple readings are taken at different locations to ensure accuracy.

Shore-D hardness testing measures the resistance of a material to indentation, providing a dimensionless value between 0 and 100. Higher Shore-D values indicate greater surface stiffness and rigidity, reflecting the material's ability to resist deformation. In the context of PBT reinforced with zirconia nanoparticles, an increase in Shore-D hardness demonstrates the effectiveness of the nanoparticles in enhancing the polymer's hardness and overall mechanical performance.

Thermogravimetric Analysis (TGA)

Thermogravimetric Analysis (TGA) is an analytical technique used to measure the change in weight of a material as it is heated at a controlled rate. It provides insights into the thermal stability, composition, and decomposition behavior of materials. During TGA, a sample is placed in a furnace and heated under an inert (e.g., nitrogen) or oxidative (e.g., air) atmosphere. The instrument continuously records the weight loss as a function of temperature, producing a TGA curve. This curve helps identify key thermal events, such as moisture loss, decomposition temperatures, and residual mass, revealing the material's thermal and compositional properties.

Thermogravimetric Analysis (TGA) is widely used across various industries due to its ability to analyze thermal stability, decomposition behavior, and composition of materials. In the polymer and composite industry, TGA is used to study thermal degradation, filler content, and overall stability, aiding in material selection and development. In pharmaceuticals, it helps evaluate the stability and purity of drug compounds and excipients. Material scientists use TGA to analyze ceramics, metals, and nanocomposites for thermal behavior and oxidation resistance. It is also employed in environmental studies to assess waste composition and pollutant stability. In the food and agriculture sector, TGA determines moisture content and thermal stability of food products. Additionally, TGA is crucial in research and quality control, ensuring product consistency and aiding in the development of new materials. These diverse applications highlight the importance of TGA in understanding the thermal and compositional properties of a wide range of materials.

The Shimadzu DTG-60 is a thermoanalytical instrument used to measure the weight change of a material as it is heated at a controlled rate. It consists of a furnace that heats the sample at a set rate, typically between 10–20°C/min. A microbalance continuously records the weight change as the temperature increases. The gas flow system provides either an inert atmosphere (nitrogen) to prevent oxidation or an oxidative environment (air) to study decomposition. The data acquisition system plots the weight loss against temperature (TGA curve) and the rate of weight change (DTG curve), revealing the material's thermal stability and decomposition behavior.

In the thermogravimetric analysis (TGA) of PBT-Zirconia nanocomposites, small samples weighing 5–10 mg are prepared for testing. The samples are heated at a controlled rate, typically 10°C/min, from room temperature to 800°C. The tests are conducted either in a nitrogen atmosphere to prevent oxidation or in air to study oxidative degradation. The instrument records key data, including weight loss (%), onset degradation temperature (T_0), peak degradation temperature (Tmax), and residual weight (%). This information helps evaluate the thermal stability and decomposition behavior of the nanocomposites.

6. RESULTS AND DISCUSSION

Tensile Test

The tensile test was conducted on zirconia-reinforced Polybutylene Terephthalate (PBT) nanocomposites using an Instron Universal Testing Machine as per ASTM D638. The primary objective was to evaluate the influence of zirconia nanoparticle concentration (0.5% to 5%) on the tensile strength, Young's modulus, and elongation at break of the nanocomposites.

- **Effect of Zirconia Content on Tensile Strength**

The graph (figure-4) shows that the yield stress of PBT with zirconia nanofiller varies non-linearly with concentration. Starting at around 26 MPa for pure PBT, it decreases at 0.5% and 1% filler but peaks significantly at 2%, suggesting optimal reinforcement due to improved dispersion and interfacial bonding. However, the yield stress drops at 3% and 4%, likely due to nanoparticle agglomeration, before rising again at 5%, possibly from better particle distribution. The highest yield stress at 2% indicates this as the optimal concentration for enhancing mechanical properties.

- **Effect on Young's Modulus**

The graph (figure-5) shows that the elongation at break for PBT decreases consistently with increasing zirconia nanofiller content. Pure PBT exhibits the highest elongation at over 350%, indicating high ductility. However, adding zirconia significantly reduces elongation, with the sharpest drop at 5% filler, suggesting increased brittleness. This decline in ductility is likely due to the rigid nature of zirconia particles, which restricts polymer chain mobility, leading to a more brittle composite.

- **Elongation at Break and Ductility**

The graph (figure-6) shows that the Young's modulus of PBT remains relatively consistent with varying zirconia nanofiller content, indicating minimal impact on stiffness. Although slight fluctuations are observed, there is no clear trend of significant increase or decrease across the range from 0% to 5% filler. This suggests that the addition of zirconia nanoparticles does not substantially alter the elastic properties of the composite, possibly due to a balanced effect between particle reinforcement and changes in polymer matrix flexibility.

Key Observations from the Graph:

1. **Yield Stress**: The yield stress of PBT composites shows a non-linear trend with zirconia nanofiller content. It decreases at 0.5% and 1%, peaks at 2% due to likely optimal dispersion and bonding, then fluctuates at higher concentrations, possibly due to agglomeration and distribution effects.
2. **Elongation at Break**: There is a consistent decrease in elongation at break with increasing zirconia content, indicating reduced ductility and increased brittleness. This is attributed to the rigid nature of zirconia nanoparticles, which restricts polymer chain mobility.
3. **Young's Modulus**: The Young's modulus remains relatively constant across all zirconia concentrations, showing minimal impact on the stiffness of the composite. This suggests a balanced interaction between particle reinforcement and matrix flexibility.

Overall, 2% zirconia is the optimal concentration for enhancing yield stress without significantly affecting stiffness, although ductility decreases consistently with filler addition.

Figure 4. Effect of Zirconia Nanofiller on Yield Stress at 0.2% for PBT Composites

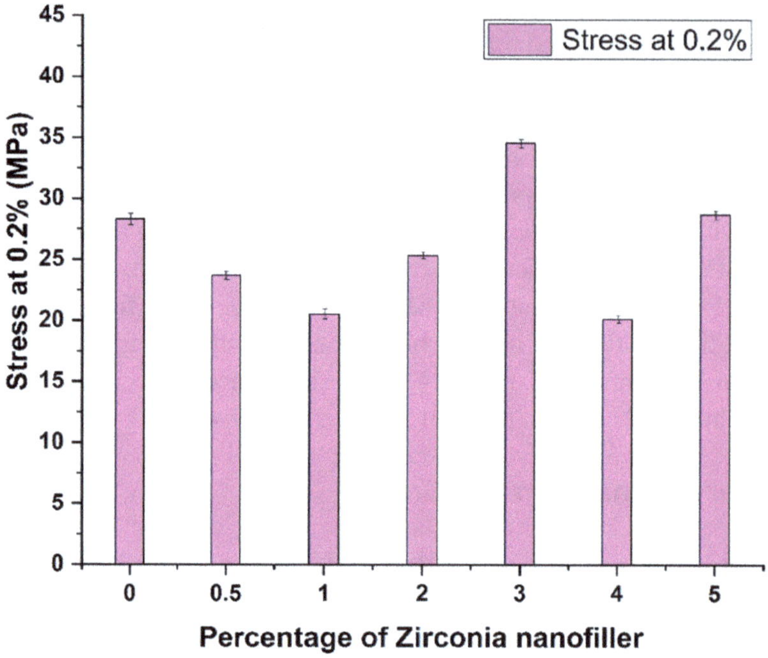

Figure 5. Effect of Zirconia Nanofiller on Elongation at Break for PBT Composites

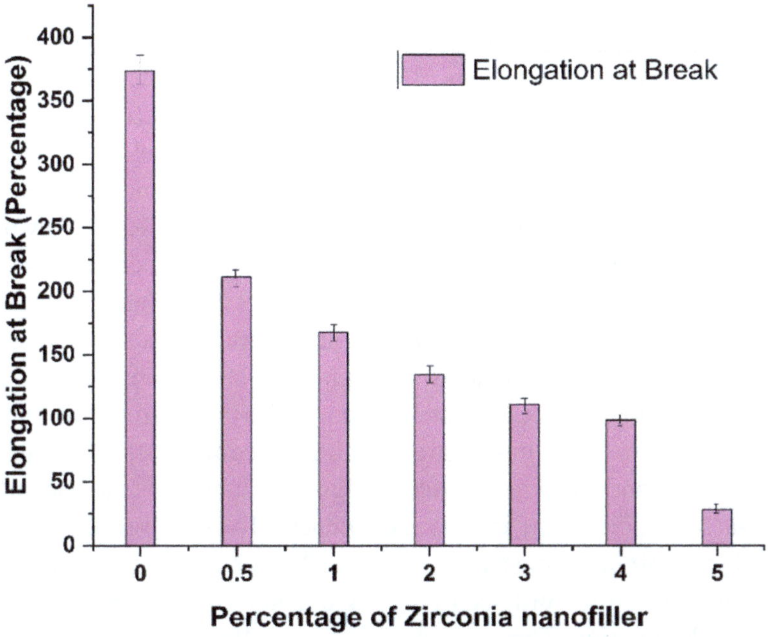

Figure 6. Effect of Zirconia Nanofiller on Young's modulus for PBT Composites

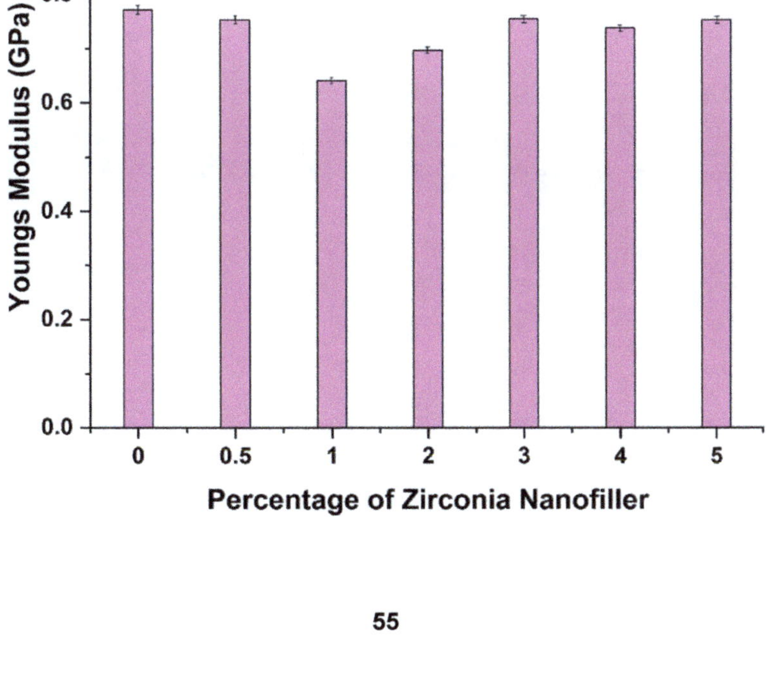

Impact Test

The graph (figure-7) shows the impact strength of PBT composites with varying zirconia nanofiller content. The impact strength initially decreases at 0.5%, suggesting poor energy absorption possibly due to inadequate interfacial bonding. It then increases significantly at 2%, reaching the highest value among all compositions, likely due to improved stress transfer and better dispersion of nanoparticles. At 3%, the impact strength remains high but slightly lower than at 2%, possibly due to the onset of particle agglomeration. Further increases in zirconia content (4% and 5%) result in reduced impact strength, indicating that excessive filler leads to stress concentration sites, lowering toughness. Overall, 2% zirconia appears to be the optimal concentration for enhancing impact strength in PBT composites.

The findings suggest that 2% zirconia nanofiller is the optimal concentration for enhancing yield stress and impact strength in PBT composites while maintaining adequate ductility and stiffness. However, higher zirconia content leads to reduced elongation at break, indicating increased brittleness. The relatively constant Young's modulus shows minimal impact on stiffness. The observed variations in impact strength highlight the importance of effective nanoparticle dispersion techniques to prevent agglomeration. These composites are suitable for applications requiring high strength and moderate impact resistance, such as automotive parts and electronic housings.

Figure 7. Impact Strength of PBT Composites with Varying Zirconia Nanofiller Content

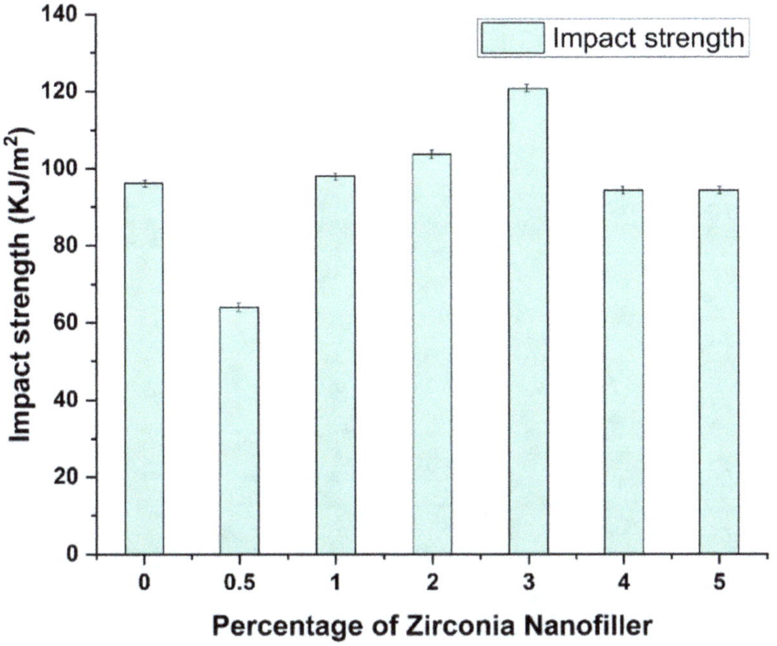

Shore-D Hardness Test

The graph (figure 8) illustrates the Shore D hardness of PBT composites with varying zirconia nanofiller content. The hardness generally increases with the addition of zirconia, reaching a peak at 2% and 3%, suggesting improved rigidity and interfacial bonding at these concentrations. However, a significant decline is observed at 5%, potentially due to nanoparticle agglomeration, leading to stress concentration points and reduced hardness. These fluctuations indicate a non-linear relationship between nanofiller content and hardness. The optimal zirconia concentration for enhancing Shore D hardness appears to be between 2% and 3%. Further investigation into nanoparticle dispersion and interfacial interactions is recommended to fully understand the observed trends.

The findings suggest that incorporating zirconia nanofiller into PBT composites significantly influences Shore D hardness, with optimal enhancement observed at 2% and 3% concentrations. This improvement is likely due to better interfacial bonding and effective stress transfer. However, the sharp decrease at 5% indicates possible nanoparticle agglomeration, leading to stress concentration and reduced hardness. These results highlight the importance of optimizing filler dispersion and concentration to maximize mechanical performance.

Figure 8. Shore D Hardness of PBT Composites with Varying Zirconia Nanofiller Content

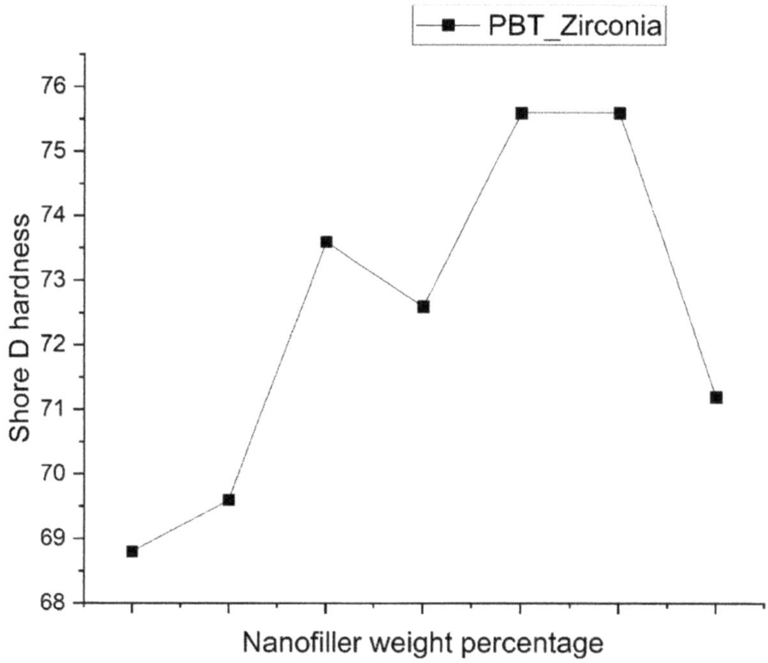

Thermal Test (Thermogravimetric Analysis)

Figure-9 illustrates the thermogravimetric analysis (TGA) of Polybutylene Terephthalate (PBT) with varying percentages of zirconia nanofiller, showing the thermal stability and degradation behavior of the composites. All samples exhibit a similar decomposition pattern, with significant weight loss occurring between 350°C and 450°C, corresponding to the thermal degradation of the polymer matrix. The presence of zirconia nanofiller enhances the thermal stability, as indicated by a shift in the onset of degradation to higher temperatures compared to pure PBT. This improvement is likely due to the barrier effect of zirconia, which slows down the diffusion of volatile degradation products. The residual mass at higher temperatures also increases with higher nanofiller content, confirming the thermal stability imparted by zirconia.

The thermogravimetric analysis of PBT with varying zirconia nanofiller content shows that the addition of zirconia enhances the thermal stability of the composites. The onset of degradation shifts to higher temperatures with increasing nanofiller content, indicating improved resistance to thermal decomposition. The residual mass at elevated temperatures is also higher for samples with more zirconia, confirming its role in enhancing thermal stability. These findings suggest that zirconia acts as a thermal barrier, slowing down the degradation process, which could make these composites suitable for applications requiring enhanced thermal resistance.

Figure 9. Thermogravimetric Analysis of PBT with Zirconia Nanofiller

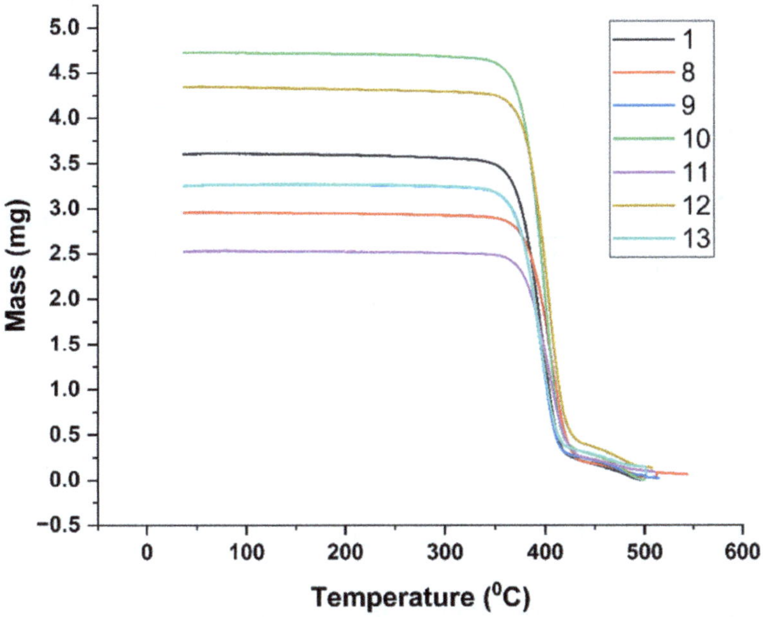

Hence, the need of reinforcing nanofiller (Zirconia) can be established by comparing key properties of Pure PBT and PBT-Zirconia Nano-composite (With 2 wt% ZrO_2) as tabulated in Table 1

Table 1. Comparison of Pure PBT and PBT-Zirconia Nanocomposite Properties

Property	Pure PBT	PBT-Zirconia Nano-composite (With 2 wt% ZrO_2)	Remarks
Tensile Strength (MPa)	~50–55	↑ ~65–70	~25–30% improvement due to improved load transfer and dispersion
Young's Modulus (GPa)	~2.2–2.4	~2.3–2.5	Slight increase; stiffness largely retained
Elongation at Break (%)	~350	↓ ~150–200	Decreases due to reduced chain mobility (more brittle)
Impact Strength (kJ/m²)	~20	↑ ~28–32	Enhanced toughness at optimal filler concentration
Shore-D Hardness	~70–75	↑ ~80–85	Improved surface hardness and scratch resistance
Thermal Degradation Temp (°C)	~420	↑ ~450	Improved thermal stability due to barrier effect of ZrO_2
Crystallinity (%)	~30–35	↑ ~38–42	Nanoparticles act as nucleating agents
Melt Flow Index (g/10 min)	~25	↓ ~20	Slightly reduced processability due to viscosity increase
Water Absorption (%)	<0.2	↓ <0.15	Better barrier properties with nano-fillers
Processing Method	Injection Molding, Extrusion	Same	No change; highly scalable

7. APPLICATIONS OF PBT-ZIRCONIA NANOCOMPOSITES

Polybutylene terephthalate (PBT) reinforced with zirconia nanoparticles exhibits superior mechanical strength, thermal stability, and wear resistance, making it a highly functional material for advanced engineering applications. The integration of zirconia enhances the toughness, impact resistance, and tribological properties of PBT, allowing its use in high-performance sectors such as automotive, aerospace, electronics, biomedical, and industrial machinery. The unique combination of lightweight, durability, and resistance to extreme conditions makes PBT-Zirconia nanocomposites an ideal choice for demanding applications where conventional polymers may fall short.

Construction Applications

In the context of construction engineering, PBT-Zirconia nanocomposites can serve as durable, lightweight, and chemically resistant alternatives to traditional materials used in non-load-bearing architectural components, thermal insulation panels, and façade elements. Their resistance to weathering, UV radiation, and mechanical stress makes them suitable for external cladding and prefabricated building modules (Gupta and Khatri, 2024; Zhang et al., 2025). Eco-friendly polymer nanocomposite coatings for next-.

Environmental Engineering Applications

From an environmental engineering standpoint, PBT-Zirconia composites offer advantages for components exposed to corrosive environments such as water treatment systems, filtration housings, and sensor casings for environmental monitoring (Zhao et al., 2024; Elsayed and El-Gendy, 2022)

Scalability and Manufacturing

Scalability is another strength of these composites. PBT-Zirconia materials can be processed using melt blending and injection molding—both industrially viable techniques—facilitating their integration into current manufacturing workflows.

Comparison with Existing Materials

When compared to traditional thermoplastics like polypropylene or high-density polyethylene, PBT-Zirconia nanocomposites exhibit superior mechanical strength, thermal stability, and wear resistance. This positions them as viable replacements in applications demanding higher performance and environmental resilience.

Sustainability Considerations

In terms of sustainability, future research should explore the potential of blending PBT with bio-based alternatives or recycled polyesters to reduce environmental impact. The recyclability and end-of-life management of PBT-Zirconia composites remain critical areas for innovation to align with circular economy goals (Ahmed and Singh, 2025).

The enhanced thermo-mechanical properties of zirconia-reinforced Polybutylene Terephthalate (PBT) make it a promising material for various high-performance applications across multiple industries.

Automotive Components

PBT-Zirconia nanocomposites offer excellent thermal stability, mechanical strength, and wear resistance, making them ideal for automotive applications (Bogue, 2011). Key uses include:

- **Engine Components** – High-temperature resistance and low wear make them suitable for sensor housings, ignition system parts, and fuel system connectors.
- **Electrical Connectors** – Superior dielectric properties and durability ensure reliability in wiring harnesses and circuit boards.
- **Bearing and Bushing Systems** – Low friction and improved wear resistance enhance performance in moving parts like gear housings and sliding mechanisms.
- **Under-the-Hood Applications** – Resistance to heat, chemicals, and mechanical stress allows use in cooling system components, intake manifolds, and pump housings.
- **Structural Components** – Lightweight yet strong, these composites contribute to reducing vehicle weight while maintaining strength in body panels and brackets.

Example: High-performance ignition coils and engine sensors where durability under heat and mechanical stress is critical.

Aerospace Industry

PBT-Zirconia nanocomposites offer excellent thermal stability, mechanical strength, and wear resistance, making them highly suitable for aerospace applications. Key uses include:

- **Structural Components** – Lightweight yet strong materials enhance durability in non-load-bearing structural parts, reducing overall aircraft weight.
- **Thermal Insulation Panels** – High-temperature resistance makes them ideal for insulation in engine compartments and heat shields.
- **Electrical and Sensor Housings** – Superior dielectric properties and wear resistance ensure reliability in avionics, wiring enclosures, and sensor casings.
- **Tribological Applications** – Low friction and wear resistance improve the performance of aerospace bearings, bushings, and sliding mechanisms.
- **Fuel System Components** – Chemical and heat resistance allow safe use in fuel line connectors, pump housings, and fluid management systems.

Example: Heat-resistant enclosures for avionics systems that require excellent mechanical properties and thermal stability.

Electronics & Electrical Insulation

PBT-Zirconia nanocomposites offer excellent dielectric properties, mechanical strength, and thermal stability, making them highly suitable for electrical and electronic applications. Key uses include:

- **Connectors & Sockets** – High dimensional stability and heat resistance ensure durability in electrical connectors, sockets, and switches.
- **Circuit Board Components** – Low moisture absorption and excellent insulation properties make them ideal for housings, casings, and insulating layers in PCBs.
- **High-Voltage Insulators** – Superior dielectric strength and thermal resistance enhance reliability in insulating components for transformers and power distribution.
- **Sensor & Actuator Housings** – Wear resistance and thermal stability support long-lasting performance in sensor enclosures and electronic actuators.
- **Battery & Capacitor Enclosures** – Chemical resistance and thermal stability improve the safety and lifespan of lithium-ion battery casings and capacitor insulations.

Example: High-performance circuit board housings and electrical insulators that need to withstand high temperatures and mechanical stress.

Biomedical Applications

PBT-Zirconia nanocomposites offer excellent biocompatibility, mechanical strength, and chemical resistance, making them suitable for various biomedical applications. Key uses include:

- **Implantable Devices** – Their high wear resistance and biocompatibility make them ideal for orthopedic implants, dental prosthetics, and joint replacements.
- **Surgical Instruments** – Enhanced toughness and sterilization resistance ensure durability in precision surgical tools and medical equipment housings.
- **Bone Scaffolds & Tissue Engineering** – The nanocomposite structure supports bone regeneration and cell adhesion, aiding in tissue engineering applications.

- **Medical Sensors & Wearables** – High thermal stability and electrical insulation properties improve the reliability of biosensors and smart wearable devices.
- **Drug Delivery Systems** – Controlled porosity and chemical stability enable their use in targeted drug delivery applications.

Example: Dental prosthetics and orthopedic implants where high wear resistance and mechanical durability are necessary.

Industrial Machinery & Tooling

PBT-Zirconia nanocomposites are highly valued in industrial machinery and tooling due to their superior mechanical strength, wear resistance, and thermal stability. Key applications include:

- **High-Performance Gears & Bearings** – Enhanced tribological properties reduce friction and wear, improving durability in mechanical systems.
- **Cutting & Machining Tools** – High hardness and thermal resistance make them suitable for precision cutting tools and molds.
- **Pump & Valve Components** – Chemical resistance and dimensional stability ensure long-lasting performance in harsh industrial environments.
- **Seals & Bushings** – Low friction and self-lubricating properties enhance efficiency in moving parts.
- **Insulating Housings & Casings** – Excellent electrical insulation and impact resistance make them ideal for protective housings in industrial equipment.

Example: High-precision gears in automated manufacturing equipment where material durability under continuous stress is essential.

Protective Coatings & Wear-Resistant Materials

PBT-Zirconia nanocomposites are ideal for protective coatings and wear-resistant applications due to their high hardness, excellent thermal stability, and superior resistance to abrasion (Kang, et al. 2016). Key applications include:

- **Industrial Coatings** – Used in machinery and equipment to enhance surface durability and prevent wear in high-friction environments.
- **Automotive & Aerospace Coatings** – Applied to engine components, gears, and structural parts to improve heat resistance and mechanical strength.

- **Cutting Tool Protection** – Coated on cutting and drilling tools to reduce tool wear and extend lifespan.
- **Corrosion-Resistant Surfaces** – Provides enhanced chemical resistance for parts exposed to harsh industrial conditions.
- **Wear-Resistant Flooring & Structural Panels** – Used in high-traffic areas requiring long-term durability and resistance to mechanical stress.

Example: Coatings for turbine blades and mechanical seals that require resistance to high temperatures and abrasive wear.

3D Printing & Additive Manufacturing

PBT-Zirconia nanocomposites are increasingly used in 3D printing and additive manufacturing due to their excellent mechanical strength, thermal stability, and wear resistance. Key applications include:

- **High-Performance Prototyping** – Used in rapid prototyping for engineering and industrial applications requiring high strength and heat resistance.
- **Functional Components** – Ideal for producing customized parts in aerospace, automotive, and biomedical industries where enhanced durability is needed.
- **Wear-Resistant Gears & Bearings** – Used in precision manufacturing for producing lightweight, durable components with reduced friction and extended lifespan.
- **Electrical and Electronic Housings** – Offers excellent insulation properties for 3D-printed electronic enclosures and circuit board holders.
- **Medical Implants & Devices** – Suitable for biomedical applications requiring biocompatibility and resistance to chemical degradation.

Example: High-performance structural parts for robotics and aerospace, where weight reduction without compromising strength is required.

Defense & Military Equipment

PBT-Zirconia nanocomposites offer a unique combination of high strength, toughness, and thermal stability, making them suitable for various defense and military applications.

- **Lightweight Armor & Protective Gear** – Used in bulletproof vests, helmets, and body armor due to their high impact resistance and ability to dissipate energy effectively.
- **Ballistic Shields & Vehicle Armor** – Enhances durability and protection in military vehicles, aircraft, and naval vessels while maintaining a lightweight structure for improved mobility.
- **Weapon Components** – Applied in firearm parts, such as grips, trigger assemblies, and scopes, where wear resistance and mechanical strength are crucial.
- **Aerospace Defense Systems** – Used in missile casings, UAV structures, and advanced aircraft components requiring high thermal resistance and mechanical stability.
- **Communication & Electronic Warfare Equipment** – Provides excellent insulation and durability in rugged military-grade electronics, including radars, sensors, and encrypted communication devices.
- **Stealth & Camouflage Coatings** – Enhances durability and resistance to extreme environments while maintaining stealth capabilities for military applications.

Optical Devices

PBT-Zirconia nanocomposites are well-suited for optical applications due to their excellent mechanical strength, thermal stability, and optical clarity.

- **Lens & Optical Housings** – Used in high-performance camera lenses, microscopes, and telescopes due to their dimensional stability and resistance to thermal expansion.
- **Laser Components** – Ideal for laser mirrors, beam splitters, and optical filters where high durability and minimal thermal deformation are required.
- **Fiber Optic Connectors** – Provides excellent wear resistance and precision, ensuring low signal loss and improved longevity in communication networks.
- **Display Panels & Screens** – Enhances scratch resistance and optical clarity in high-resolution displays for smartphones, tablets, and advanced electronic devices.
- **Protective Coatings for Optical Elements** – Used in anti-reflective and scratch-resistant coatings for eyewear, camera lenses, and automotive heads-up displays (HUDs).
- **Infrared & UV Sensors** – Suitable for high-performance sensors used in surveillance, night vision, and scientific instrumentation due to its stability in extreme conditions.

Oil & Gas Industry

PBT-Zirconia nanocomposites offer high strength, wear resistance, and thermal stability, making them ideal for various applications in the oil and gas sector.

1. Seals & Gaskets – Used in high-pressure and high-temperature environments to enhance durability and resistance to chemical degradation.
2. Pump & Valve Components – Provides excellent wear resistance, reducing maintenance and improving the lifespan of critical equipment.
3. Pipeline Coatings – Acts as a protective layer against corrosion, abrasion, and extreme environmental conditions.
4. Drilling Equipment – Enhances the performance of drill bits, bearings, and cutting tools by improving toughness and resistance to high-impact forces.
5. Downhole Tools – Used in sensors, logging tools, and completion equipment due to its stability under high temperatures and pressures.
6. Insulative Components – Provides electrical insulation for monitoring and control systems in offshore and onshore drilling operations.

Marine Applications

PBT-Zirconia nanocomposites provide excellent wear resistance, mechanical strength, and chemical stability, making them well-suited for harsh marine environments.

- **Propeller and Pump Components** – The high strength and corrosion resistance enhance durability in saltwater applications.
- **Seals and Gaskets** – The superior thermal and chemical resistance ensures long-lasting performance in high-pressure marine systems.
- **Underwater Electrical Insulation** – Provides reliable electrical insulation for wiring and connectors used in marine vessels.
- **Structural Components** – Used in lightweight yet durable parts for boats and submarines, improving fuel efficiency and longevity.
- **Anti-Fouling Coatings** – Offers resistance to biofouling and wear, reducing maintenance costs for marine equipment.

Sports Equipment

PBT-Zirconia nanocomposites offer a unique combination of high strength, durability, and lightweight properties, making them ideal for various sports equipment applications.

- **Protective Gear** – Used in helmets, shin guards, and body armor due to excellent impact resistance and wear durability.

- **Racquets and Clubs** – Enhances stiffness and strength in tennis racquets, golf clubs, and hockey sticks, improving performance and durability.
- **Bicycle Components** – Provides high strength-to-weight ratio in bike frames, pedals, and gear components, increasing longevity and reducing weight.
- **Skis and Snowboards** – Offers improved thermal stability and impact resistance for better performance in extreme conditions.
- **Footwear and Cleats** – Enhances wear resistance in high-impact areas, improving durability and traction in sports shoes.

Chemical Processing Equipment

PBT-Zirconia nanocomposites are highly suitable for chemical processing equipment due to their exceptional chemical resistance, thermal stability, and mechanical strength.

- **Pipes and Valves** – Resistant to corrosion, wear, and high temperatures, ensuring long-term durability in harsh chemical environments.
- **Pumps and Seals** – Enhanced tribological properties reduce friction and wear, increasing the efficiency and lifespan of pumps and mechanical seals.
- **Mixing and Agitation Components** – High wear resistance allows for use in chemical mixers and agitators, minimizing material degradation over time.
- **Chemical Storage Tanks and Linings** – Provides excellent chemical resistance, preventing material breakdown and contamination of stored chemicals.
- **Heat Exchanger Components** – Thermal stability ensures reliable performance in high-temperature and corrosive environments.

Renewable Energy Systems

PBT-Zirconia nanocomposites play a crucial role in renewable energy systems due to their high strength, thermal stability, and resistance to wear and environmental degradation.

- **Solar Panel Components** – Used in mounting structures, connectors, and housing for solar inverters due to their UV resistance and durability.
- **Wind Turbine Parts** – Suitable for bearings, gears, and electrical insulation in wind turbines, enhancing wear resistance and longevity.
- **Hydroelectric Power Systems** – Ideal for turbine blades, seals, and pump components, offering excellent corrosion resistance in water-intensive environments.
- **Battery and Energy Storage Casings** – Provides robust protection for lithium-ion and solid-state battery enclosures, improving safety and longevity.

- **Hydrogen Fuel Cell Components** – Used in bipolar plates and insulating components, benefiting from high thermal stability and chemical resistance.

PBT-Zirconia nanocomposites offer a unique combination of mechanical strength, thermal stability, wear resistance, and electrical insulation, making them highly versatile across multiple industries. From automotive and aerospace to biomedical and renewable energy systems, these materials enhance the durability and performance of critical components. Their ability to withstand harsh environments, resist corrosion, and improve wear properties ensures long-term reliability in demanding applications. As research and innovation continue, the potential of PBT-Zirconia nanocomposites will expand further, paving the way for next-generation advanced materials in diverse engineering and industrial domains.

8. CONCLUSION

The study on the thermo-mechanical properties of Polybutylene Terephthalate (PBT) reinforced with zirconia nanoparticles demonstrates that the inclusion of zirconia significantly enhances the mechanical strength, toughness, hardness, and thermal stability of the polymer. The elongation at break decreases with increasing zirconia content, indicating reduced ductility, while the relatively stable Young's modulus suggests that stiffness is not significantly impacted. Impact strength peaks at 2% zirconia, revealing improved toughness at this concentration. Shore D hardness shows an increase up to 2% filler loading, followed by a decline at higher percentages, indicating an optimal range for maximizing hardness. Thermogravimetric analysis confirms enhanced thermal stability with zirconia addition, as evidenced by the shift in degradation temperatures. Overall, the optimal performance is observed around 2% zirconia nanofiller, achieving a balance between mechanical reinforcement and thermal stability.

These improved properties make zirconia-reinforced PBT suitable for advanced engineering applications, including automotive components such as gears and connectors, electronic housings requiring high dimensional stability, and aerospace parts where enhanced thermal resistance and mechanical strength are crucial. Additionally, its superior impact strength and hardness at optimal filler loading make it ideal for structural components in demanding environments, highlighting its potential as a high-performance material.

9. FUTURE SCOPE

The incorporation of zirconia nanoparticles into Polybutylene Terephthalate (PBT) has demonstrated significant enhancements in thermo-mechanical properties, yet further research is needed to optimize its performance for broader applications. One key area of future study involves improving the dispersion of zirconia nanoparticles within the polymer matrix. Advanced surface functionalization techniques or coupling agents can help minimize agglomeration, ensuring better load transfer and mechanical reinforcement. Additionally, hybrid nanocomposites combining zirconia with graphene, carbon nanotubes, or other ceramic fillers could be explored to achieve superior strength, thermal stability, and wear resistance.

Long-term durability and aging studies are crucial to understanding the stability of PBT-ZrO_2 composites under real-world environmental conditions, including humidity, UV exposure, and thermal cycling. Investigating alternative processing techniques such as in-situ polymerization, electrospinning, or 3D printing could also lead to enhanced material performance while maintaining structural integrity. Beyond mechanical and thermal improvements, the electrical conductivity, dielectric behavior, and gas barrier properties of zirconia-reinforced PBT should be evaluated to expand its applications in electronics, sensors, and high-performance packaging.

Sustainability and recyclability are also essential considerations, particularly as industries seek environmentally friendly alternatives. Future research can focus on the recyclability of PBT-ZrO_2 composites and explore bio-based polymer matrices to reduce environmental impact. Additionally, tailoring the composite properties for specific applications, such as automotive components, aerospace structures, biomedical implants, and protective coatings, will be critical for their commercial adoption. By addressing these challenges and opportunities, PBT-ZrO_2 nanocomposites can be further developed into high-performance materials suitable for next-generation engineering applications.

ACKNOWLEDGEMENTS

The authors want to acknowledge SERB, GOI for funding the project through Core Research Grant.

The authors are also grateful to the Central Instrumentation Facility, Birla Institute of Technology Mesra, Ranchi for providing facilities to conduct various tests.

The authors also appreciate the encouragement provided and sincerely acknowledges the comments and suggestions of the editors and reviewers that have been instrumental for improving and upgrading the paper in its present form.

REFERENCES

Ahmed, R., & Singh, A. (2025). Mechanical durability and recyclability of montmorillonite-reinforced polyester nanocomposites. *Heliyon, 11*(4), e01403. DOI: 10.1016/j.heliyon.2025.e01403

Alhotan, A., Yates, J., Zidan, S., Haider, J., & Silikas, N. (2021). Assessing fracture toughness and impact strength of PMMA reinforced with nano-particles and fibre as advanced denture base materials. *Materials (Basel), 14*(15), 4127. DOI: 10.3390/ma14154127 PMID: 34361320

Barbaros, I., Yang, Y., Safaei, B., Yang, Z., Qin, Z., & Asmael, M. (2022). State-of-the-art review of fabrication, application, and mechanical properties of functionally graded porous nanocomposite materials. *Nanotechnology Reviews, 11*(1), 321–371. DOI: 10.1515/ntrev-2022-0017

Barletta, M., Aversa, C., Ayyoob, M., Gisario, A., Hamad, K., Mehrpouya, M., & Vahabi, H. (2022). Poly (butylene succinate)(PBS): Materials, processing, and industrial applications. *Progress in Polymer Science, 132*, 101579. DOI: 10.1016/j.progpolymsci.2022.101579

Bas, Y., & Dike, A. S. (2024). Comprehensive characterizations of organoclay types based on their integration to poly (butylene terephthalate) nanocomposites. *Journal of Nanoparticle Research, 26*(3), 60. DOI: 10.1007/s11051-024-05972-x

Bikiaris, D. (2011). Can nanoparticles really enhance thermal stability of polymers? Part II: An overview on thermal decomposition of polycondensation polymers. *Thermochimica Acta, 523*(1-2), 25-45 Can nanoparticles really enhance thermal stability of polymers? Part II: An overview on thermal decomposition of polycondensation polymers. *Thermochimica Acta, 523*(1-2), 25–45. DOI: 10.1016/j.tca.2011.06.012

Bogue, R. (2011). Nanocomposites: A review of technology and applications. *Assembly Automation*, *31*(2), 106–112. DOI: 10.1108/01445151111117683

Chan, K., & Zinchenko, A. (2023). Design and synthesis of functional materials by chemical recycling of waste polyethylene terephthalate (PET): Opportunities and challenges. *Journal of Cleaner Production*, *139828*, 139828. Advance online publication. DOI: 10.1016/j.jclepro.2023.139828

Chen, Y., Zhang, Y., & Li, L. (2023). Biochar-derived nanocomposites for sustainable environmental remediation: A review. *Environmental Nanotechnology, Monitoring & Management*, *20*, 100673. DOI: 10.1016/j.enmm.2023.100673

Colombo, M. A., Díaz, F. R., Kodali, D., Rangari, V., Güven, O., & Moura, E. A. (2023). Influence of Reinforcing Efficiency of Clay on the Mechanical Properties of Poly (butylene terephthalate) Nanocomposite. *Ceramics*, *6*(1), 58–73. DOI: 10.3390/ceramics6010005

Elsayed, M. A., & El-Gendy, A. S. (2022). Development of cellulose nanofiber-reinforced PET membranes for petroleum wastewater treatment. *Journal of Hazardous Materials*, *430*, 128415. DOI: 10.1016/j.jhazmat.2022.128415

Fukushima, K., Fina, A., Geobaldo, F., Venturello, A., & Camino, G. (2012). Properties of poly (lactic acid) nanocomposites based on montmorillonite, sepiolite and zirconium phosphonate. *Express Polymer Letters*, *6*(11), 914–926. Advance online publication. DOI: 10.3144/expresspolymlett.2012.97

Gad, M. M., Abualsaud, R., Alqarawi, F. K., Emam, A. N. M., Khan, S. Q., Akhtar, S., Mahrous, A. A., & Al-Harbi, F. A. (2021). Translucency of nanoparticle-reinforced PMMA denture base material: An in-vitro comparative study. *Dental Materials Journal*, *40*(4), 972–978. DOI: 10.4012/dmj.2020-296 PMID: 33883327

Gupta, D. K., Kumar, S., & Wani, M. Y. (2024). MOF magic: Zirconium-based frameworks in theranostic and bio-imaging applications. *Journal of Materials Chemistry. B*, *12*(11), 2691–2710. Advance online publication. DOI: 10.1039/D3TB02562D PMID: 38419476

Gupta, P., & Khatri, V. (2024). Eco-friendly polymer nanocomposite coatings for next-generation fire retardants in building materials. *Polymers*, *16*(14), 2045. DOI: 10.3390/polym16142045 PMID: 39065362

He, X. F., Nie, G. Z., Zhang, S., Xie, Y. Q., Ge, T. J., & Liu, P. H. (2024). Preparation and properties of modified zirconium dioxide reinforced poly (butylene adipate-co-terephthalate). *Journal of Thermoplastic Composite Materials, 08927057241230877*(10), 3238–3251. Advance online publication. DOI: 10.1177/08927057241230877

Hong, R. Y., & Chen, Q. (2015). Dispersion of inorganic nanoparticles in polymer matrices: challenges and solutions. *Organic-Inorganic Hybrid Nanomaterials*, 1-38 https://doi.org/DOI: 10.1007/12_2014_286

Idumah, C. I. (2023). Thermal expansivity of polymer nanocomposites and applications. *Polymer-Plastics Technology and Materials*, *62*(9), 1178–1203. DOI: 10.1080/25740881.2023.2204952

Kanda, M., Puggal, S., Dhall, N., & Sharma, A. (2018). Recent developments in the fabrication, characterization, and properties enhancement of polymer nanocomposites: A critical review. *Materials Today: Proceedings*, *5*(14), 28243–28252. DOI: 10.1016/j.matpr.2018.10.069

Kang, D. J., Park, G. U., Im, H. G., Park, H. Y., & Jin, J. (2016). Silica nanoparticle-embedded urethane acrylate nanohybrid thermosets for photo-patternable transparent hard coating. *Polymer*, *105*, 19–24. DOI: 10.1016/j.polymer.2016.10.008

Khan, Z. I., Habib, U., Mohamad, Z. B., Rahmat, A. R. B., & Abdullah, N. A. S. B. (2022). Mechanical and thermal properties of sepiolite strengthened thermoplastic polymer nanocomposites: A comprehensive review. *Alexandria Engineering Journal*, *61*(2), 975–990. DOI: 10.1016/j.aej.2021.06.015

Kumar, A. P., Depan, D., Tomer, N. S., & Singh, R. P. (2009). Nanoscale particles for polymer degradation and stabilization—Trends and future perspectives. *Progress in Polymer Science*, *34*(6), 479–515. DOI: 10.1016/j.progpolymsci.2009.01.002

Kumar, K., Ghosh, P. K., Kumar, A., & Singh, O. (2021). Enhanced Thermomechanical Properties of ZrO 2 Particle Reinforced Epoxy Nanocomposite. *Journal of Materials Engineering and Performance*, *30*(1), 145–153. DOI: 10.1007/s11665-020-05350-3

Li, T., Sun, H., Lei, F., Li, D., Leng, J., Chen, L., Huang, Y., & Sun, D. (2019). High performance linear low density polyethylene nanocomposites reinforced by two-dimensional layered nanomaterials. *Polymer*, *172*, 142–151. DOI: 10.1016/j.polymer.2019.03.072

Mendes, L. C., Mariano, D. M., Freitas, D. F. S., Albitres, G. A. V., Tavares, M. I. B., & Garcia, E. E. (2024). Ecofriendly composites based on poly (lactic acid) with nano-zirconium phosphate and nano-zinc oxide/zirconium phosphate: Physico-chemical and aging characteristics. *Journal of Thermal Analysis and Calorimetry*, *149*(17), 9297–9306. DOI: 10.1007/s10973-024-13541-6

Mishra, T., Mandal, P., Rout, A. K., & Sahoo, D. (2022). A state-of-the-art review on potential applications of natural fiber-reinforced polymer composite filled with inorganic nanoparticle. *Composites Part C: Open Access*, *9*, 100298. DOI: 10.1016/j.jcomc.2022.100298

Mulla, M. Z., Rahman, M. R. T., Marcos, B., Tiwari, B., & Pathania, S. (2021). Poly Lactic Acid (PLA) Nanocomposites: Effect of inorganic nanoparticles reinforcement on its performance and food packaging applications. *Molecules (Basel, Switzerland)*, *26*(7), 1967. DOI: 10.3390/molecules26071967 PMID: 33807351

Naeem, S., Gilani, S. Q. Z., Baheti, V., Wiener, J., Militky, J., Javed, S., ... & ul Hassan, S. Z. (2018). Electrical Conductivity of PLA Films Reinforced With Carbon Nano Particles from Waste Acrylic Fibers. In *Advances in Natural Fibre Composites: Raw Materials, Processing and Analysis* (pp. 205-222). Springer International Publishing DOI: 10.1007/978-3-319-64641-1_18

Naffakh, M., Díez-Pascual, A. M., Marco, C., Ellis, G. J., & Gomez-Fatou, M. A. (2013). Opportunities and challenges in the use of inorganic fullerene-like nanoparticles to produce advanced polymer nanocomposites. *Progress in Polymer Science*, *38*(8), 1163–1231. DOI: 10.1016/j.progpolymsci.2013.04.001

Nikam, P. C., Rao, A. R., & Shertukde, V. V. (2021). Enhancement of thermo-mechanical and chemical resistance properties of polyurethane composite reinforced with hydrophobic nano-silica and scrape PET derived bis (2-hydroxyethyl terephthalate). *Materials Today. Communications*, *29*, 102788. DOI: 10.1016/j.mtcomm.2021.102788

Ogbonna, V. E., Popoola, A. P. I., Popoola, O. M., & Adeosun, S. O. (2022). A review on the recent advances on improving the properties of epoxy nanocomposites for thermal, mechanical, and tribological applications: Challenges and recommendations. *Polymer-Plastics Technology and Materials*, *61*(2), 176–195. DOI: 10.1080/25740881.2021.1967391

Owen, M. M., Wong, L. S., Achukwu, E. O., Saharudin, M. S., & Akil, H. M. (2024). Enhanced mechanical, thermal, and morphological properties of waste PET plastics reinforced with coated biodegradable kenaf fibers for infrastructure applications. *Construction & Building Materials, 442*, 137659. DOI: 10.1016/j.conbuildmat.2024.137659

Pathak, R., Punetha, M., Bhatt, S., Pillai, S. A., Dhapola, P. S., & Punetha, V. D. (2024). Organic and inorganic nanofillers for polymer nanocomposites: Trends, opportunities, and challenges. *Advances in Functionalized Polymer Nanocomposites*, 1-34 DOI: 10.1016/B978-0-443-18860-2.00003-7

Paul, D. R., & Robeson, L. M. (2008). Polymer nanotechnology: Nanocomposites. *Polymer, 49*(15), 3187–3204. DOI: 10.1016/j.polymer.2008.04.017

Petousis, M., Moutsopoulou, A., Korlos, A., Papadakis, V., Mountakis, N., Tsikritzis, D., Ntintakis, I., & Vidakis, N. (2023). The effect of nano zirconium dioxide (ZrO2)-optimized content in polyamide 12 (PA12) and polylactic acid (PLA) matrices on their thermomechanical response in 3D printing. *Nanomaterials (Basel, Switzerland), 13*(13), 1906. DOI: 10.3390/nano13131906 PMID: 37446421

Samieifakhr, M., & Shojaei, A. (2024). Improved crystallization behavior and enhanced impact strength of melt processed poly (ethylene terephthalate)/UiO-66 nanocomposites. *Polymer, 290*, 126593. DOI: 10.1016/j.polymer.2023.126593

Schmidt, H., Jonschker, G., Goedicke, S., & Mennig, M. (2000). The sol-gel process as a basic technology for nanoparticle-dispersed inorganic-organic composites. *Journal of Sol-Gel Science and Technology, 19*(1-3), 39–51. DOI: 10.1023/A:1008706003996

Sudhakar, Y. N., Selvakumar, M., & Bhat, D. K. (2021). Investigations on thermo-mechanical properties of organically modified polymer clay nanocomposites for packaging application. *Polymers & Polymer Composites, 29*(8), 1191–1199. DOI: 10.1177/0967391120960645

Utracki, L. A., Sepehr, M., & Boccaleri, E. (2007). Synthetic, layered nanoparticles for polymeric nanocomposites (PNCs). *Polymers for Advanced Technologies, 18*(1), 1–37. DOI: 10.1002/pat.852

Vaia, R. A., & Maguire, J. F. (2007). Polymer nanocomposites with prescribed morphology: Going beyond nanoparticle-filled polymers. *Chemistry of Materials, 19*(11), 2736–2751. DOI: 10.1021/cm062693+

Verma, D., Goh, K. L., & Vimal, V. (2022). Interfacial studies of natural fiber-reinforced particulate thermoplastic composites and their mechanical properties. *Journal of Natural Fibers, 19*(6), 2299–2326. DOI: 10.1080/15440478.2020.1808147

Visakh, P. M. (2015). Polyethylene terephthalate: blends, composites, and nano-composites–state of art, new challenges, and opportunities. *Poly (Ethylene Terephthalate) Based Blends, Composites and Nanocomposites*, 1-14 https://doi.org/ DOI: 10.1016/B978-0-323-31306-3.00001-4

Zhang, M., Zhou, J., Zhang, Q., Zhang, C., & Song, W. (2025). Construction of flame-retardant and water-resistant polybutylene terephthalate composites by reactive extrusion. *Frontiers in Materials, 11*, 1536536. DOI: 10.3389/fmats.2025.1536536

Zhao, W., Yao, M., Liu, L., & Wang, H. (2024). Preparation and characterization of PBT@PP-CNC@cellulose wood pulp paper double-layers fuel filtration materials. *Journal of Materials Science, 59*(6), 4123–4138. DOI: 10.1007/s10853-023-09100-3

Chapter 3
Innovative Energy-Efficient Materials for Sustainable Construction

Salim Barbhuiya
https://orcid.org/0000-0002-4325-281X
University of East London, UK

Bibhuti Das
NITK Surathkal, India

Dibyendu Adak
https://orcid.org/0000-0003-1494-3010
NIT Meghalaya, Shillong, India

ABSTRACT

The construction industry plays a pivotal role in global energy consumption and environmental impact, making the adoption of energy-efficient materials essential for sustainable development. This chapter explores the latest advancements in energy-efficient construction materials, examining their properties, applications, and contributions to reducing energy use and carbon emissions. Key materials discussed include advanced insulation solutions such as aerogels and spray foam, high-performance glazing like Low-E and smart windows, and sustainable concrete alternatives such as geopolymer and recycled aggregate concrete. Natural and renewable resources, including bamboo, hempcrete, and straw bale, are also highlighted for their ecological benefits and performance. The chapter delves into innovative technologies, such as phase-change materials, nanotechnology, and 3D-printed components, that enhance energy performance in modern buildings. It further reviews energy efficiency standards and certifications like LEED, BREEAM,

DOI: 10.4018/979-8-3373-2414-2.ch003

and Passive House, offering guidance on best practices for material selection. Additionally, it addresses challenges in material adoption, including cost, supply chain issues, and regulatory barriers, while forecasting future trends in material science and sustainable construction practices.

1. INTRODUCTION TO ENERGY-EFFICIENT MATERIALS

Energy-efficient materials are designed to reduce thermal transmittance and optimize building energy use (Cabeza et al., 2010; Ascione et al., 2016), contributing significantly to indoor environmental quality and reduced operational energy demand. These materials are commonly used in construction, manufacturing, and product design to enhance sustainability and lower energy costs. Examples include high-performance insulation, reflective roofing, low-emissivity glass, and phase-change materials. The significance of energy-efficient materials lies in their ability to enhance energy conservation, reduce carbon emissions, and promote sustainability. In the construction sector, they help maintain indoor temperatures, reducing the need for heating and cooling systems. This leads to lower energy bills and decreased dependence on fossil fuels. In industrial applications, such materials improve efficiency by reducing energy waste during production processes (Zhang et al. 2017; Mondal et al. 2019). Moreover, using energy-efficient materials supports environmental conservation by minimizing resource depletion and greenhouse gas emissions (Worrell et al. 2016; Gilbert et al. 2017; Zhang et al. 2022). They also contribute to improving comfort levels in buildings, enhancing durability, and increasing property value. As global energy demands rise, incorporating these materials into everyday applications is crucial for achieving energy efficiency, reducing operational costs, and supporting green building initiatives. Their role in sustainable development makes them essential for creating eco-friendly and resilient infrastructure.

Sustainable construction relies on the use of eco-friendly materials to minimize environmental impact while enhancing energy efficiency, durability, and resource conservation. The choice of materials plays a crucial role in reducing carbon footprints, improving building performance, and promoting sustainability throughout a structure's life cycle. Energy-efficient materials, such as insulated concrete forms, low-emissivity glass, and reflective roofing, help regulate indoor temperatures, reducing energy consumption for heating and cooling (Jelle et al. 2015; Ling et al. 2024; Salah et al. 2024). Recycled and renewable materials, including reclaimed wood, recycled steel, and bamboo, decrease waste and reliance on non-renewable resources. Geopolymer concretes have shown significant reductions in embodied carbon, with studies estimating CO_2 reductions between 40–80% (Provis & van Deventer, 2009; Zeyad, 2019). Water-efficient materials, such as permeable concrete and rainwater-harvesting systems, contribute to better water management, while non-toxic, low-VOC (volatile organic compound) materials improve indoor air quality (Xu et al. 2023; Supraja et al. 2024; Xu et al. 2025). Additionally, smart materials like self-healing concrete and phase-change materials enhance building longevity and adaptability to climate conditions. By integrating sustainable materials, construction projects not only reduce their environmental impact but also improve energy efficiency, lower maintenance costs, and create healthier living spaces. As urbanization grows, adopting sustainable materials is essential for building resilient, eco-friendly infrastructure that supports a greener future. Conceptually, sustainable construction involves a cradle-to-cradle philosophy. Empirical research, such as that by Gursel et al. (2014), has quantified life cycle energy savings across various material categories.

Energy-efficient construction is gaining momentum worldwide as governments, industries, and consumers prioritize sustainability. One major trend is the adoption of green building certifications like LEED, BREEAM, and Passive House, which set energy efficiency standards for construction projects (Altomonte & Schiavon, 2013; Awadh, 2017; Mihai et al. 2017). Innovative energy-efficient materials, such as aerogel insulation, smart glass, and phase-change materials, are being integrated into modern buildings to reduce energy consumption. Additionally, net-zero energy buildings (NZEBs), which generate as much energy as they consume, are becoming more common, utilizing solar panels, wind energy, and energy storage solutions (Deng et al. 2014; Wells et al. 2018; Jaysawal et al. 2022). Another significant trend is the use of prefabrication and modular construction, which minimizes waste and improves energy efficiency during building assembly. Smart building technologies, including IoT-based energy management systems, optimize heating, cooling, and lighting, further reducing energy use. With stricter regulations and increased awareness, the global construction industry is shifting towards sustainable, energy-efficient solutions, reducing environmental impact and long-term operational costs. Empirical analyses have demonstrated the impact of these materials on energy performance. For example, Ascione et al. (2016) report energy demand reductions of up to 40% when advanced insulation materials are incorporated in retrofits

The motivation behind writing this chapter is to highlight the critical role of energy-efficient materials and sustainable construction in addressing environmental challenges. With rising global energy demands and climate concerns, it is essential to explore innovative solutions that reduce energy consumption, minimize carbon footprints, and enhance building performance. This chapter aims to educate readers on the significance of using eco-friendly materials, emerging trends, and best practices in sustainable construction. By promoting awareness and encouraging the adoption of energy-efficient strategies, this chapter seeks to contribute to the development of greener, more resilient infrastructure for a sustainable future.

2. PRINCIPLES OF ENERGY EFFICIENCY IN CONSTRUCTION MATERIALS

2.1 Thermal Insulation and Heat Retention

Energy efficiency in construction materials is primarily achieved through thermal insulation and heat retention, which help regulate indoor temperatures and reduce energy consumption. These principles ensure that buildings maintain comfortable conditions with minimal reliance on heating and cooling systems, lowering energy costs and environmental impact. Thermal insulation refers to the ability of a material to resist heat flow. High-performance insulating materials, such as fiberglass, mineral wool, polyurethane foam, and aerogels, reduce heat transfer between the interior and exterior of a building (Woltman et al. 2017; Flynn et al. 2017; Berardi, 2018; Liu et al. 2022; Nandy et al. 2023). Proper insulation in walls, roofs, and floors minimizes energy loss, maintaining stable indoor temperatures regardless of external weather conditions. Heat retention involves storing and gradually releasing heat to maintain a balanced indoor climate. Materials with high thermal mass, such as concrete, brick, and rammed earth, absorb heat during the day and release it at night, reducing temperature fluctuations (Samadianfard & Toufigh, 2020; Ben-Alon & Rempel, 2023; Kaitouni et al. 2024). In a study by Kosny et al. (2011), wall systems with high thermal mass materials reduced peak indoor temperatures by over 5°C during hot climates. Phase-change materials (PCMs) further enhance heat retention by storing excess heat and releasing it when temperatures drop. By integrating thermal insulation and heat retention strategies, buildings achieve higher energy efficiency, improved comfort, and lower carbon emissions. These principles are fundamental to designing sustainable structures that optimize energy use and contribute to a greener future.

2.2 Material Embodied Energy

Material embodied energy refers to the total energy consumed throughout a material's life cycle, including extraction, processing, transportation, installation, maintenance, and disposal. It plays a crucial role in the energy efficiency of construction materials, as high embodied energy contributes to greater carbon emissions and environmental impact. To improve sustainability, construction materials should have low embodied energy while maintaining durability and functionality. Natural and renewable materials such as bamboo, straw, and timber have significantly lower embodied energy than traditional materials like concrete and steel (Li et al. 2021; Gan et al. 2022; Netto et al. 2024). Recycled and reclaimed materials, such as recycled steel, reclaimed wood, and crushed concrete, also help reduce embodied energy by minimizing the need for raw material extraction and processing.

Efficient manufacturing processes further contribute to lower embodied energy. Techniques like geopolymer concrete production and advanced prefabrication methods reduce energy-intensive steps, making materials more eco-friendly (Li et al. 2021; Liu et al. 2021; Qaidi et al. 2022; Ahmed et al. 2022). Additionally, local sourcing of materials reduces transportation energy, further cutting down emissions. By selecting materials with low embodied energy, builders and designers can significantly reduce a building's overall environmental footprint. This approach promotes sustainable construction, enhances resource efficiency, and supports global efforts to mitigate climate change while maintaining high performance and durability in building structures.

Table 1 outlines the embodied energy of various energy-efficient materials used in sustainable construction. Embodied energy measures the total energy required for material extraction, processing, and transportation, expressed in megajoules per kilogram (MJ/kg). Lower values indicate lower environmental impact. Materials like rammed earth, straw bale, and hempcrete have minimal embodied energy, making them highly sustainable. Recycled materials, such as steel and concrete, reduce resource depletion. Embodied energy values in Table 1 are compiled from empirical life-cycle assessment studies (Hammond & Jones, 2008; Dixit, 2017), ensuring comparative accuracy. Bamboo and cross-laminated timber (CLT) offer renewable alternatives with good structural properties. Additionally, aerated concrete and cork provide excellent insulation. Using these materials helps reduce carbon footprints and improve energy efficiency in buildings.

Table 1. Comparative Analysis of Sustainable Building Materials: Embodied Energy, Benefits, and Applications

Material	Embodied Energy (MJ/kg)	Energy Category	Key Sustainability Benefits	Common Uses	Comparative Notes
Rammed Earth	0.7 – 1.2	Very Low	Low energy, locally sourced, recyclable	Walls, foundations	One of the lowest energy materials; best suited for thermal mass and local builds
Straw Bale	0.2 – 0.5	Very Low	Renewable, carbon-sequestering	Insulation, wall infill	Exceptional for insulation with negative carbon potential
Recycled Steel	8 – 12	High	Reduces mining, highly durable	Structural framing, roofing	High energy cost but excellent durability and recyclability

continued on following page

Table 1. Continued

Material	Embodied Energy (MJ/kg)	Energy Category	Key Sustainability Benefits	Common Uses	Comparative Notes
Bamboo	0.5 – 1.2	Very Low	Fast-growing, high strength-to-weight ratio	Flooring, panels, structures	Strong alternative to timber; sustainable with rapid regeneration
Hempcrete	0.7 – 1.5	Low	Absorbs CO_2, breathable, lightweight	Walls, insulation	Carbon-negative and ideal for passive design strategies
Cross-Laminated Timber	2.5 – 3.5	Medium	Renewable, carbon storage, lightweight	Structural panels, floors	Balances structural integrity with moderate embodied energy
Recycled Concrete	0.9 – 1.4	Low	Diverts waste, lower carbon footprint	Pavements, foundations	Strong sustainability profile for reuse in large-scale infrastructure
Aerated Concrete	1.0 – 3.0	Low–Medium	Lightweight, good insulation properties	Walls, partitions	Trades higher energy input for thermal performance benefits
Cork	1.1 – 1.4	Low	Sustainable harvesting, excellent insulation	Flooring, wall panels	Renewable and resilient, especially suitable for interior finishes
Recycled Glass	3.5 – 5.0	Medium–High	Reuses waste, reduces mining impact	Countertops, insulation panels	Energy-intensive to process but a good way to divert glass waste

2.3 Lifecycle Assessment of Construction Materials

Lifecycle Assessment (LCA) is a systematic approach used to evaluate the environmental impact of construction materials from their extraction to disposal. It provides a comprehensive analysis of the energy, emissions, and resource consumption associated with a material throughout its life cycle stages, which include raw material extraction, manufacturing, transportation, construction, usage, and end-of-life disposal or recycling. LCA helps identify materials with lower environmental footprints, guiding sustainable construction practices. Raw material extraction is often energy-intensive, especially for materials like cement and steel, leading to high carbon emissions. Manufacturing processes also contribute to environmental impact through energy consumption, pollution, and waste generation.

Transportation and construction stages further add to the material's footprint, emphasizing the importance of using locally sourced materials and efficient construction techniques like prefabrication. During the usage phase, materials with high durability and energy-efficient properties, such as insulated panels and thermal mass materials, reduce operational energy consumption (Grazieschi et al. 2021; Disckson & Pavia, 2021; Zilberberg et al. 2021). At the end-of-life stage, materials that can be reused, recycled, or biodegraded contribute to a circular economy and lower environmental impact. Lifecycle assessments cited are aligned with ISO 14040/14044 standards (ISO, 2006) and draw from comparative LCA studies such as those by Cabeza et al. (2014). By integrating LCA into material selection, the construction industry can promote energy efficiency, reduce carbon emissions, and enhance sustainability, ultimately supporting environmentally responsible building practices.

2.4 Integrating Energy Transition Theory and Circular Economy Principles

Contemporary efforts to improve energy efficiency in construction materials should be situated within broader frameworks of socio-technical systems change and circular economy strategies. Energy systems transition theory, especially the Multi-Level Perspective (MLP), offers valuable insights into how innovation in materials can scale from niche adoption to widespread transformation. This perspective, developed by scholars such as Geels (2002) and Schot & Kanger (2018), highlights the interplay between technological innovations, policy landscapes, and user practices in shaping sustainable transitions.

In the context of building materials, circular economy (CE) principles emphasize the importance of material reuse, design for disassembly, and minimizing lifecycle impacts. Rather than focusing solely on energy performance, CE advocates a regenerative approach — keeping materials in use at their highest value, reducing embodied energy, and closing resource loops (Kirchherr et al. 2017; Blomsma & Brennan 2017). For example, using cross-laminated timber (CLT) or recycled aggregates aligns not only with low-carbon goals but also with CE priorities by extending material lifespans and reducing raw material extraction. At the systems level, policies that encourage product-as-a-service models or extended producer responsibility (Stahel, 2016) can accelerate the adoption of circular materials in construction.

Bridging these theoretical domains enriches the assessment of sustainable materials beyond technical performance, integrating environmental, economic, and systemic transition perspectives that are essential for achieving climate targets in the built environment.

3. TYPES OF ENERGY-EFFICIENT CONSTRUCTION MATERIALS

3.1 Insulating Materials

Insulating materials are essential in construction for reducing heat transfer, improving energy efficiency, and maintaining indoor comfort. These materials minimize energy loss, lowering heating and cooling costs. Common insulating materials include fiberglass, mineral wool, polyurethane foam, cellulose, and aerogels. Each offers varying levels of thermal resistance (R-value), durability, and environmental impact. Natural insulation options, such as sheep wool, cork, and hemp, provide eco-friendly alternatives with sustainable benefits. Proper insulation in walls, roofs, and floors enhances building performance, reduces carbon emissions, and contributes to sustainable construction. Choosing the right insulating material is crucial for optimizing energy efficiency and occupant comfort.

3.1.1 Fibreglass Insulation

Fibreglass insulation is a widely used material for improving energy efficiency and thermal comfort in residential, commercial, and industrial buildings. Made from fine glass fibres, it is available in batts, rolls, and loose-fill forms, offering flexibility for different applications. This insulation effectively reduces heat transfer, keeping interiors warm in winter and cool in summer, which lowers energy costs and improves overall building performance (Al-Homoud, 2005; Shekarchian et al. 2012; Lee & Lim, 2016; Choudhary & Eastes, 2024). One of the key advantages of fiberglass insulation is its affordability and ease of installation. It is non-combustible, moisture-resistant, and does not support mold growth, making it a durable and safe option. Additionally, it has excellent soundproofing qualities, reducing noise transmission between rooms and from external sources. Fibreglass insulation is also an environmentally friendly choice, as it often contains recycled glass and reduces the need for excessive heating and cooling, thereby lowering carbon emissions. However, proper handling is essential, as fine glass fibres can irritate the skin, eyes, and respiratory system. Wearing protective gear during installation minimizes these risks. Used in walls, attics, floors, and HVAC systems, fibreglass insulation plays a crucial role in sustainable building practices (Porter, 2020; Alassaf, 2024). Its combination of affordability, efficiency, and environmental benefits makes it a popular choice for modern construction and retrofitting projects.

In Fig. 1, the fiberglass grid is the 7th layer in the insulation system, playing a crucial role in reinforcing the thermal insulating layer. It prevents cracks by distributing stress caused by thermal expansion, moisture, and mechanical impacts. The grid enhances adhesion between the insulation (5th layer) and external coating (8th layer), ensuring structural stability. Being moisture-resistant, it prevents water infiltration while maintaining durability under temperature variations. Its non-combustibility improves fire safety by reducing flame spread. Lightweight and flexible, it is easy to install and adapts to complex surfaces, making it an essential component for long-term insulation performance.

Figure 1. Thermal insulation material installed at the external side of the building element

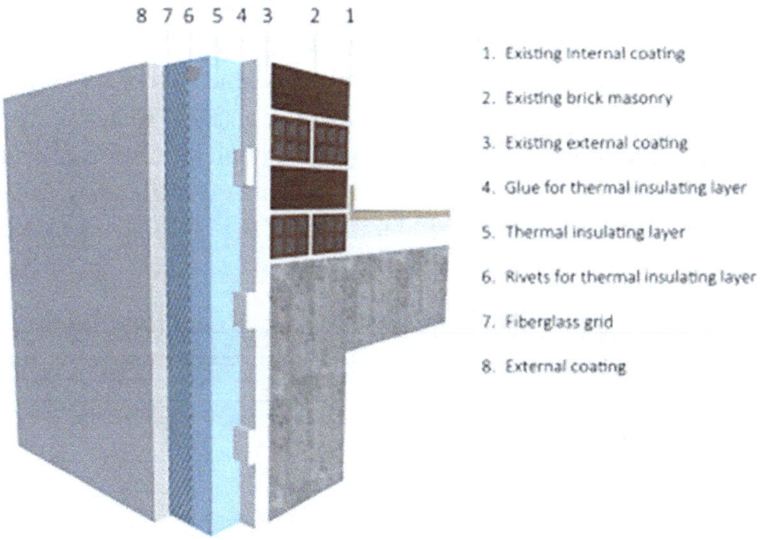

Source: Kylili and Fokaides (2017)

3.1.2 Spray Foam Insulation

Spray foam insulation is a highly effective insulating material that expands upon application, filling gaps and sealing surfaces to improve energy efficiency. It is composed of polyurethane or isocyanate-based materials that react when mixed, creating a dense, insulating foam. Spray foam comes in two main types: open-cell and closed-cell. Open-cell foam is softer, more flexible, and allows some air movement, making it ideal for soundproofing and insulating interior walls (Das et al. 2023; Fu et al. 2024; Li et al. 2024). Closed-cell foam is denser, providing superior thermal resistance, moisture control, and structural reinforcement, making it suitable for exterior walls, roofs, and basements (Amran et al. 2022; Ali et al. 2024). One of the key advantages of spray foam insulation is its ability to create an airtight seal, significantly reducing energy loss by minimizing air infiltration. This leads to lower heating and cooling costs and improved indoor comfort. Additionally, closed-cell spray foam acts as a moisture barrier, preventing mold growth and enhancing the durability of structures. Spray foam insulation is widely used in residential, commercial, and industrial buildings due to its longevity and effectiveness. However, installation requires professional expertise due to the chemicals involved. While the initial cost is higher than traditional insulation materials, its long-term energy savings and durability make it a cost-effective solution for modern insulation needs.

Some joints of the building envelope where polyurethane filling may be used, as shown in Fig. 2, include the tightening of joints between the external wall and the window, partition walls and inserted ceilings, and partition walls and doors. These areas require proper sealing to prevent air leakage, enhance thermal insulation, and improve structural integrity. Polyurethane foam is an ideal material for these applications as it expands to fill gaps, ensuring airtight and moisture-resistant joints. Its use in these critical locations helps improve energy efficiency, reduce noise transmission, and enhance the overall durability of the building envelope.

Figure 2. Some joints of the building envelope where polyurethane filling may be used

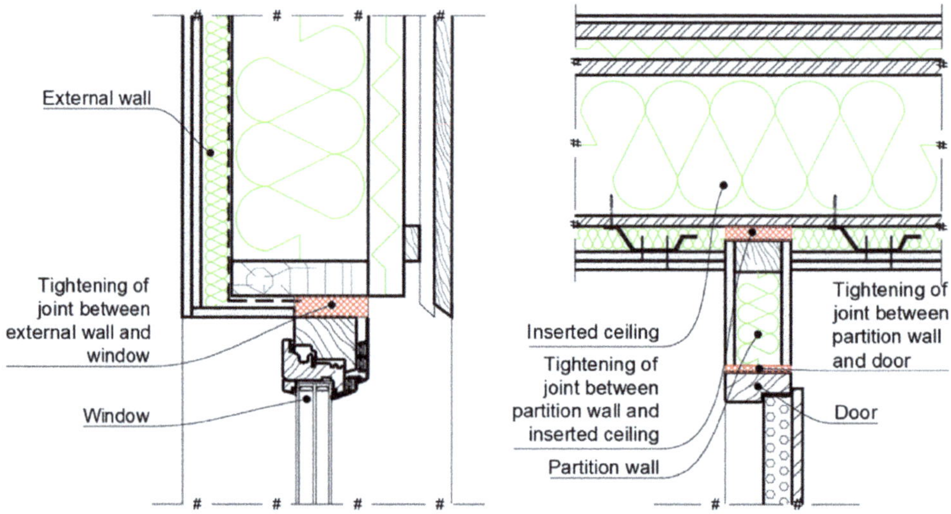

Source: Hallik et al. (2019)

3.1.3 Aerogels and Vacuum Insulated Panels

Aerogels and vacuum-insulated panels (VIPs) are advanced insulation materials known for their exceptional thermal performance and space efficiency. Aerogels are ultra-light materials derived from silica, with up to 99% porosity, making them one of the best thermal insulators available. Their extremely low thermal conductivity, around 0.013 W/m·K, allows for excellent insulation in thin layers (Cuce et al. 2014; Liu et al. 2019; Sambucci et al. 2023). Aerogels are also moisture-resistant, fireproof, and highly durable, making them suitable for building insulation, aerospace, and industrial applications. Studies by Baetens et al. (2011) and Buratti et al. (2012) demonstrate aerogel panels can achieve thermal conductivity values as low as 0.013 W/m·K in real-world building envelopes. Though highly efficient, aerogels can be brittle and expensive, limiting their widespread use in standard construction.

Vacuum Insulated Panels (VIPs) consist of a porous core material enclosed in a vacuum-sealed barrier, significantly reducing heat transfer. With a thermal conductivity as low as 0.004 W/m·K, VIPs provide up to ten times better insulation than conventional materials like fiberglass or foam (Biswas et al. 2018; Boafo et al. 2021). They are ideal for space-constrained applications, such as refrigeration, walls, and high-performance building envelopes. However, VIPs have a fragile vacuum seal, which, if compromised, reduces their efficiency. Both aerogels and VIPs offer cutting-edge insulation solutions, particularly in energy-efficient buildings, aerospace, and industrial applications. While cost and handling challenges exist, their superior performance makes them essential for high-tech and sustainable construction projects.

Figure 3 illustrates the manufacturing process of Vacuum Insulated Panels (VIPs) with aerogel composite cores, which occur in three main stages. Initially, aerogel composite cores are prepared as the primary insulation material. These cores are placed inside barrier envelopes designed to enclose and protect the core while ensuring an airtight seal. Next, the enclosed aerogel cores undergo an outgassing process in a vacuum chamber, where trapped gases and moisture are removed to prevent air pockets that could compromise insulation performance. This step ensures long-term stability and efficiency by reducing the risk of internal pressure build-up. Following this, the prepared panels are placed in a heat-sealing device within the vacuum chamber, where air is removed using a vacuum pump. The panel is then heat-sealed, creating a completely airtight and moisture-resistant enclosure. This process results in a Vacuum Insulated Panel (VIP) with high thermal resistance, ensuring superior insulation performance for applications requiring energy efficiency.

Figure 3. Preparation schematic of VIPs with aerogel composite cores

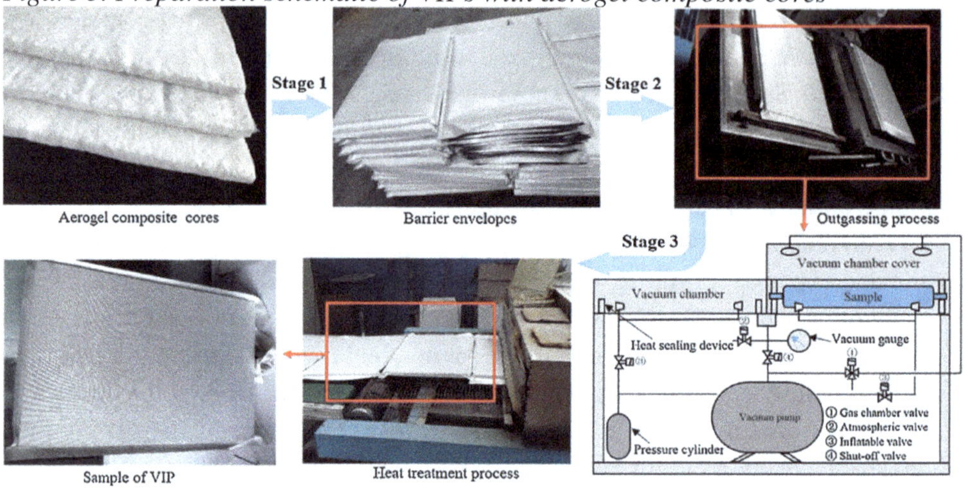

Source: Liang et al. (2017)

3.2 High-Performance Glazing and Windows

High-performance glazing and windows enhance energy efficiency, comfort, and sustainability in buildings. These advanced window systems incorporate technologies like low-emissivity (Low-E) coatings, multi-pane glass, gas fills (argon or krypton), and insulated frames to minimize heat transfer. They reduce energy consumption by improving thermal insulation, reducing heat gain in summer, and retaining warmth in winter. High-performance glazing also improves sound insulation, blocks harmful UV rays, and enhances indoor comfort. Common applications include residential, commercial, and industrial buildings seeking energy savings and sustainability certifications. Investing in these windows leads to lower utility bills, reduced carbon footprint, and improved overall building performance.

3.2.1 Low-E Glass

Low-emissivity (Low-E) glass is a technologically advanced glazing solution designed to improve energy efficiency by minimizing heat transfer while maintaining visible light transmission. It features a microscopically thin, transparent coating of metal or metal oxide that reduces infrared radiation, helping to regulate indoor temperatures. This coating reflects heat into the building during winter and prevents excessive heat from entering during summer, reducing heating and cooling costs. Low-E glass is available in two primary types: hard-coat (pyrolytic) and soft-coat (sputtered). Hard-coat Low-E glass is more durable and suitable for single-pane applications, while soft-coat Low-E glass offers superior performance and is commonly used in double or triple-glazed units (Duan et al. 2020; Somasundaram et al. 2020). One of its key advantages is enhanced thermal insulation, significantly improving the window's U-value, which measures heat loss. It also reduces ultraviolet (UV) radiation, preventing furniture, carpets, and artworks from fading. Despite these benefits, Low-E glass maintains high visibility and natural light transmission, ensuring bright indoor spaces. Used in residential, commercial, and industrial buildings, Low-E glass is an essential component of energy-efficient windows and facades, contributing to sustainability and reduced carbon emissions. While slightly more expensive than standard glass, its long-term energy savings make it a cost-effective investment.

Figure 4 illustrates the structure and manufacturing process of laminated Low-E glass. It consists of a top clear glass layer, a polyvinyl butyral (PVB) interlayer and a Low-E coated glass layer. The PVB interlayer enhances adhesion, impact resistance, and sound insulation. The assembly is placed in a heating chamber at 140-160°C, where heat and pressure fuse the layers, creating a durable, energy-efficient glazing material. The sputtered Low-E coating minimizes heat transfer while allowing natural light, improving thermal performance. This laminated Low-E glass is widely used in architectural applications for energy savings, safety, and UV protection.

Figure 4. Schematic representation of laminated Low-E type coated glass structure

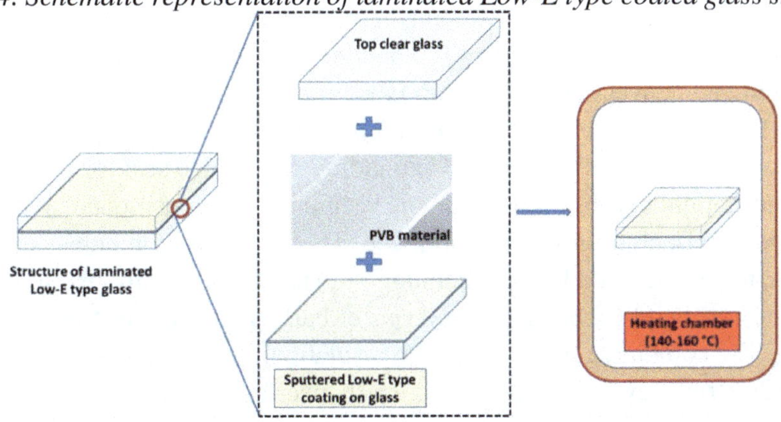

Source: Nur-E-Alam (2024)

3.2.2 Triple-Glazed Windows

Triple-glazed windows are advanced energy-efficient window systems that consist of three panes of glass separated by insulating gas-filled spaces, typically argon or krypton. These windows provide superior thermal insulation, significantly reducing heat transfer and improving indoor comfort (Zeng et al. 2019; Liu et al. 2021; Wang et al. 2024; Nsaif et al. 2024). Compared to double-glazed windows, triple-glazing offers better energy efficiency, making it ideal for extreme climates where maintaining indoor temperatures is crucial. One of the key benefits of triple-glazed windows is their low U-value, which measures heat loss. With an additional pane and insulating layer, they prevent cold air from entering and warm air from escaping, reducing heating and cooling costs. They also incorporate Low-E (low-emissivity) coatings, further enhancing their ability to reflect heat while allowing natural light transmission. Beyond thermal performance, triple glazing enhances sound insulation, making it effective for homes and buildings in noisy environments (Liu, 2015; Rezaei et al. 2017). It also improves security, as the extra glass pane increases resistance to breakage. Additionally, the insulation helps prevent condensation, reducing moisture-related issues and mold growth. Although more expensive than double-glazing, triple-glazed windows offer long-term savings through reduced energy bills. They are widely used in passive houses and sustainable building projects, contributing to lower carbon footprints and enhanced indoor comfort.

Figure 5 compares two configurations of Air-Vacuum Layered Triple-Glazed Windows (IVW), optimized for different thermal insulation needs. In (1a) IVW cold, the vacuum layer faces the outdoor (cold side), minimizing heat loss by reducing conduction and convection. This setup is ideal for cold climates, preventing heat escape while maintaining indoor warmth. The structure includes three glass sheets, an air layer, a support pillar array, and an indium alloy edge seal for durability and airtight insulation. In (1b) IVW warm, the vacuum layer is positioned on the indoor (warm side), reflecting infrared heat inside. This configuration enhances thermal retention, making it suitable for maintaining warm indoor conditions. Both designs use desiccants, butyl rubber seals, and spacer bars to ensure moisture control and longevity. By combining air and vacuum insulation, these windows provide superior energy efficiency and comfort, making them effective for buildings in extreme temperature environments.

Figure 5. Schematic diagram of air-vacuum layered triple glazed window in which vacuum layer is facing the EN-ISO ambient conditions

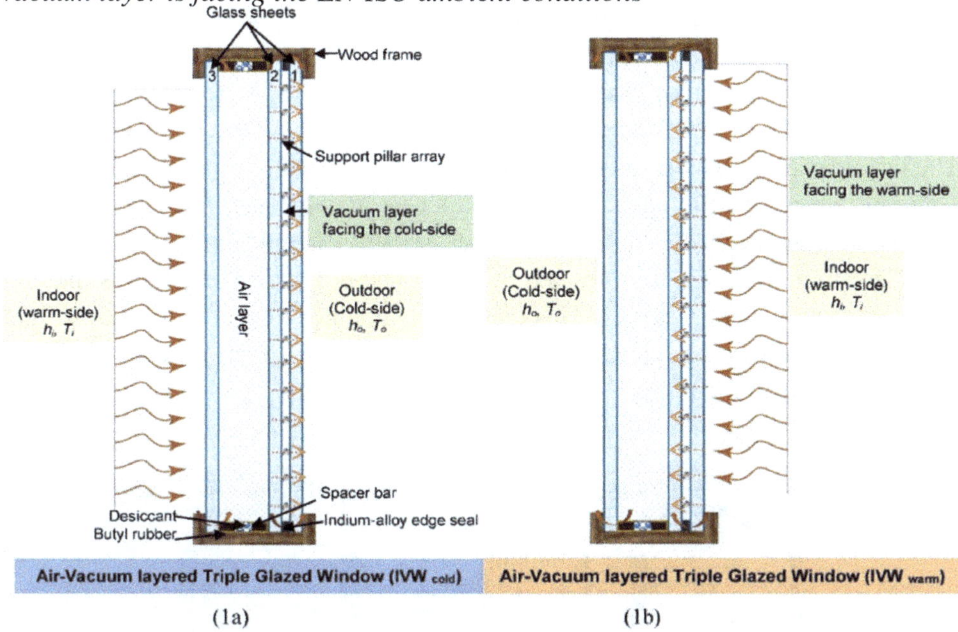

Source: Fang et al. (2020)

3.2.3 Smart Windows and Electrochromic Glass

Smart windows and electrochromic glass are advanced glazing technologies that dynamically adjust transparency and tint to optimize energy efficiency, privacy, and comfort. These windows use electrical, thermal, or chemical triggers to control light and heat transmission, reducing reliance on artificial lighting and HVAC systems. Electrochromic glass is a widely used smart window technology that changes tint when voltage is applied (Huang et al. 2012; Cannavale et al. 2020; Brzezicki, 2021). It consists of multiple thin-film layers that alter their optical properties in response to electrical stimuli, switching between clear and tinted states. This process controls solar heat gain, glare, and visibility, enhancing indoor comfort while lowering cooling and heating costs. Electrochromic windows are particularly beneficial in smart buildings, offices, and residential spaces, where they improve sustainability and occupant well-being. Other smart window technologies include thermochromic and photochromic glass, which respond to temperature and sunlight, respectively (Cao et al. 2018; Xu et al. 2022; Zhang et al. 2023; Gao et al. 2025). Liquid crystal and suspended particle devices (SPD) offer rapid adjustments for privacy control. Although more expensive than conventional windows, smart windows provide long-term energy savings, reduced carbon footprints, and enhanced architectural aesthetics. Their integration with automated building systems and IoT makes them a crucial component of modern energy-efficient designs, improving both functionality and sustainability in buildings and vehicles.

Figure 6 illustrates the working mechanism of the Suspended Particle Device (SPD) smart glass, which dynamically adjusts transparency by applying voltage. The left side shows the opaque state (SPD switch off, 0V, 5% transparency), where randomly oriented SPD particles scatter light, reducing visibility. The right side depicts the transparent state (SPD switch on, 110V, 55% transparency), where SPD particles align, allowing light to pass through. The polymer matrix holds the particles between two glass panes. This technology enables instant control over light transmission, enhancing privacy, glare reduction, and energy efficiency in buildings and automotive applications.

Figure 6. Suspended particle device type smart window (left: off state; right: on state)

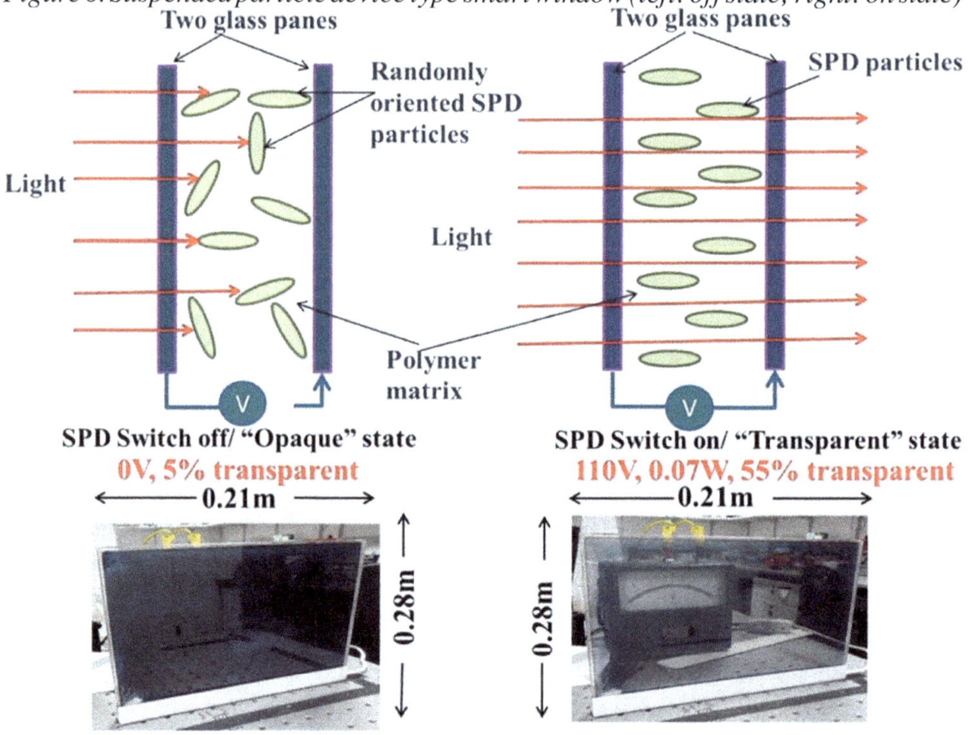

Source: Ghosh et al. (2016)

Figure 7 illustrates the function of an electrochromic double-pane window in controlling sunlight and solar heat. The top diagram represents the window in its clear state, allowing maximum sunlight and solar heat to enter. The bottom diagram shows the window in its tinted state, where the electrochromic coating darkens to reduce sunlight and block solar heat, enhancing energy efficiency. The window has multiple glass surfaces, with the electrochromic layer applied to the outer pane. This technology helps regulate indoor temperatures, reducing cooling costs in hot climates and improving comfort by minimizing glare and heat gain inside the building.

Figure 7. Electrochromic glass technology in clear and tinted states

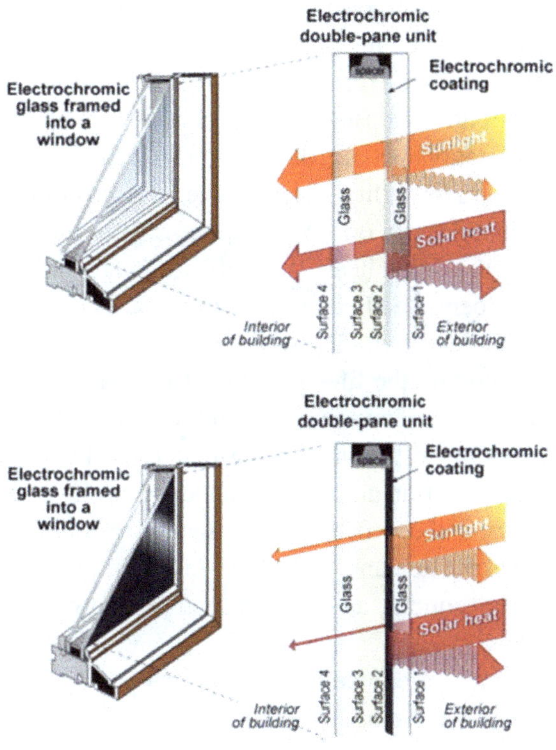

Source: Sbar et al. (2012)

3.3 Reflective and Cool Roofing Materials

Reflective and cool roofing materials help reduce heat absorption, lowering indoor temperatures and improving energy efficiency. These materials, including cool roof coatings, reflective shingles, and metal roofing, are designed to reflect more sunlight and emit absorbed heat efficiently. They reduce the urban heat island effect, decrease cooling costs, and enhance indoor comfort. Cool roofs are commonly made from materials like white membranes, specially coated tiles, or reflective metal panels. By minimizing heat retention, they extend roof lifespan and reduce thermal stress. Widely used in residential, commercial, and industrial buildings, cool roofing supports sustainability efforts and helps combat climate change.

3.3.1 Cool Roof Coatings

Cool roof coatings are specialized reflective materials applied to rooftops to reduce heat absorption and improve energy efficiency. These coatings, typically made from acrylic, silicone, or elastomeric materials, work by reflecting more sunlight and emitting absorbed heat, keeping buildings cooler. This technology is particularly beneficial in hot climates, where excessive heat retention can lead to increased indoor temperatures and higher air-conditioning costs (Dabaieh et al. 2015; Pisello, 2017; Sedaghat et al. 2023; Aggarwal et al. 2024; Ling et al. 2024). A primary advantage of cool roof coatings is their ability to lower energy consumption, reducing electricity bills and decreasing strain on HVAC systems. By minimizing heat buildup, they also help mitigate the urban heat island effect, where cities experience higher temperatures due to concentrated heat-absorbing surfaces (Santamouris, 2013; Liu & Morawska, 2020; Wai et al. 2025). Additionally, cool roof coatings contribute to sustainability by extending the lifespan of roofing materials, reducing the need for frequent replacements, and lowering carbon footprints. These coatings come in various formulations to suit different roofing materials, including metal, asphalt, and concrete. Proper application and maintenance ensure long-term benefits, making them a cost-effective and eco-friendly solution for residential, commercial, and industrial buildings. As the demand for energy-efficient building solutions grows, cool roof coatings are becoming an essential component of modern construction and retrofitting projects.

Figure 8 illustrates the difference between an uncoated roof and a cool-coated roof in terms of heat absorption and reflection. On the left, the uncoated roof absorbs more solar radiation, leading to higher temperatures inside the building, as indicated by the red thermometer. On the right, the cool-coated roof, enhanced with NIR (Near-Infrared) reflective pigments, reflects more sunlight and emits absorbed heat efficiently, keeping the building cooler (shown by the blue thermometer). This demonstrates how cool roof coatings reduce indoor temperatures, lower energy consumption for cooling, and contribute to mitigating the urban heat island effect in hot climates.

Figure 8. Cool Roof Coatings on Heat Reflection and Temperature Reduction

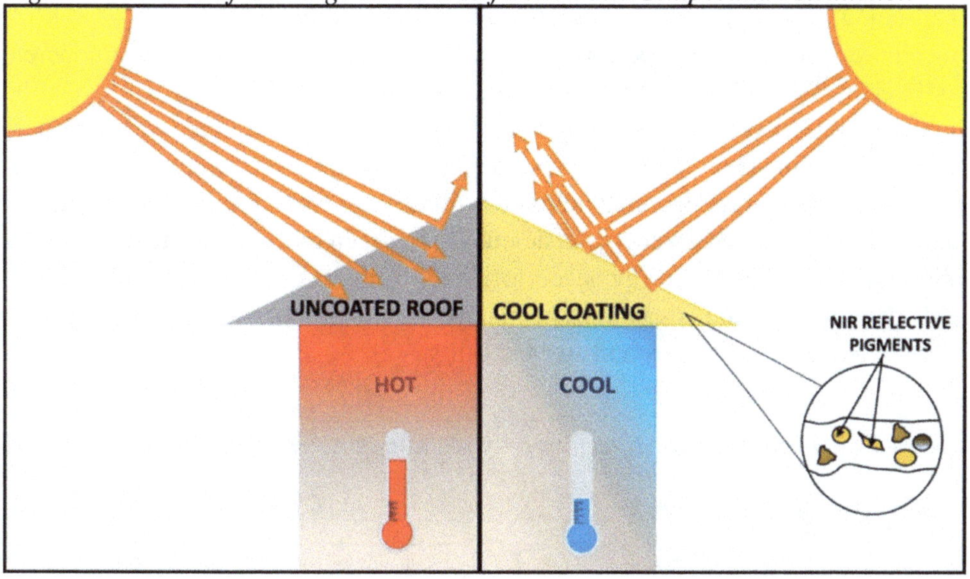

Source: Rosati et al. (2021)

3.3.2 Green Roof Systems

Green roof systems, also known as living roofs, are rooftops covered with vegetation and a growing medium over a waterproofing layer. These systems provide numerous environmental, economic, and social benefits, making them an essential feature in sustainable urban development. One of the primary advantages of green roofs is their ability to reduce urban heat island effects by absorbing heat and providing natural insulation. They help regulate indoor temperatures, reducing the need for air conditioning in summer and heating in winter, which leads to lower energy consumption and utility costs (He et al. 2021; Tseng et al. 2022; Shi et al. 2022; Mihalakakou et al. 2023). Additionally, green roofs improve air quality by capturing pollutants and filtering carbon dioxide, contributing to a healthier urban environment. Another key benefit is stormwater management. Green roofs absorb and retain rainwater, reducing runoff and lowering the risk of urban flooding. This also helps ease the burden on municipal drainage systems. Furthermore, they enhance biodiversity by providing habitat for birds, bees, and other pollinators. Green roof systems come in two types: extensive (lightweight with low-maintenance vegetation) and intensive (heavier with trees, shrubs, and even urban farms). Whether for commercial or residential buildings, green roofs enhance sustainability, promote well-being, and contribute to a greener, more resilient urban future.

Figure 9 compares the performance of PV (photovoltaic) panels on a concrete roof versus a green roof in terms of energy efficiency and heat regulation. On the left, the PV Concrete Roof shows a higher PV surface temperature, which reduces the efficiency of solar energy conversion. On the right, the PV Green Roof benefits from evapotranspiration, which helps cool the environment, reducing solar surface heat and improving green energy efficiency. The green roof system creates a cooler microclimate, enhancing PV panel performance by maintaining lower temperatures. This demonstrates the advantages of integrating green roofs with solar panels for increased energy efficiency and sustainability.

Figure 9. Photovoltaic-green roof vs photovoltaic-concrete roof

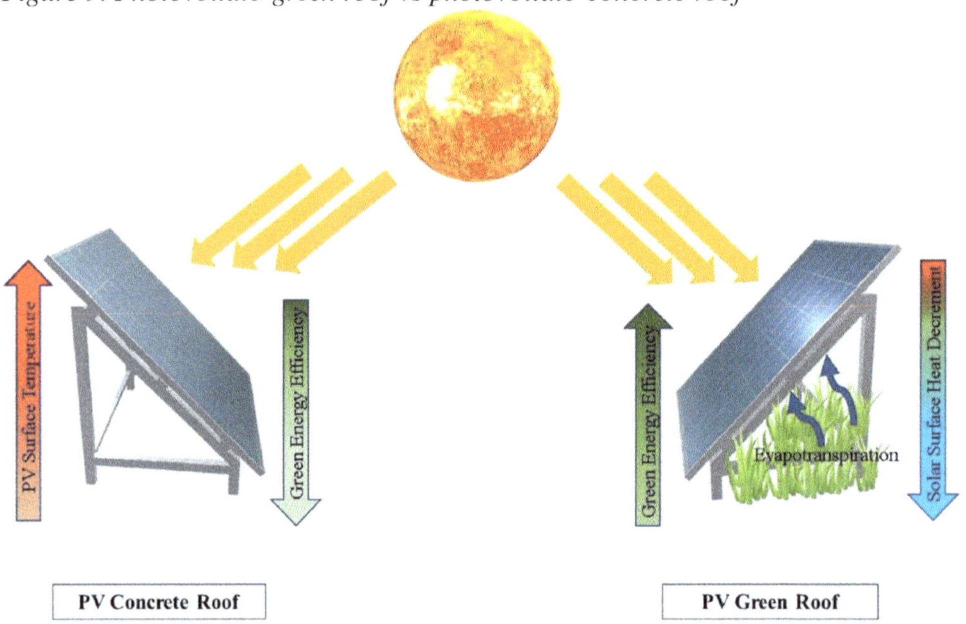

Source: Talwar et al. (2023)

3.4 Sustainable Concrete Alternatives

Sustainable concrete alternatives reduce environmental impact while maintaining strength and durability. Options include fly ash and slag-based concrete, geopolymer concrete, and carbon-capturing concrete. Fly ash and slag replace traditional Portland cement, cutting carbon emissions. Geopolymer concrete uses industrial byproducts, requiring less energy and producing fewer emissions. Carbon-capturing concrete incorporates CO_2 during production, reducing atmospheric greenhouse gases. Other eco-friendly choices include recycled aggregate concrete and hempcrete, which offer insulation benefits and lower embodied carbon. These alternatives help decrease the construction industry's carbon footprint, conserve natural resources, and support green building initiatives while ensuring structural integrity and longevity in modern construction.

3.4.1 Fly Ash and Slag Cement

Fly ash and slag cement are widely used as sustainable alternatives to traditional Portland cement, contributing to environmentally friendly and durable concrete production. Both materials are industrial by-products that help reduce carbon emissions and enhance concrete properties. Fly ash is a fine powder produced from coal combustion in power plants. It is rich in silica and alumina, making it an excellent pozzolanic material. When mixed with cement, fly ash reacts with calcium hydroxide to form additional cementitious compounds, improving strength, durability, and resistance to sulphate attack (Papadakis, 2000; Chindaprasirt et al. 2007; Wang et al. 2017; Elahi et al. 2021; Harilal et al. 2021). Using fly ash reduces landfill waste, decreases cement demand, and lowers CO_2 emissions, making it a sustainable choice in construction. Slag cement, or ground granulated blast furnace slag (GGBFS), is a by-product of the steel industry. It is formed by rapidly cooling molten iron slag and grinding it into a fine powder. Slag cement enhances concrete durability, workability, and resistance to chemical attacks while reducing heat generation during hydration. Its use minimizes industrial waste and significantly reduces the environmental impact of cement production. Both fly ash and slag cement contribute to sustainable construction by conserving natural resources, reducing energy consumption, and enhancing the performance of concrete structures (Berndt, 2009; Xu & Shi, 2018 ; Song et al. 2021). Their adoption supports green building initiatives and lowers the carbon footprint of the construction industry.

Table 2 compares fly ash and slag cement, highlighting their sustainability benefits, chemical properties, and performance in concrete. Both materials are industrial by-products, reducing landfill waste and lowering CO_2 emissions. While fly ash is a pozzolanic material requiring calcium hydroxide for strength development, slag cement is hydraulic and gains strength faster. Both improve durability, workability and reduce heat of hydration, making them ideal for mass concrete structures. Fly ash is widely used in bricks and soil stabilization, whereas slag cement is common in marine and high-performance concrete. Their adoption supports eco-friendly construction and green building initiatives.

Table 2. Fly Ash and Slag Cement: Sustainable Alternatives for Green Concrete Production

Feature	Fly Ash	Slag Cement (GGBFS)	Key Facts
Source	By-product of coal combustion in power plants	By-product of iron and steel production (blast furnace slag)	Utilizes industrial waste, reducing landfill disposal
Chemical Composition	Rich in silica (SiO_2) and alumina (Al_2O_3)	Rich in calcium, silica, and alumina	Both materials contribute to pozzolanic and hydraulic reactions
Cement Type	Pozzolanic material	Latent hydraulic material	Fly ash requires calcium hydroxide to react, while slag reacts with water
Reaction with Cement	Reacts with calcium hydroxide to form additional cementitious compounds	Reacts with water and hydrates like Portland cement	Slag cement provides faster strength gain than fly ash
Strength Development	Improves long-term strength	Enhances both early and long-term strength	Slag is beneficial for high-strength concrete applications
Durability Benefits	Improves resistance to sulphate attack and alkali-silica reaction	Increases resistance to chloride attack and chemical deterioration	Both improve concrete longevity
Workability	Improves workability, reduces water demand	Improves workability, reduces bleeding	Better workability means easier placement of concrete
Heat of Hydration	Lowers heat of hydration, useful for mass concrete	Significantly reduces heat generation, preventing thermal cracks	Beneficial in large structures like dams and bridges
Environmental Impact	Reduces CO_2 emissions and landfill waste	Lowers cement demand and industrial waste	Helps in sustainable construction by reducing carbon footprint
Common Applications	Used in concrete, bricks, and soil stabilization	Used in marine structures, pavements, and high-performance concrete	Both materials are widely used in green building projects

3.4.2 Geopolymer Concrete

Geopolymer concrete (GPC) is an innovative and eco-friendly alternative to conventional Portland cement concrete. It is made by utilizing industrial by-products like fly ash, slag, or metakaolin, which react with alkaline activators such as sodium hydroxide (NaOH) and sodium silicate (Na_2SiO_3) to form a strong, durable binder. This chemical reaction, known as geopolymerization, eliminates the need for traditional cement, significantly reducing carbon dioxide (CO_2) emissions (Elmesalami & Celik, 2022; Garces et al. 2022; Freire et al. 2022; Danish et al. 2022; Mushtaq et al. 2025). One of the main advantages of geopolymer concrete is its superior strength, durability, and resistance to harsh environments, including sulphate attacks, acid exposure, and high temperatures. It also exhibits low shrinkage, high early strength, and excellent bonding properties (Zhang et al. 2018; Asayesh et al. 2021; Zhang et al. 2022). Due to these benefits, GPC is widely used in infrastructure projects, precast elements, marine structures, and high-temperature applications. Additionally, geopolymer concrete contributes to sustainability by utilizing industrial waste, reducing landfill disposal, and lowering the carbon footprint of construction activities. Despite its benefits, challenges such as the availability of raw materials, standardization of mix designs, and curing methods limit its widespread adoption. With continued research and technological advancements, geopolymer concrete has the potential to revolutionize the construction industry by offering a sustainable and high-performance alternative to conventional cement-based concrete.

Table 3 compares Geopolymer Concrete (GPC) with Conventional Concrete in terms of strength, durability, emissions, and cost. GPC exhibits 30-60% higher compressive strength (40-80 MPa) than conventional concrete (25-50 MPa), along with better flexural strength and lower shrinkage. It significantly reduces CO_2 emissions by 60-80%, making it an environmentally friendly alternative. Additionally, GPC shows higher resistance to sulphate and acid attacks, enhancing durability. While initial costs maybe 10-20% higher, its long-term benefits, including reduced maintenance and superior performance, justify its use in sustainable construction. GPC's adaptability to heat or ambient curing further improves its practicality.

Table 3. Comparison of Geopolymer Concrete and Conventional Concrete: Properties and Performance

Property	Geopolymer Concrete	Conventional Concrete	Improvement (%)
Compressive Strength (MPa)	40 - 80 MPa	25 - 50 MPa	30 - 60% higher
CO_2 Emission Reduction	60 - 80%	-	Significant reduction
Setting Time	30 - 120 min	45 - 240 min	Faster in some cases
Durability (Sulphate & Acid Resistance, Chloride Permeability)	High	Moderate	50 - 70% better
Shrinkage (%)	0.02 - 0.04%	0.05 - 0.1%	50 - 60% lower
Curing Requirements	Heat or Ambient	Water curing	More adaptable
Flexural Strength (MPa)	5 - 10 MPa	3 - 6 MPa	30 - 50% higher
Initial Cost ($/m^3)	120 - 180	90 - 150	10 - 20% higher (varies)

3.4.3 Recycled Aggregate Concrete

Recycled Aggregate Concrete (RAC) is an eco-friendly construction material that incorporates recycled aggregates from demolished concrete structures, reducing reliance on natural resources. The coarse and fine aggregates used in RAC are obtained by crushing old concrete, which is then processed and cleaned to remove impurities. This approach aligns with sustainable construction practices, minimizing construction waste and landfill use while conserving natural stone and sand.

One of the key benefits of RAC is its contribution to environmental sustainability. Using recycled materials significantly reduces the carbon footprint of concrete production (Xiao et al. 2018; Bostanci et al. 2018; Xiao et al. 2022). However, RAC generally exhibits slightly lower strength and durability than traditional concrete due to the presence of old mortar attached to the recycled aggregates, which affects bond strength and porosity. This can be mitigated by improving mix design, adding supplementary cementitious materials like fly ash or silica fume, and enhancing curing methods. RAC is widely used in non-structural applications, road bases, pavement construction, and precast concrete elements. Recent advancements in processing and mix optimization have made it feasible for structural applications, promoting circular economy principles (de Andrade Salgado & de Andrade Silva, 2022; Zandifaez et al. 2023; Yu et al. 2024; Liu et al. 2024). While challenges remain in standardization and performance consistency, RAC offers a sustainable alternative for modern construction, reducing waste and conserving natural resources.

Table 4 compares Recycled Aggregate Concrete (RAC) with Conventional Concrete (NAC) in terms of strength, durability, density, and sustainability. RAC generally exhibits 10-20% lower compressive strength and higher water absorption (50-100%), making it more porous and slightly less durable. However, it offers 10-30% CO_2 emission reduction and 10-25% cost savings, making it a more sustainable and economical choice. While RAC has slightly higher shrinkage and lower flexural strength, improvements in mix design and additives can enhance its performance. Despite some limitations, RAC is a viable alternative for non-structural and structural applications, supporting sustainable construction and reducing environmental impact.

Table 4. Comparison of Recycled Aggregate Concrete (RAC) and Conventional Concrete: Properties and Performance

Property	Recycled Aggregate Concrete (RAC)	Conventional Concrete (NAC)	Change (%)
Compressive Strength (MPa)	20 - 50 MPa	25 - 60 MPa	10 - 20% lower
Water Absorption (%)	3 - 7%	1 - 3%	50 - 100% higher
Density (kg/m^3)	2200 - 2400	2300 - 2500	~5% lower
Durability (Chloride & Sulphate Resistance)	Moderate	High	Lower resistance
Flexural Strength (MPa)	2 - 6 MPa	3 - 7 MPa	~10% lower
Shrinkage (%)	0.05 - 0.15%	0.03 - 0.10%	20 - 50% higher
CO_2 Emission Reduction	10 - 30%	-	Lower environmental impact
Cost Reduction (%)	10 - 25%	-	More economical

3.5 Natural and Renewable Materials

Natural and renewable materials are sustainable resources used in construction, design, and manufacturing to reduce environmental impact. Examples include bamboo, cork, hemp, wool, straw, and reclaimed wood, which are biodegradable, rapidly renewable, and require minimal processing. These materials offer excellent insulation, durability, and aesthetic appeal while reducing reliance on synthetic or non-renewable resources. They also promote healthier indoor air quality by minimizing toxic emissions. Used in flooring, insulation, furniture, and structural components, natural materials contribute to eco-friendly design and sustainable building certifications. Their widespread adoption supports resource conservation, carbon sequestration, and the transition to a circular, low-carbon economy.

3.5.1 Straw Bale and Hempcrete

Straw bale and hempcrete are two sustainable building materials gaining popularity due to their environmental benefits, insulation properties, and affordability. Both materials provide an eco-friendly alternative to traditional construction methods, reducing carbon footprints and promoting energy efficiency. Straw bale construction utilizes tightly packed bales of straw as insulation and structural support within walls. This method is known for its high thermal resistance, reducing energy costs for heating and cooling. Straw is an agricultural by-product, making it a renewable and low-cost material. Additionally, properly constructed straw bale walls offer good fire resistance due to the lack of oxygen in densely packed bales (Koh & Kraniotis,

2021; Tlaiji et al. 2022 ; Sun et al. 2023). However, care must be taken to prevent moisture infiltration, which can lead to mold or decay.

Hempcrete is a bio-composite material made from the inner woody core of the hemp plant (shiv), lime, and water. It provides excellent insulation, breathability, and resistance to mold and pests (Abdellatef et al. 2020; Asghari & Memari, 2024; Muhit et al. 2024). Unlike concrete, hempcrete is lightweight and has a lower carbon footprint, as hemp plants absorb CO_2 during growth. Although not a load-bearing material, hempcrete is used as an insulating infill in timber-framed buildings. Its thermal mass helps regulate indoor temperatures, improving energy efficiency. Both straw bale and hempcrete support sustainable construction by using natural, renewable resources, reducing reliance on synthetic materials, and enhancing building performance.

The image in Fig. 10 presents cast hemp concrete cubes measuring 100 mm in size. The cube on the left (a) is composed of 90% hydrated lime and fly ash, while the cube on the right (b) contains an additional component, calcium chloride ($CaCl_2$). The inclusion of $CaCl_2$ likely influences the hydration process, setting time, and mechanical properties of the concrete. The surface textures appear relatively uniform with some visible roughness, possibly due to the hemp fibres. The colour variation may indicate differences in composition, curing, or reaction with atmospheric moisture, affecting overall strength and durability characteristics.

Figure 10. Cast hemp concrete cubes of size 100 mm formed from (a) 90% hydrated lime + fly ash; (b) 90% hydrated lime + fly ash + $CaCl_2$

Source: Jami et al. (2019)

3.5.2 Bamboo and Timber

Bamboo is gaining popularity as a sustainable construction material due to its eco-friendliness, strength, and rapid renewability. Unlike traditional timber, bamboo grows exceptionally fast, with some species reaching maturity in 3 to 5 years. This makes it a highly renewable resource compared to hardwood trees, which can take decades to mature. One of bamboo's most significant advantages is its strength-to-weight ratio, which rivals that of steel. Its natural flexibility makes it an excellent material for earthquake-resistant structures, while its high compressive strength allows it to be used in load-bearing applications. Additionally, bamboo's hollow structure makes it lightweight and easy to transport, reducing the carbon footprint associated with construction (Zhang et al. 2021; Liu et al. 2023; Zhao et al. 2023). Bamboo is also an environmentally friendly material because it absorbs more carbon dioxide and produces more oxygen than most trees. Furthermore, it regenerates naturally without replanting, making it a highly sustainable choice. When used in construction, bamboo can be treated to resist pests and moisture, increasing its lifespan (Manandhar et al. 2019; Sain et al. 2024). From housing and bridges to flooring and furniture, bamboo offers a sustainable alternative to conventional materials. As global demand for eco-friendly construction rises, bamboo stands out as a viable solution for green building, reducing environmental impact while promoting sustainability.

Figure 11 showcases two bamboo-based construction techniques. The first image illustrates a typical all-bamboo roof, consisting of a bamboo frame that supports layers of bamboo basketry and bamboo tiles. The basketry provides an initial layer, while the tiles act as the outer protective covering, offering durability against weather conditions. The second image presents a bamboo-reinforced clay wall, where a bamboo frame supports woven bamboo basketry, which is then coated with clay. This method enhances structural stability, insulation, and earthquake resistance. Both techniques highlight bamboo's adaptability as a sustainable, lightweight, and eco-friendly material in traditional and modern construction.

Figure 11. Traditional bamboo-bearing structures

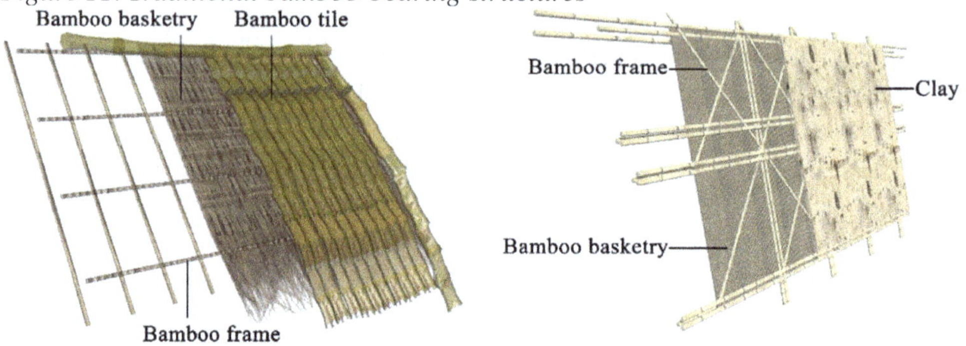

(A) Typical all-bamboo roof

(B) Bamboo-reinforced clay wall

Source: Ding and Xian (2024)

Timber is a highly sustainable construction material due to its renewability, carbon sequestration properties, and energy efficiency. Sourced from responsibly managed forests, timber is a renewable resource that can be replenished through sustainable forestry practices. When trees grow, they absorb carbon dioxide from the atmosphere, storing it within the wood even after being used in construction. This makes timber a carbon-negative material, helping to mitigate climate change.

Compared to concrete and steel, timber requires significantly less energy to produce and process, reducing its overall environmental impact (Skullestad et al. 2016; Morris et al. 2021; Abed et al. 2022). Its lightweight nature makes transportation more energy-efficient, further lowering carbon emissions. Additionally, timber provides excellent natural insulation, reducing heating and cooling costs in buildings and contributing to energy efficiency (Asdrubali et al. 2017; Pajek et al. 2017; Busselli et al. 2021). Modern engineered wood products, such as cross-laminated timber (CLT) and glue-laminated timber (glulam), enhance the strength and versatility of timber, making it suitable for large-scale structures. Timber is also biodegradable and can be repurposed or recycled at the end of its life cycle, reducing construction waste. As the demand for sustainable building materials increases, timber stands out as a viable alternative to traditional materials, offering durability, aesthetic appeal, and a lower environmental footprint in eco-friendly construction projects.

Figure 12 presents two aspects of timber construction. Image (A) shows a workshop setting where workers are assembling a timber structure, likely a roof truss or prefabricated timber frame. The use of clamps and precision tools suggests careful craftsmanship and structural integrity in timber-based construction. Image (B) displays a close-up of cross-laminated timber (CLT) panels, which are composed of multiple layers of wood arranged perpendicularly for enhanced strength and stability. CLT is widely used in modern sustainable construction due to its durability, load-bearing capacity, and eco-friendly properties, making it a viable alternative to traditional concrete and steel materials.

Figure 12. (A) Small-scale production of nailed stacked plank panel system and (B) completed nailed stacked plank panel for use in housing service core module

(A) (B)

Source: Jones and Brischke (2017)

3.5.3 Cork and Natural Fibre Insulation

Cork and natural fibre insulation are sustainable and eco-friendly alternatives to traditional synthetic insulation materials. Derived from renewable sources, these materials provide excellent thermal and acoustic insulation while reducing environmental impact (Malanho et al. 2021; Novais et al. 2021; Ouakarrouch et al. 2022). Cork is harvested from the bark of the cork oak tree, primarily found in Portugal and Spain. The bark regenerates after harvesting, making cork an exceptionally sustainable resource. Cork insulation is naturally resistant to moisture, mold, and pests, and it has excellent thermal performance due to its cellular structure. Additionally, it provides soundproofing properties and fire resistance without the need for chemical additives.

Natural fibre insulation includes materials like hemp, wool, flax, cotton, and wood fibre. These fibres are biodegradable, breathable, and have low embodied energy compared to synthetic insulations like fiberglass or foam. Wool, for example, has excellent moisture-wicking properties and maintains its insulating capacity even when damp. Hemp insulation is durable, mold-resistant, and has a low carbon footprint. Both cork and natural fibre insulation contribute to healthier indoor air quality as they do not release harmful VOCs (volatile organic compounds). Their ability to regulate humidity and provide long-term efficiency makes them ideal choices for sustainable and energy-efficient construction. As the demand for green building materials grows, cork and natural fibres stand out as viable, eco-conscious solutions.

4. INNOVATIVE TECHNOLOGIES IN ENERGY-EFFICIENT MATERIALS

Innovative technologies are transforming energy-efficient materials, driving sustainability in construction. Advanced materials such as aerogels, phase-change materials (PCMs), and bio-based composites enhance insulation, reducing energy consumption in buildings. AI-driven material discovery accelerates the development of eco-friendly alternatives, optimizing thermal properties and durability. 3D printing and nanotechnology enable precision manufacturing, minimizing material waste while improving energy performance. Digital twin technology simulates material behaviour, ensuring optimal selection for sustainable projects. Additionally, self-healing concrete and photocatalytic coatings contribute to longevity and air purification. These advancements support the transition toward greener construction, reducing carbon footprints and improving building efficiency worldwide.

4.1 Phase-Change Materials (PCMs)

Phase-change materials (PCMs) are an innovative solution for improving energy efficiency in buildings by regulating temperature through latent heat storage. These materials absorb, store, and release thermal energy as they transition between solid and liquid phases, helping maintain indoor comfort while reducing heating and cooling demands (Zhou et al. 2012; Sharma et al. 2015; Abuelnuor et al. 2018; Suresh et al. 2022). PCMs are categorized into organic (paraffins, fatty acids), inorganic (salt hydrates, metal alloys), and eutectic (a combination of organic and inorganic) types. Organic PCMs offer chemical stability and non-corrosiveness, while inorganic PCMs provide higher thermal conductivity and energy storage capacity. Eutectic PCMs combine the advantages of both, enhancing performance and reliability. Studies have quantified thermal lag improvement and peak temperature reduction using PCMs in building walls (Sharma et al., 2009; Kalnæs & Jelle, 2015).

Applications of PCMs in sustainable construction include integration into building walls, ceilings, and flooring to regulate temperature fluctuations passively. Encapsulated PCMs are also used in thermal insulation panels, concrete, and plaster to improve energy efficiency (Huang et al. 2023; Shoeibi et al. 2024). Smart glazing with PCM coatings enhances window insulation, reducing winter heat loss and minimizing summer cooling needs. By reducing reliance on HVAC systems, PCMs contribute to lower energy consumption and carbon emissions. Their ability to enhance thermal comfort makes them crucial in green building design, supporting the transition toward energy-efficient and sustainable construction practices.

4.2 Self-Healing and Smart Materials

Self-healing and smart materials represent a groundbreaking advancement in sustainable construction, enhancing durability, efficiency, and maintenance reduction. These materials can repair damage autonomously or respond to environmental changes, extending their lifespan and reducing resource consumption. Self-healing materials, such as self-healing concrete, contain embedded microcapsules of healing agents (e.g., bacteria, polymers, or minerals) that activate when cracks form (Qian et al. 2021; Hossain et al. 2022; Nguyen et al. 2023; Zhang et al. 2024). These agents fill and repair the cracks, preventing further deterioration and reducing maintenance costs. This innovation significantly enhances infrastructure longevity, minimizing the need for repairs and lowering carbon emissions associated with reconstruction.

Smart materials, including thermochromic and photochromic coatings, adapt to environmental conditions by changing their properties in response to stimuli such as temperature, light, or moisture. Shape-memory alloys and polymers can revert to a predefined shape when exposed to specific triggers, enhancing structural resilience (Molod et al. 2022; Raza et al. 2022; Tabrizikahou et al. 2022). Additionally, piezoelectric materials convert mechanical stress into electrical energy, offering the potential for self-powered sensors in buildings (Abid et al. 2021; Wang et al. 2022; Moussa et al. 2022). By integrating self-healing and smart materials into construction, energy efficiency improves, material waste decreases, and long-term sustainability is enhanced. These innovations contribute to the development of resilient, low-maintenance structures that align with green building practices and the circular economy.

4.3 Nanotechnology in Building Materials

Nanotechnology is revolutionizing the construction industry by enhancing the performance, durability, and sustainability of building materials. By manipulating materials at the nanoscale, scientists and engineers are developing innovative solutions that improve strength, energy efficiency, and environmental resistance. One of the most significant applications of nanotechnology is in nanoparticles and nanocomposites. Nano-enhanced concrete, infused with carbon nanotubes or nanosilica, offers increased strength, reduced permeability, and self-healing properties. This leads to more durable structures with lower maintenance requirements. Similarly, nanocoating, such as hydrophobic and self-cleaning coatings, repels water and pollutants, extending the lifespan of facades and glass surfaces.

Thermal insulation benefits greatly from nanotechnology, with aerogels and nano-enhanced insulation materials providing superior heat resistance while maintaining lightweight properties (Adhikary et al. 2021; Wu et al. 2023; Yang et al.

2024). Nanophotonic materials used in smart windows regulate light transmission, improving energy efficiency by reducing heating and cooling demands. Additionally, antimicrobial and air-purifying materials utilizing nanoparticles of titanium dioxide (TiO_2) help break down pollutants and kill bacteria upon exposure to sunlight, enhancing indoor air quality (Singh et al. 2021; Akin & Mutlu Danaci, 2021; Castro-Hoyos et al. 2022). Piezoelectric nanomaterials can even harvest mechanical energy from building vibrations, contributing to self-sustaining energy systems. By integrating nanotechnology into building materials, the construction industry is advancing toward more resilient, energy-efficient, and eco-friendly structures, supporting sustainable development and green building initiatives.

Nanotechnology is revolutionizing building materials by enhancing strength, durability, energy efficiency, and sustainability. Table 5 highlights the diverse applications and benefits of nanotechnology in building materials. Nano-silica strengthens concrete, while carbon nanotubes enhance flexibility and conductivity (Du et al. 2014; Abhilash et al. 2021; Zhang et al. 2023; Jayakumari et al. 2023). Nano-TiO_2 provides self-cleaning properties, and nano-clay improves fire resistance. Other nanomaterials like graphene, alumina, copper, and aerogel contribute to durability, antimicrobial properties, insulation, and energy efficiency. Nano-silica improves concrete strength and reduces permeability, while carbon nanotubes (CNTs) reinforce cement and coatings for better flexibility and conductivity. Nano-titanium dioxide (TiO_2) enables self-cleaning surfaces and air purification, and nano-clay enhances fire resistance. Graphene strengthens concrete and improves thermal conductivity, while nano-alumina increases wear resistance. Nano-copper and nano-zinc oxide offer antimicrobial protection in paints and coatings. Aerogels provide superior insulation, and quantum dots enhance energy-efficient lighting and smart windows. These advancements lead to sustainable, high-performance buildings with improved safety and functionality.

Table 5. Applications and Benefits of Nanotechnology in Building Materials

Nanomaterial	Application in Building Materials	Benefits	Examples
Nano-Silica	Concrete, coatings	Increases strength, and durability, reduces permeability	High-performance concrete, self-cleaning coatings
Carbon Nanotubes (CNTs)	Reinforcement in cement, coatings	Enhances mechanical strength, flexibility, conductivity	CNT-reinforced concrete, smart coatings
Nano-Titanium Dioxide (TiO_2)	Self-cleaning surfaces, UV protection	Provides photocatalytic properties for air purification and self-cleaning	Self-cleaning glass, pollution-reducing paints

continued on following page

Table 5. Continued

Nanomaterial	Application in Building Materials	Benefits	Examples
Nano-Clay	Fire-resistant coatings, composites	Improves fire resistance, barrier properties, and durability	Fire-retardant coatings, reinforced polymers
Nano-Graphene	Concrete, insulation, coatings	Increases strength, thermal conductivity, and electrical properties	Graphene-enhanced cement, conductive coatings
Nano-Alumina	Structural materials, coatings	Enhances wear resistance, thermal stability, and hardness	Corrosion-resistant coatings, reinforced composites
Nano-Copper	Antibacterial coatings, paints	Provides antimicrobial properties, prevents biofilm formation	Antimicrobial hospital surfaces, hygiene coatings
Nano-Zinc Oxide	UV-blocking coatings, paints	Provides UV resistance, antimicrobial effects	UV-resistant exterior paints, antibacterial coatings
Aerogel Nanoparticles	Insulation materials	Ultra-lightweight, excellent thermal insulation	High-performance insulation panels, fireproof coatings
Nano-Iron Oxide	Energy-efficient windows, pigments	Enhances UV absorption, thermal regulation	Smart glass, heat-reflective coatings
Quantum Dots	Smart lighting, energy-efficient windows	Enhances optical properties, improves energy efficiency	Solar window coatings, LED lighting

4.4 3D-Printed Sustainable Materials

3D printing is revolutionizing sustainable construction by enabling the creation of eco-friendly, efficient, and customizable building materials. This technology minimizes waste, reduces construction time, and allows for innovative material compositions that enhance energy efficiency and sustainability (Chen et al. 2022; Nguyen-Van et al. 2023; Ahmed, 2023). One of the most significant advantages of 3D-printed materials is their ability to use recycled and bio-based resources. Materials such as recycled concrete, biodegradable bioplastics, and composite mixtures of clay, sand, and natural fibres are increasingly being used in 3D-printed structures. Additionally, geopolymer cement, a low-carbon alternative to traditional Portland cement, reduces CO_2 emissions while maintaining durability.

3D printing allows for precise material usage, significantly reducing construction waste compared to traditional methods. Complex, high-performance architectural designs can be printed with optimized material distribution, improving structural integrity while using fewer resources. Furthermore, insulating materials can be integrated directly into printed walls, enhancing energy efficiency in buildings. The adaptability of 3D printing also supports modular and prefabricated construction, making it easier to design for disassembly and reuse. This aligns with circular economy principles, promoting sustainability. By integrating 3D-printed sustainable materials, the construction industry moves toward eco-friendly, cost-effective, and resilient infrastructure, reducing environmental impact while meeting the growing demand for green buildings.

5. ENERGY EFFICIENCY STANDARDS AND CERTIFICATIONS

Energy efficiency standards and certifications ensure buildings and materials meet specific sustainability and performance criteria, reducing energy consumption and environmental impact. Major certifications include LEED (Leadership in Energy and Environmental Design), BREEAM (Building Research Establishment Environmental Assessment Method), ENERGY STAR, and Passive House. These standards promote efficient insulation, renewable energy use, smart building technologies, and low-carbon materials. Governments and organizations adopt these certifications to lower operational costs, enhance indoor comfort, and reduce greenhouse gas emissions. By following energy efficiency standards, the construction industry supports sustainable development, resource conservation, and the creation of high-performance, eco-friendly buildings worldwide.

5.1 LEED (Leadership in Energy and Environmental Design)

LEED (Leadership in Energy and Environmental Design) is a globally recognized green building certification system developed by the U.S. Green Building Council (USGBC). It provides a framework for designing, constructing, and operating environmentally responsible and resource-efficient buildings. LEED certification applies to various building types, including commercial, residential, and institutional structures, promoting sustainable practices across the construction industry. LEED evaluates buildings based on multiple criteria, including energy efficiency, water conservation, indoor environmental quality, sustainable materials, and site development (Lee et al. 2008; Azhar et al. 2011; Asdrubali et al. 2015; Leite Ribeiro et al. 2025). The certification is awarded at four levels—Certified, Silver, Gold, and Platinum—depending on the number of sustainability points earned. However, studies suggest limitations in LEED's actual impact on energy use. Newsham et al. (2009)

found variability in LEED-certified buildings' energy performance, suggesting a need for post-occupancy verification.

Key benefits of LEED-certified buildings include reduced energy and water consumption, lower operational costs, improved occupant health, and enhanced property value. The system encourages the use of renewable energy sources, efficient insulation, low-carbon materials, and smart building technologies to minimize environmental impact. Additionally, LEED promotes waste reduction and encourages recycling during construction and throughout a building's lifecycle. With growing concerns about climate change and resource depletion, LEED plays a crucial role in advancing sustainable construction. Governments, businesses, and individuals worldwide adopt LEED standards to create greener, healthier, and more efficient buildings, contributing to long-term environmental and economic sustainability.

5.2 BREEAM (Building Research Establishment Environmental Assessment Method)

BREEAM (Building Research Establishment Environmental Assessment Method) is one of the world's leading sustainability assessment and certification systems for buildings. Developed by the Building Research Establishment (BRE) in the UK, BREEAM evaluates the environmental, social, and economic sustainability of a building throughout its lifecycle, from design to operation. BREEAM certification assesses buildings based on several key categories, including energy efficiency, water conservation, material sustainability, waste management, pollution reduction, and occupant well-being (Grzegorzewska & Kirschke, 2021; Nicolini, 2022; Ferrari et al. 2022). The certification is awarded at five levels—Pass, Good, Very Good, Excellent, and Outstanding—based on the sustainability performance of the project. Dwaikat & Ali (2018) analyzed 30 case studies showing that BREEAM buildings achieved, on average, 25% operational energy savings compared to baseline buildings.

One of BREEAM's primary goals is to encourage low-carbon and energy-efficient designs, promoting the use of renewable energy sources, sustainable building materials, and efficient resource management. It also supports the implementation of smart technologies, such as automated energy monitoring and climate-responsive designs, to enhance building performance. BREEAM-certified buildings benefit from lower operating costs, improved market value, reduced environmental impact, and healthier indoor environments. As sustainability becomes a priority in the construction industry, BREEAM plays a crucial role in guiding developers, architects, and engineers toward more sustainable and energy-efficient building solutions, supporting global efforts to combat climate change, and promoting green infrastructure.

5.3 Energy Star and Passive House Standards

Energy Star and Passive House standards are two widely recognized certifications promoting energy efficiency in buildings. While both aim to reduce energy consumption and environmental impact, they differ in approach and scope. Energy Star is a U.S. government-backed certification managed by the Environmental Protection Agency (EPA). It evaluates buildings and appliances based on their energy performance compared to industry standards. Energy Star-certified buildings consume at least 15-30% less energy than conventional buildings by integrating high-efficiency HVAC systems, insulation, lighting, and water management solutions (An & Pivo, 2020; Brolinson et al. 2023). This certification focuses on improving operational energy efficiency, reducing carbon footprints, and lowering utility costs for businesses and homeowners.

Passive House (Passivhaus) is a rigorous building standard developed in Germany that emphasizes ultra-low energy consumption and superior indoor comfort (Jochem, 2009; Han et al. 2022; Riffat et al. 2024). It relies on high-performance insulation, airtight construction, advanced ventilation systems, and minimal heating and cooling needs. Passive House buildings use up to 90% less energy than traditional structures by optimizing heat retention, thermal mass, and passive solar gains (Sadineni et al. 2011; Proietti et al. 2013; Was et al. 2020). While Energy Star focuses on improving efficiency across various buildings and appliances, Passive House takes a holistic design approach to achieve near-zero energy demand. Both standards play a critical role in promoting sustainable construction, reducing energy consumption, and lowering greenhouse gas emissions globally.

5.4 Regional and Local Regulatory Frameworks

Regional and local regulatory frameworks play a crucial role in ensuring energy efficiency and sustainability in the construction industry. These regulations set guidelines, codes, and policies that mandate the use of energy-efficient materials, technologies, and building practices to reduce environmental impact and optimize resource use. One of the most widely implemented frameworks is the International Energy Conservation Code (IECC), which provides minimum efficiency standards for residential and commercial buildings in many regions (Golbazi & Aktas, 2018). Additionally, countries and municipalities adopt green building codes, such as California's Title 24 Energy Code in the U.S. and the European Energy Performance of Buildings Directive (EPBD), to enforce strict energy efficiency measures (Sun et al. 2016).

Local governments often introduce incentives and penalties to encourage compliance with energy efficiency standards. These may include tax credits, rebates, or subsidies for buildings that achieve certifications like LEED, BREEAM, or Passive House. Some cities have implemented net-zero building mandates, requiring new constructions to generate as much energy as they consume. By enforcing regional and local regulatory frameworks, governments can reduce carbon emissions, lower energy costs, and promote sustainable urban development. Compliance with these frameworks ensures that buildings are designed for long-term efficiency, resilience, and environmental responsibility, benefiting both communities and the planet.

6. BRIDGING THE POLICY AND STRATEGIC GAPS IN ENERGY-EFFICIENT MATERIALS

While the chapter provides a thorough overview of technical innovations and advanced materials in sustainable construction, the integration of governance mechanisms, market structures, and long-term resilience strategies is essential to foster real-world implementation and appeal to a broader policy-oriented audience. This section addresses these dimensions.

6.1 Governance Mechanisms: Policy Levers and Institutional Support

Governance frameworks play a crucial role in promoting the adoption of energy-efficient materials. Public policy can drive change through regulatory mandates, incentives, and long-term planning instruments. For instance, building energy codes like the International Energy Conservation Code (IECC) and the European Energy Performance of Buildings Directive (EPBD) establish minimum performance benchmarks. These codes are often updated periodically to reflect new technologies and climate goals. Additionally, financial instruments such as tax credits, rebates, and green procurement standards (e.g., mandating the use of certified materials in public buildings) can stimulate demand. Government-backed pilot projects and public-private partnerships (PPPs) also help de-risk the use of novel materials by demonstrating feasibility and scalability. Strengthening institutional capacity to enforce compliance and providing training to local authorities are essential for ensuring uniform implementation.

6.2 Market Structures: Incentives, Innovation, and Investment

To accelerate adoption, the market ecosystem must support both supply and demand for energy-efficient materials. This includes aligning incentives across the value chain, from manufacturers to end users. Carbon pricing mechanisms, such as emissions trading systems or embodied carbon taxes, can make low-impact materials more competitive. Similarly, environmental product declarations (EPDs) and material passports enable transparency, influencing procurement and consumer decisions. Investment in local production facilities and circular supply chains can mitigate bottlenecks while creating green jobs. Furthermore, integrating ESG (Environmental, Social, and Governance) criteria into financial evaluations incentivizes builders and developers to prioritize sustainability. Programs such as green bonds, climate funds, and concessional loans for high-efficiency retrofitting can mobilize capital toward scalable projects.

6.3 Long-Term Resilience: Climate Adaptation and Lifecycle Durability

In a rapidly changing climate, long-term resilience must be a core design principle. Energy-efficient materials should be evaluated not only on their environmental performance but also on their ability to withstand stressors like extreme weather, flooding, and heatwaves. Materials such as fire-resistant insulation, moisture-regulating bio-composites, and thermal-mass-rich components are crucial in building adaptive capacity. Lifecycle assessment tools should be integrated with future climate projections to guide material choices. Moreover, aligning construction practices with circular economy principles (reuse, recyclability, modularity) enhances resource efficiency and resilience. Urban-level planning, including climate risk zoning and material flow analysis, can support systemic resilience strategies.

7. CHALLENGES AND BARRIERS TO THE ADOPTION OF ENERGY-EFFICIENT MATERIALS

The adoption of energy-efficient materials in construction faces several challenges and barriers. High initial costs of advanced materials, such as aerogels and phase-change materials, discourage widespread use despite long-term savings. This trend is corroborated by NIBS (2016), which reports that advanced materials like VIPs cost up to 10–15 times more per m^2 than conventional insulation. Limited awareness and technical knowledge among builders and developers hinder proper implementation. Regulatory inconsistencies across regions create confusion in compliance. Additionally, supply chain limitations and lack of local availability increase costs and delays. Resistance to change, traditional construction practices, and the slow adoption of new technologies further impede progress. Survey-based studies (Janda, 2013; Darko et al., 2017) show that knowledge gaps among professionals remain one of the top three barriers to green material adoption globally. Overcoming these barriers requires incentives, education, and stronger policies to promote energy-efficient construction globally.

7.1 Cost Implications and Economic Viability

The use of energy-efficient materials in construction involves both cost implications and economic viability. While these materials often require a higher initial investment, they provide substantial long-term financial and environmental benefits. Advanced materials such as high-performance insulation, low-emissivity glass, and phase-change materials generally have elevated production costs and may require specialized labour for installation (Alassaf, 2024; Yaras et al. 2024). This can discourage widespread adoption, particularly in regions where financial incentives are limited or where traditional construction methods dominate. Despite the upfront costs, energy-efficient materials significantly reduce operational expenses over a building's lifespan. By minimizing heating, cooling, and electricity demands, these materials lower utility bills, making them cost-effective in the long run (Han et al. 2010; Elaouzy et al. 2022; Farghali et al. 2023). Additionally, buildings incorporating energy-efficient materials tend to have higher market value, appealing to environmentally conscious buyers and investors. Governments and municipalities often provide financial incentives, including tax credits and rebates, to encourage their use, further improving their economic feasibility. Beyond financial savings, energy-efficient materials enhance building durability and require less maintenance, reducing overall lifecycle costs. They also contribute to sustainability goals by decreasing carbon footprints and promoting energy conservation. While initial costs can be a barrier, the long-term benefits of reduced energy consumption, lower maintenance costs, and increased property value make energy-efficient materials a financially viable and sustainable choice.

7.2 Availability and Supply Chain Issues

The availability of energy-efficient materials and supply chain challenges play a crucial role in the adoption of sustainable construction practices. One major issue is the limited production capacity of advanced materials like aerogels, phase-change materials, and high-performance insulation. These materials require specialized manufacturing processes, which are not widely available in all regions, leading to higher costs and longer lead times. Supply chain disruptions also impact the accessibility of energy-efficient materials. Factors such as global trade restrictions, transportation delays, raw material shortages, and geopolitical instability can create bottlenecks, affecting the timely delivery of sustainable construction materials (Enshassi et al. 2009; Chang et al. 2011; Ndukwe et al. 2020; Luo et al. 2020).

Additionally, fluctuations in the availability of raw materials, such as recycled steel, bamboo, and bio-based insulation, further complicate supply chain stability. Regional disparities in material distribution mean that developing countries often struggle to access high-performance sustainable materials at competitive prices. This limits the widespread implementation of energy-efficient construction, as builders may rely on locally available, but less sustainable, alternatives. To address these issues, increased investment in local production facilities, material innovation, and government incentives is essential. Strengthening supply chains through diversification, sustainable sourcing, and improved logistics will enhance the availability of energy-efficient materials, making sustainable construction more accessible and cost-effective globally.

7.3 Technical Limitations and Performance Concerns

Technical limitations and performance concerns are significant barriers to the widespread adoption of energy-efficient materials in construction. One of the primary challenges is material compatibility and integration with existing building systems. Many energy-efficient materials, such as aerogels, phase-change materials, and vacuum insulation panels, require specialized installation techniques, making them difficult to incorporate into traditional construction methods. Improper installation can reduce their effectiveness, leading to suboptimal thermal performance and increased maintenance costs. Durability and long-term performance are also concerns. Some energy-efficient materials degrade faster under extreme weather conditions, reducing their lifespan. For example, bio-based insulation like hempcrete and straw bales may be susceptible to moisture damage if not properly treated (Thomson & Walker, 2014; Rahim et al. 2016; Andersen & Rasmussen, 2025). Similarly, phase-change materials may lose effectiveness over time due to material fatigue or leakage, affecting their ability to regulate indoor temperatures efficiently (Da Cunha & de Aguiar, 2020; Wang et al. 2022). Another issue is cost-performance trade-offs. While some high-tech materials provide superior energy efficiency, their benefits may not always justify the higher costs compared to conventional alternatives. Additionally, builders and contractors may lack sufficient expertise in handling these materials, leading to improper application and diminished energy savings. Overcoming these limitations requires ongoing research, improved material formulations, better training for construction professionals, and rigorous testing standards to ensure long-term performance and reliability.

7.4 Regulatory and Policy Challenges

Regulatory and policy challenges significantly impact the adoption of energy-efficient materials in the construction industry. One of the primary issues is the lack of uniform standards and codes across different regions. While some countries have strict building energy codes, others have outdated or inconsistent regulations, making it difficult to implement energy-efficient practices on a global scale (Björklund et al. 2023; Rodriguez et al. 2024; Kirme & Kapse, 2024). This inconsistency creates uncertainty for developers, architects, and builders, slowing down innovation and adoption. Another major challenge is compliance enforcement. Even when regulations exist, inadequate enforcement mechanisms often result in non-compliance. Many jurisdictions lack the resources to conduct proper inspections or penalize non-adherence, leading to the continued use of inefficient materials and construction methods.

Financial constraints also play a role. Governments may not provide sufficient incentives, subsidies, or tax benefits to encourage the use of energy-efficient materials. Without financial support, many builders opt for cheaper, less sustainable alternatives, prioritizing short-term cost savings over long-term efficiency. The slow process of policy updates is another barrier. As new technologies emerge, policies need to evolve quickly to accommodate advancements in energy-efficient materials. However, bureaucratic delays and resistance to change often result in outdated regulations that fail to support modern sustainable construction practices. Addressing these challenges requires stronger enforcement, financial incentives, and continuous policy updates to accelerate the transition to energy-efficient buildings.

Table 6 presents a clear linkage between key challenges of the linear construction model and circular economy solutions. It demonstrates how high-impact issues—such as carbon emissions, waste generation, and resource depletion—can be addressed through material innovation and regenerative strategies. Each challenge is paired with practical solutions like recycled aggregates, bio-based insulation, and geopolymer concrete, showing their potential to reduce environmental impact and enhance building performance. The table also emphasizes design thinking for durability, adaptability, and reuse, underscoring a shift toward systems that prioritize lifecycle efficiency. This synthesis strengthens the case for circularity as a viable path for sustainable construction.

Table 6. Linking Linear Energy Paradigm Challenges to Circular Economy Solutions in Construction

Linear Paradigm Challenge	Circular Economy Solution	Material / Strategy
High carbon emissions from cement production	Use of low-carbon alternatives and industrial byproducts	Geopolymer concrete, Fly ash and slag cement
Construction and demolition waste	Material reuse and recycling	Recycled aggregate concrete, reclaimed wood, recycled steel
Resource depletion (e.g., sand, limestone)	Renewable and rapidly regenerating resources	Bamboo, straw bale, hempcrete
Energy inefficiency and heat loss in buildings	Passive design and high-performance thermal insulation	Aerogels, spray foam, triple-glazed Low-E windows, phase-change materials
Fossil fuel-based heating and cooling	Thermal energy storage and adaptive insulation	Phase-change materials, smart glazing, green roofs
Short material lifespan and high maintenance	Durable, self-healing, and low-maintenance materials	Self-healing concrete, nanotechnology-enhanced coatings
Embodied energy and emissions in long transport	Use of local and prefabricated materials	Rammed earth, CLT (cross-laminated timber), modular construction
Toxic emissions and poor indoor air quality	Use of low-VOC, biodegradable, and non-toxic natural materials	Cork, wool, hemp insulation, lime-based composites
Inflexibility in design for end-of-life reuse	Design for disassembly and material circularity	3D-printed modular panels, recyclable bioplastics, prefabricated elements

8. INSTITUTIONAL, FINANCIAL, AND BEHAVIOURAL DRIVERS FOR THE ADOPTION OF ENERGY-EFFICIENT MATERIALS

The successful adoption of energy-efficient materials in sustainable construction is not solely determined by technological advancements or environmental urgency. It also hinges on a complex set of institutional, financial, and behavioural drivers that shape decision-making across the building lifecycle. Understanding and integrating these dimensions is essential for accelerating the transition to circular and low-carbon construction practices. Government institutions, urban planning bodies, and international regulatory frameworks play a pivotal role in mainstreaming energy-efficient materials. National building codes, such as the International Energy Conservation Code (IECC), and directives like the EU's Energy Performance of Buildings Directive (EPBD), establish performance baselines that influence market demand. Furthermore, green public procurement (GPP) policies and municipal incentives (e.g., density bonuses, expedited permitting) can encourage the use of certified sustainable materials. Institutional leadership in education, standardization, and enforcement ensures that innovation is not only developed but also applied at scale.

Despite their long-term cost savings, energy-efficient materials often face up-front cost barriers that deter adoption. Financial instruments such as green bonds, tax incentives, low-interest loans, and energy performance contracting can mitigate these barriers. Organizations like the International Finance Corporation (IFC) and Green Climate Fund support financing schemes that de-risk sustainable construction investment. Additionally, the internalization of lifecycle costs through tools like Total Cost of Ownership (TCO) or Life Cycle Cost Analysis (LCCA) can shift investor focus from first-cost to long-term value.

User perception, contractor preferences, and cultural norms significantly affect material choices. Risk aversion, lack of awareness, or perceived complexity of new materials often result in the preference for traditional, familiar options. Behavioural interventions—such as awareness campaigns, professional training, and demonstration projects—can shift mindsets and build confidence in sustainable alternatives. Incorporating human-centered design principles and fostering community co-creation can further enhance the social acceptability of energy-efficient solutions.

A thriving innovation ecosystem—comprising research institutions, startups, material manufacturers, industry associations, and policy makers—is critical for scaling material innovation. Collaborative platforms that connect these stakeholders facilitate knowledge exchange, reduce market fragmentation, and accelerate adoption. Multi-actor partnerships (e.g., university-industry consortia or public-private partnerships) can also help align innovation with regulatory, financial, and social readiness. For example, living labs and testbeds enable real-world piloting of advanced materials in diverse contexts, helping to build the case for broader implementation.

Energy-efficient materials intersect with broader urban systems, including energy infrastructure, water systems, and transportation. Embracing a circular economy approach—in which materials are designed for reuse, recyclability, and minimal embodied carbon—demands intersectoral coordination. For instance, construction waste can be valorized through industrial symbiosis, supporting energy recovery or feedstock for other industries. Recognizing these interdependencies and fostering systems-level thinking will be key to achieving net-zero and regenerative construction goals.

9. EXPANDING THE LANDSCAPE: POLICY, EQUITY, AND DIGITAL INNOVATION IN ENERGY-EFFICIENT CONSTRUCTION

Achieving widespread adoption of energy-efficient materials in construction not only requires technological innovation but also supportive regulatory frameworks, social equity considerations, and integration of emerging digital technologies. These elements are critical for accelerating sustainable development and ensuring that the transition to greener construction practices is inclusive and scalable.

While building standards like LEED, BREEAM, and Passive House have set benchmarks for sustainability, recent developments in global policy provide additional momentum. The EU Taxonomy for Sustainable Activities, for example, defines clear criteria for what qualifies as an environmentally sustainable construction project. This classification guides investors toward green construction and promotes the use of energy-efficient materials through targeted funding and sustainable finance. Governments at all levels are introducing subsidies, tax incentives, and grant programs to encourage the adoption of advanced insulation, renewable materials, and sustainable concrete. However, disparities in regulatory enforcement and outdated building codes in some regions hinder consistent progress. Stronger governance, transparent compliance mechanisms, and regular policy updates are necessary to drive adoption and investment in sustainable construction technologies.

The environmental benefits of energy-efficient construction must be matched by attention to social equity. Energy poverty, where households cannot afford adequate heating or cooling, affects millions worldwide and is particularly acute in under-resourced communities. Buildings constructed with energy-efficient materials can significantly reduce energy costs, thereby mitigating energy poverty and improving public health outcomes. Equitable access to green technologies and affordable housing solutions—such as prefabricated buildings using natural insulation like hempcrete or straw bale—can bridge socioeconomic divides. Policies must ensure that low-income and marginalized populations are not excluded from the benefits of energy-efficient design. This includes providing financial assistance, community-focused development models, and capacity-building programs for local labour and builders.

Emerging digital technologies are transforming how energy-efficient materials are designed, implemented, and managed:

- Artificial Intelligence (AI) is being deployed to optimize material selection, simulate energy performance, and manage building operations for maximum efficiency. AI-driven tools also enable predictive maintenance and lifecycle cost analysis.
- Blockchain technology offers transparent and immutable records of material sourcing, enabling supply chain traceability, which is essential for circular construction practices and sustainability certifications.
- Internet of Things (IoT) devices embedded in building envelopes can monitor temperature, humidity, and structural performance in real-time. These smart systems facilitate adaptive energy use and help identify materials that can be recovered, reused, or recycled at the end of a building's life.

Together, these innovations support a circular economy model in construction—minimizing waste, maximizing material reuse, and extending the lifecycle of resources.

10. FUTURE TRENDS AND DEVELOPMENTS

10.1 Emerging Materials and Technologies

Emerging materials and technologies in energy-efficient construction are revolutionizing the industry by reducing energy consumption, enhancing durability, and improving sustainability. These innovations are key to building environmentally friendly structures that minimize carbon footprints while optimizing resource efficiency. Aerogels and Vacuum Insulation Panels (VIPs) are among the most advanced insulation materials available (Kan et al. 2023; Kaushik et al. 2024). With exceptional thermal resistance and minimal thickness, they provide superior insulation, making them ideal for energy-efficient buildings with space constraints. Their use significantly reduces heat transfer, improving overall energy savings.

Phase-change materials (PCMs) help regulate indoor temperatures by absorbing and releasing heat as they change phases. These materials reduce the need for artificial heating and cooling, leading to lower energy consumption and improved thermal comfort. PCMs are increasingly used in walls, ceilings, and flooring systems to enhance passive energy efficiency (Saffari et al. 2022; Arumugam et al. 2022; Yang et al. 2023). Self-healing concrete incorporates bacteria or microcapsules that repair cracks automatically, increasing structural durability and reducing maintenance costs (Amran et al. 2022; Zhang et al. 2024; Xiang et al. 2024). This innovation extends the lifespan of concrete structures, making them more sustainable and cost-effective over time. Cool roof coatings and smart glass technologies also contribute to energy efficiency. Cool roofs reflect solar heat, reducing cooling loads, while smart glass adjusts its transparency based on sunlight intensity, optimizing lighting and temperature control. With advancements like 3D-printed sustainable materials using recycled and eco-friendly substances, construction is moving towards more efficient, resource-conscious solutions that promote long-term sustainability.

10.2 Integration of Renewable Energy Systems with Building Materials

The integration of renewable energy systems with building materials is a transformative approach to sustainable construction, allowing buildings to generate clean energy while enhancing efficiency. This strategy reduces reliance on fossil fuels, lowers operational costs, and minimizes environmental impact. One of the most innovative solutions is building-integrated photovoltaics (BIPV), where solar panels are embedded into roofs, facades, and windows, seamlessly blending with architectural designs (Liu et al. 2021; Martín-Chivelet et al. 2022; Taser et al. 2023; Aguacil et al. 2024). Unlike traditional solar panels, BIPV materials serve dual purposes—providing structural function while generating renewable energy. Transparent solar glass, for example, allows natural light while producing electricity, making it ideal for windows and skylights.

Another advancement is solar shingles and photovoltaic roofing tiles, which replace conventional roofing materials while harnessing solar energy. These solutions improve aesthetics, reduce installation complexity, and enhance building efficiency. Thermal energy systems, such as solar water heating panels integrated into building exteriors, efficiently utilize solar radiation to provide hot water and space heating (Zhang et al. 2022; Vassiliades et al. 2022). Additionally, phase-change materials (PCMs) and thermally active building systems help store and distribute renewable energy more effectively. The integration of wind turbines, kinetic energy flooring, and bio-based energy solutions further expands renewable energy potential in construction. By embedding these technologies into building materials, structures become more energy-efficient, resilient, and environmentally sustainable, supporting global decarbonization efforts.

10.3 The Role of AI and Digital Tools in Material Innovation

The construction industry is rapidly evolving, with artificial intelligence (AI) and digital tools playing a crucial role in the development of energy-efficient materials for sustainable construction. AI-driven material innovation is transforming how buildings are designed, constructed, and maintained by optimizing material properties, reducing waste, and improving energy efficiency. Machine learning algorithms analyse vast datasets to identify novel material compositions with enhanced thermal insulation, durability, and carbon footprint reduction (Li et al. 2022; Wang et al. 2022; Tien et al. 2022; Fu et al. 2022). AI-powered simulations enable researchers to predict material behaviour under various environmental conditions, accelerating the development of eco-friendly alternatives such as bio-based composites, aerogels, and phase-change materials.

Digital twin technology, combined with AI, allows architects and engineers to model, test, and optimize material performance before implementation. Advanced robotics and 3D printing facilitate the precise manufacturing of complex, energy-efficient building components, minimizing resource waste. Furthermore, AI-driven supply chain management ensures the sustainable sourcing and efficient use of construction materials. These innovations contribute to green building certifications, reducing reliance on traditional high-carbon materials like cement and steel.

By integrating AI and digital tools, the construction industry can advance toward net-zero energy buildings, promoting a more sustainable and resilient built environment. The synergy between AI and material science is revolutionizing the future of energy-efficient construction.

10.4 Circular Economy and Material Reuse

The circular economy plays a vital role in the development of energy-efficient materials for sustainable construction by promoting resource efficiency, reducing waste, and extending material lifecycles. Traditional construction methods rely heavily on virgin resources, leading to high energy consumption and environmental degradation. In contrast, a circular approach focuses on material reuse, recycling, and repurposing to create energy-efficient building components that minimize carbon footprints and improve sustainability.

Reclaimed materials such as recycled concrete, repurposed steel, and salvaged wood are increasingly used in green building projects (Bertino et al. 2021; Ghobadi & Sepasgozar et al. 2023; Kitek Kuzman et al. 2024). Advanced processing techniques, including AI-driven material analysis and digital tracking systems, enhance the efficiency of recovering and reintegrating these materials. Additionally, innovative materials such as recycled insulation, bio-based composites, and aerogels derived from industrial byproducts contribute to energy efficiency while supporting circular economy principles. Modular construction and design-for-disassembly further facilitate material reuse by enabling easy deconstruction and repurposing of building components (Olawumi et al. 2022; Pan et al. 2022). Digital platforms and blockchain technology also enhance supply chain transparency, ensuring that recycled and reused materials meet high-performance standards. By integrating circular economy strategies with material innovation, the construction industry can significantly reduce waste, lower embodied carbon, and support the transition toward a more sustainable and resilient built environment.

11. CONCLUSIONS

11.1 Summary of Key Points

This chapter synthesizes critical insights into the development and application of energy-efficient materials as a cornerstone of sustainable construction practices. The deployment of advanced insulation systems—such as aerogels, spray foam, and phase-change materials (PCMs)—demonstrably enhances thermal regulation, leading to significant reductions in heating and cooling energy demands. Concurrently, high-performance glazing technologies, including low-emissivity (Low-E) glass and electrochromic smart windows, serve to optimize solar heat gain and maintain indoor thermal comfort, thereby improving overall energy efficiency. Sustainable concrete alternatives, notably geopolymer and recycled aggregate concrete, present low-carbon substitutes to traditional Portland cement, contributing to both emissions mitigation and resource conservation. Natural and renewable materials such as bamboo, hempcrete, and straw bale further support ecological objectives, offering favourable insulation performance and a reduced environmental footprint.

Technological innovations—particularly nanotechnology, self-healing materials, and additive manufacturing (3D printing)—advance the durability, adaptability, and lifecycle performance of construction materials while simultaneously minimizing waste. The adoption of green building standards, including LEED (Leadership in Energy and Environmental Design), BREEAM (Building Research Establishment Environmental Assessment Method), and the Passive House standard, institutionalizes best practices and incentivizes the integration of energy-efficient solutions. Nevertheless, several barriers persist. These include elevated material costs, inconsistent regulatory frameworks, insufficient market awareness, and supply chain constraints. Future trajectories in the field are likely to emphasize the integration of renewable energy systems—such as building-integrated photovoltaics (BIPV)—and the use of AI-driven design tools to enhance material selection and system performance. Collectively, these findings underscore the central role of energy-efficient materials in reducing environmental impact, enhancing energy performance, and advancing sustainable, climate-resilient infrastructure.

11.2 The Future of Energy-Efficient Construction

The trajectory of energy-efficient construction is increasingly shaped by the convergence of material innovation, digital technologies, and sustainability imperatives. As global climate policy frameworks demand rapid decarbonization of the built environment, materials such as aerogels, PCMs, and smart glazing are poised to become foundational elements in high-performance building envelopes. Their superior insulation properties, responsiveness to environmental conditions, and durability enable substantial reductions in operational energy use. The institutionalization of green certification systems—particularly LEED, BREEAM, and Passive House—continues to exert a transformative influence on design and construction practices by mandating energy-efficient material usage and comprehensive lifecycle assessments. These standards encourage the adoption of geopolymer concrete, recycled aggregates, and bio-based insulations that support circular economy objectives and enhance building sustainability.

Simultaneously, emerging technologies such as additive manufacturing (3D printing), artificial intelligence (AI)-based material optimization, and nanoscale engineering are revolutionizing material functionality and construction methodologies. Smart materials capable of self-healing, thermoregulation, or energy harvesting are expected to play an increasingly significant role in net-zero energy building (NZEB) strategies. In parallel, the integration of renewable energy systems—including solar-integrated façades, photovoltaic roof tiles, and hybrid thermal-electric systems—will further support decentralized, clean energy generation at the building scale. Despite these promising developments, challenges persist. These include high initial investment requirements, material availability disparities, and regulatory fragmentation. Addressing these issues will necessitate a coordinated effort involving research institutions, policymakers, and industry stakeholders. Strategic investment in research and development, coupled with supportive policy environments and educational outreach, will be instrumental in realizing the full potential of energy-efficient construction in the coming decades.

11.3 Final Thoughts on Sustainable Building Practices

Sustainable building practices represent a critical response to the intertwined challenges of climate change, resource scarcity, and urbanization. The incorporation of energy-efficient materials—such as aerogels, PCMs, and high-performance glazing systems—enables substantial reductions in building energy use while enhancing thermal comfort and indoor air quality. Furthermore, alternatives to conventional construction materials, including geopolymer concrete, recycled aggregates, and natural fiber insulations, contribute to the decarbonization of the construction sector by lowering embodied carbon and promoting resource circularity.

Despite demonstrable benefits, the implementation of these materials is hindered by persistent challenges. High capital costs, fragmented regulatory landscapes, technical skill gaps, and logistical barriers continue to constrain widespread adoption. These challenges underscore the need for a holistic strategy involving governmental incentives, targeted policy reforms, public-private collaboration, and enhanced knowledge dissemination within the construction sector.

The future of sustainable construction will be increasingly defined by the integration of intelligent materials and renewable energy systems, supported by advancements in digital fabrication and AI-enabled design tools. Such integration not only facilitates energy efficiency but also aligns with broader sustainable development goals. Green building standards, notably LEED and BREEAM, play an essential role in codifying best practices, guiding material selection, and encouraging innovation in design. As the global built environment expands to accommodate population growth and urban development, the systematic adoption of energy-efficient and sustainable building practices will be imperative. These practices offer a viable pathway toward climate-resilient infrastructure that meets both environmental and socioeconomic needs, contributing meaningfully to the creation of a more sustainable future.

The paper identifies key findings in the development and application of energy-efficient materials for sustainable construction. Advanced insulation materials, including aerogels, spray foam, and phase-change materials, significantly enhance thermal efficiency, reducing heating and cooling demands. High-performance glazing, such as Low-E glass and smart windows, optimizes energy use by regulating heat transfer and improving indoor comfort. Sustainable concrete alternatives, including geopolymer and recycled aggregate concrete, offer lower carbon emissions and reduced reliance on non-renewable materials. Renewable and natural resources, such as bamboo, hempcrete, and straw bale, provide environmentally friendly solutions with strong insulation properties.

Innovative technologies like nanotechnology, self-healing materials, and 3D printing contribute to energy efficiency by enhancing durability and reducing construction waste. The adoption of green building certifications, including LEED, BREEAM, and Passive House, encourages sustainable construction practices. Despite the benefits, barriers such as high material costs, limited awareness, and regulatory inconsistencies pose challenges to widespread implementation. Future trends highlight the integration of renewable energy systems with building materials and AI-driven material innovation to optimize performance and sustainability. The findings emphasize the importance of energy-efficient materials in reducing environmental impact, lowering energy consumption, and advancing resilient, eco-friendly construction methods to support global sustainability goals.

REFERENCES

Abdellatef, Y., Khan, M. A., Khan, A., Alam, M. I., & Kavgic, M. (2020). Mechanical, thermal, and moisture buffering properties of novel insulating hemp-lime composite building materials. *Materials (Basel), 13*(21), 5000. DOI: 10.3390/ma13215000 PMID: 33171950

Abed, J., Rayburg, S., Rodwell, J., & Neave, M. (2022). A review of the performance and benefits of mass timber as an alternative to concrete and steel for improving the sustainability of structures. *Sustainability (Basel), 14*(9), 5570. DOI: 10.3390/su14095570

Abhilash, P. P., Nayak, D. K., Sangoju, B., Kumar, R., & Kumar, V. (2021). Effect of nano-silica in concrete; a review. *Construction & Building Materials, 278*, 122347. DOI: 10.1016/j.conbuildmat.2021.122347

Abuelnuor, A. A., Omara, A. A., Saqr, K. M., & Elhag, I. H. (2018). Improving indoor thermal comfort by using phase change materials: A review. *International Journal of Energy Research, 42*(6), 2084–2103. DOI: 10.1002/er.4000

Adhikary, S. K., Ashish, D. K., & Rudžionis, Ž. (2021). Aerogel based thermal insulating cementitious composites: A review. *Energy and Building, 245*, 111058. DOI: 10.1016/j.enbuild.2021.111058

Aggarwal, C., & Molleti, S. (2024). State-of-the-art review: Effects of using cool building cladding materials on roofs. *Buildings (Basel, Switzerland), 14*(8), 2257. DOI: 10.3390/buildings14082257

Aguacil, S., Duque, S., Lufkin, S., & Rey, E. (2024). Designing with building-integrated photovoltaics (BIPV): A pathway to decarbonize residential buildings. *Journal of Building Engineering, 96*, 110486. DOI: 10.1016/j.jobe.2024.110486

Ahmed, G. H. (2023). A review of "3D concrete printing": Materials and process characterization, economic considerations and environmental sustainability. *Journal of Building Engineering, 66*, 105863. DOI: 10.1016/j.jobe.2023.105863

Ahmed, H. U., Mahmood, L. J., Muhammad, M. A., Faraj, R. H., Qaidi, S. M., Sor, N. H., Mohammed, A. S., & Mohammed, A. A. (2022). Geopolymer concrete as a cleaner construction material: An overview on materials and structural performances. *Cleaner Materials, 5*, 100111. DOI: 10.1016/j.clema.2022.100111

Akın, N., & Mutlu Danacı, H. (2021). An investigation into the architectural use of nanotechnology in the context of the titanium dioxide. *Environmental Science and Pollution Research International*, *28*(45), 64130–64136. DOI: 10.1007/s11356-021-15268-5 PMID: 34247357

Al-Homoud, M. S. (2005). Performance characteristics and practical applications of common building thermal insulation materials. *Building and Environment*, *40*(3), 353–366. DOI: 10.1016/j.buildenv.2004.05.013

Alassaf, Y. (2024). Comprehensive review of the advancements, benefits, challenges, and design integration of energy-efficient materials for sustainable buildings. *Buildings (Basel, Switzerland)*, *14*(9), 2994. DOI: 10.3390/buildings14092994

Ali, A., Issa, A., & Elshaer, A. (2024). A comprehensive review and recent trends in thermal insulation materials for energy conservation in buildings. *Sustainability (Basel)*, *16*(20), 8782. DOI: 10.3390/su16208782

Altomonte, S., & Schiavon, S. (2013). Occupant satisfaction in LEED and non-LEED certified buildings. *Building and Environment*, *68*, 66–76. DOI: 10.1016/j.buildenv.2013.06.008

Amran, M., Onaizi, A. M., Fediuk, R., Danish, A., Vatin, N. I., Murali, G., Abdelgader, H. S., Mosaberpanah, M. A., Cecchin, D., & Azevedo, A. (2022). An ultra-lightweight cellular concrete for geotechnical applications–A review. *Case Studies in Construction Materials*, *16*, e01096. DOI: 10.1016/j.cscm.2022.e01096

An, X., & Pivo, G. (2020). Green buildings in commercial mortgage-backed securities: The effects of LEED and Energy Star certification on default risk and loan terms. *Real Estate Economics*, *48*(1), 7–42. DOI: 10.1111/1540-6229.12228

Andersen, B., & Rasmussen, T. V. (2025). Biobased building materials: Moisture characteristics and fungal susceptibility. *Building and Environment*, *275*, 112720. DOI: 10.1016/j.buildenv.2025.112720

Arumugam, P., Ramalingam, V., & Vellaichamy, P. (2022). Effective PCM, insulation, natural and/or night ventilation techniques to enhance the thermal performance of buildings located in various climates–A review. *Energy and Building*, *258*, 111840. DOI: 10.1016/j.enbuild.2022.111840

Asayesh, S., Javid, A. A. S., Ziari, H., & Mehri, B. (2021). Evaluating fresh state, hardened state, thermal expansion and bond properties of geopolymers for the repairing of concrete pavements under restrained conditions. *Construction & Building Materials*, *292*, 123398. DOI: 10.1016/j.conbuildmat.2021.123398

Asdrubali, F., Baldinelli, G., Bianchi, F., & Sambuco, S. (2015). A comparison between environmental sustainability rating systems LEED and ITACA for residential buildings. *Building and Environment*, *86*, 98–108. DOI: 10.1016/j.buildenv.2015.01.001

Asdrubali, F., Ferracuti, B., Lombardi, L., Guattari, C., Evangelisti, L., & Grazieschi, G. (2017). A review of structural, thermo-physical, acoustical, and environmental properties of wooden materials for building applications. *Building and Environment*, *114*, 307–332. DOI: 10.1016/j.buildenv.2016.12.033

Asghari, N., & Memari, A. M. (2024). State of the art review of attributes and mechanical properties of hempcrete. *Biomass*, *4*(1), 65–91. DOI: 10.3390/biomass4010004

Awadh, O. (2017). Sustainability and green building rating systems: LEED, BREEAM, GSAS and Estidama critical analysis. *Journal of Building Engineering*, *11*, 25–29. DOI: 10.1016/j.jobe.2017.03.010

Azhar, S., Carlton, W. A., Olsen, D., & Ahmad, I. (2011). Building information modeling for sustainable design and LEED rating analysis. *Automation in Construction*, *20*(2), 217–224. DOI: 10.1016/j.autcon.2010.09.019

Ben-Alon, L., & Rempel, A. R. (2023). Thermal comfort and passive survivability in earthen buildings. *Building and Environment*, *238*, 110339. DOI: 10.1016/j.buildenv.2023.110339

Berardi, U. (2018). Aerogel-enhanced systems for building energy retrofits: Insights from a case study. *Energy and Building*, *159*, 370–381. DOI: 10.1016/j.enbuild.2017.10.092

Berndt, M. L. (2009). Properties of sustainable concrete containing fly ash, slag and recycled concrete aggregate. *Construction & Building Materials*, *23*(7), 2606–2613. DOI: 10.1016/j.conbuildmat.2009.02.011

Bertino, G., Kisser, J., Zeilinger, J., Langergraber, G., Fischer, T., & Österreicher, D. (2021). Fundamentals of building deconstruction as a circular economy strategy for the reuse of construction materials. *Applied Sciences (Basel, Switzerland)*, *11*(3), 939. DOI: 10.3390/app11030939

Biswas, K., Desjarlais, A., Smith, D., Letts, J., Yao, J., & Jiang, T. (2018). Development and thermal performance verification of composite insulation boards containing foam-encapsulated vacuum insulation panels. *Applied Energy*, *228*, 1159–1172. DOI: 10.1016/j.apenergy.2018.06.136

Björklund, M., von Malmborg, F., & Nordensvärd, J. (2023). Lessons learnt from 20+ years of research on multilevel governance of energy-efficient and zero-carbon buildings in the European Union. *Energy Efficiency*, *16*(8), 98. DOI: 10.1007/s12053-023-10178-6

Boafo, F. E., Kim, J. H., Ahn, J. G., Kim, S. M., & Kim, J. T. (2021). Slim curtain wall spandrel integrated with vacuum insulation panel: A state-of-the-art review and future opportunities. *Journal of Building Engineering*, *42*, 102445. DOI: 10.1016/j.jobe.2021.102445

Bostanci, S. C., Limbachiya, M., & Kew, H. (2018). Use of recycled aggregates for low carbon and cost effective concrete construction. *Journal of Cleaner Production*, *189*, 176–196. DOI: 10.1016/j.jclepro.2018.04.090

Brolinson, B., Palmer, K., & Walls, M. (2023). Does Energy Star certification reduce energy use in commercial buildings? *Journal of the Association of Environmental and Resource Economists*, *10*(1), 55–93. DOI: 10.1086/720952

Brzezicki, M. (2021). A systematic review of the most recent concepts in smart windows technologies with a focus on electrochromics. *Sustainability (Basel)*, *13*(17), 9604. DOI: 10.3390/su13179604

Busselli, M., Cassol, D., Prada, A., & Giongo, I. (2021). Timber-based integrated techniques to improve energy efficiency and seismic behaviour of existing masonry buildings. *Sustainability (Basel)*, *13*(18), 10379. DOI: 10.3390/su131810379

Cannavale, A., Ayr, U., Fiorito, F., & Martellotta, F. (2020). Smart electrochromic windows to enhance building energy efficiency and visual comfort. *Energies*, *13*(6), 1449. DOI: 10.3390/en13061449

Cao, D., Xu, C., Lu, W., Qin, C., & Cheng, S. (2018). Sunlight-driven photo-thermochromic smart windows. *Solar RRL*, *2*(4), 1700219. DOI: 10.1002/solr.201700219

Castro-Hoyos, A. M., Rojas Manzano, M. A., & Maury-Ramírez, A. (2022). Challenges and opportunities of using titanium dioxide photocatalysis on cement-based materials. *Coatings*, *12*(7), 968. DOI: 10.3390/coatings12070968

Chang, Y., Wilkinson, S., Potangaroa, R., & Seville, E. (2011). Identifying factors affecting resource availability for post-disaster reconstruction: A case study in China. *Construction Management and Economics*, *29*(1), 37–48. DOI: 10.1080/01446193.2010.521761

Chen, Y., He, S., Gan, Y., Çopuroğlu, O., Veer, F., & Schlangen, E. (2022). A review of printing strategies, sustainable cementitious materials and characterization methods in the context of extrusion-based 3D concrete printing. *Journal of Building Engineering*, *45*, 103599. DOI: 10.1016/j.jobe.2021.103599

Chindaprasirt, P., Kanchanda, P., Sathonsaowaphak, A., & Cao, H. T. (2007). Sulfate resistance of blended cements containing fly ash and rice husk ash. *Construction & Building Materials*, *21*(6), 1356–1361. DOI: 10.1016/j.conbuildmat.2005.10.005

Choudhary, M. K., & Eastes, W. (2024). Effective thermal conductivity of fiberglass insulation. *International Journal of Applied Glass Science*, *15*(3), 307–316. DOI: 10.1111/ijag.16652

Cuce, E., Cuce, P. M., Wood, C. J., & Riffat, S. B. (2014). Toward aerogel based thermal superinsulation in buildings: A comprehensive review. *Renewable & Sustainable Energy Reviews*, *34*, 273–299. DOI: 10.1016/j.rser.2014.03.017

da Cunha, S. R. L., & de Aguiar, J. L. B. (2020). Phase change materials and energy efficiency of buildings: A review of knowledge. *Journal of Energy Storage*, *27*, 101083. DOI: 10.1016/j.est.2019.101083

Dabaieh, M., Wanas, O., Hegazy, M. A., & Johansson, E. (2015). Reducing cooling demands in a hot dry climate: A simulation study for non-insulated passive cool roof thermal performance in residential buildings. *Energy and Building*, *89*, 142–152. DOI: 10.1016/j.enbuild.2014.12.034

Danish, A., Ozbakkaloglu, T., Mosaberpanah, M. A., Salim, M. U., Bayram, M., Yeon, J. H., & Jafar, K. (2022). Sustainability benefits and commercialization challenges and strategies of geopolymer concrete: A review. *Journal of Building Engineering*, *58*, 105005. DOI: 10.1016/j.jobe.2022.105005

Das, P., Nayak, P. K., Muthusamy, S., & Kesavan, R. K. (2023). Foams for thermal insulation. In *Polymeric Foams: Applications of Polymeric Foams* (Vol. 2, pp. 145–165). American Chemical Society. DOI: 10.1021/bk-2023-1440.ch007

de Andrade Salgado, F., & de Andrade Silva, F. (2022). Recycled aggregates from construction and demolition waste towards an application on structural concrete: A review. *Journal of Building Engineering*, *52*, 104452. DOI: 10.1016/j.jobe.2022.104452

Deng, S., Wang, R. A., & Dai, Y. J. (2014). How to evaluate performance of net zero energy building–A literature research. *Energy*, *71*, 1–16. DOI: 10.1016/j.energy.2014.05.007

Ding, D., & Xian, M. (2024). Research trends and design strategies for bamboo architecture: A bibliometric and systematic review. *Advances in Bamboo Science*, *8*, 100094. DOI: 10.1016/j.bamboo.2024.100094

Du, H., Du, S., & Liu, X. (2014). Durability performances of concrete with nano-silica. *Construction & Building Materials*, *73*, 705–712. DOI: 10.1016/j.conbuildmat.2014.10.014

Duan, Q., Zhao, Y., & Wang, J. (2020). Thermal performance and condensation risk of single-pane glazing with low emissivity coatings. *MRS Advances*, *5*(50), 2555–2564. DOI: 10.1557/adv.2020.160

Elahi, M. M. A., Shearer, C. R., Reza, A. N. R., Saha, A. K., Khan, M. N. N., Hossain, M. M., & Sarker, P. K. (2021). Improving the sulfate attack resistance of concrete by using supplementary cementitious materials (SCMs): A review. *Construction & Building Materials*, *281*, 122628. DOI: 10.1016/j.conbuildmat.2021.122628

Elaouzy, Y., & El Fadar, A. (2022). Energy, economic and environmental benefits of integrating passive design strategies into buildings: A review. *Renewable & Sustainable Energy Reviews*, *167*, 112828. DOI: 10.1016/j.rser.2022.112828

Elmesalami, N., & Celik, K. (2022). A critical review of engineered geopolymer composite: A low-carbon ultra-high-performance concrete. *Construction & Building Materials*, *346*, 128491. DOI: 10.1016/j.conbuildmat.2022.128491

Enshassi, A., Mohamed, S., & Abushaban, S. (2009). Factors affecting the performance of construction projects in the Gaza Strip. *Journal of Civil Engineering and Management*, *15*(3), 269–280. DOI: 10.3846/1392-3730.2009.15.269-280

Fang, Y., Memon, S., Peng, J., Tyrer, M., & Ming, T. (2020). Solar thermal performance of two innovative configurations of air-vacuum layered triple-glazed windows. *Renewable Energy*, *150*, 167–175. DOI: 10.1016/j.renene.2019.12.115

Farghali, M., Osman, A. I., Mohamed, I. M., Chen, Z., Chen, L., Ihara, I., Yap, P. S., & Rooney, D. W. (2023). Strategies to save energy in the context of the energy crisis: A review. *Environmental Chemistry Letters*, *21*(4), 2003–2039. DOI: 10.1007/s10311-023-01591-5 PMID: 37362011

Ferrari, S., Zoghi, M., Blázquez, T., & Dall'O, G. (2022). New Level(s) framework: Assessing the affinity between the main international Green Building Rating Systems and the European scheme. *Renewable & Sustainable Energy Reviews*, *155*, 111924. DOI: 10.1016/j.rser.2021.111924

Flynn, M. A., Richman, R., Gorgolewski, M., Saunders, K., & Race, C. (2017). An investigation into the hygrothermal performance of a mineral wool based externally insulated enclosure in a cold climate. *Energy Procedia, 132*, 345–350. DOI: 10.1016/j.egypro.2017.09.746

Freire, A. L., José, H. J., & Moreira, R. D. F. P. M. (2022). Potential applications for geopolymers in carbon capture and storage. *International Journal of Greenhouse Gas Control, 118*, 103687. DOI: 10.1016/j.ijggc.2022.103687

Fu, Q., Han, Z., Chen, J., Lu, Y., Wu, H., & Wang, Y. (2022). Applications of reinforcement learning for building energy efficiency control: A review. *Journal of Building Engineering, 50*, 104165. DOI: 10.1016/j.jobe.2022.104165

Fu, Y., Qiu, C., Ni, L., Ye, H., Zou, H., Luo, Y., & Liang, M. (2024). Cell structure control and performance of rigid polyurethane foam with lightweight, good mechanical, thermal insulation and sound insulation. *Construction & Building Materials, 447*, 138068. DOI: 10.1016/j.conbuildmat.2024.138068

Gan, J., Chen, M., Semple, K., Liu, X., Dai, C., & Tu, Q. (2022). Life cycle assessment of bamboo products: Review and harmonization. *The Science of the Total Environment, 849*, 157937. DOI: 10.1016/j.scitotenv.2022.157937 PMID: 35952867

Gao, M., Qian, D., Zhao, L., & Jin, R. (2025). Experimental study on the photothermal properties of thermochromic glass. *Buildings (Basel, Switzerland), 15*(2), 233. DOI: 10.3390/buildings15020233

Garces, J. I. T., Beltran, A. B., Tan, R. R., Ongpeng, J. M. C., & Promentilla, M. A. B. (2022). Carbon footprint of self-healing geopolymer concrete with variable mix model. *Cleaner Chemical Engineering, 2*, 100027. DOI: 10.1016/j.clce.2022.100027

Ghobadi, M., & Sepasgozar, S. M. (2023). Circular economy strategies in modern timber construction as a potential response to climate change. *Journal of Building Engineering, 77*, 107229. DOI: 10.1016/j.jobe.2023.107229

Ghosh, A., Norton, B., & Duffy, A. (2016). First outdoor characterisation of a PV-powered suspended particle device switchable glazing. *Solar Energy Materials and Solar Cells, 157*, 1–9. DOI: 10.1016/j.solmat.2016.05.013

Gilbert, P., Wilson, P., Walsh, C., & Hodgson, P. (2017). The role of material efficiency to reduce CO2 emissions during ship manufacture: A life cycle approach. *Marine Policy, 75*, 227–237. DOI: 10.1016/j.marpol.2016.04.003

Golbazi, M., & Aktas, C. B. (2018). Energy efficiency of residential buildings in the US: Improvement potential beyond IECC. *Building and Environment, 142*, 278–287. DOI: 10.1016/j.buildenv.2018.06.029

Grazieschi, G., Asdrubali, F., & Thomas, G. (2021). Embodied energy and carbon of building insulating materials: A critical review. *Cleaner Environmental Systems*, *2*, 100032. DOI: 10.1016/j.cesys.2021.100032

Grzegorzewska, M., & Kirschke, P. (2021). The impact of certification systems for architectural solutions in green office buildings in the perspective of occupant well-being. *Buildings (Basel, Switzerland)*, *11*(12), 659. DOI: 10.3390/buildings11120659

Hallik, J., Gustavson, H., & Kalamees, T. (2019). Air leakage of joints filled with polyurethane foam. *Buildings (Basel, Switzerland)*, *9*(7), 172. DOI: 10.3390/buildings9070172

Han, F., Wang, Y., Feist, W., Cao, X., Yu, Z., Song, B., Benli, H., & Dermentzis, G. (2022). Exploring solutions to achieve carbon neutrality in China: A comparative study of a large-scale passive house district and a green building district in Qingdao. *Energy and Building*, *268*, 112224. DOI: 10.1016/j.enbuild.2022.112224

Han, H. J., Jeon, Y. I., Lim, S. H., Kim, W. W., & Chen, K. (2010). New developments in illumination, heating and cooling technologies for energy-efficient buildings. *Energy*, *35*(6), 2647–2653. DOI: 10.1016/j.energy.2009.05.020

Harilal, M., George, R. P., Philip, J., & Albert, S. K. (2021). Binary blended fly ash concrete with improved chemical resistance in natural and industrial environments. *Environmental Science and Pollution Research International*, *28*(22), 28107–28132. DOI: 10.1007/s11356-021-12453-4 PMID: 33533001

He, Y., Lin, E. S., Tan, C. L., Yu, Z., Tan, P. Y., & Wong, N. H. (2021). Model development of Roof Thermal Transfer Value (RTTV) for green roof in tropical area: A case study in Singapore. *Building and Environment*, *203*, 108101. DOI: 10.1016/j.buildenv.2021.108101

Hossain, M. R., Sultana, R., Patwary, M. M., Khunga, N., Sharma, P., & Shaker, S. J. (2022). Self-healing concrete for sustainable buildings. A review. *Environmental Chemistry Letters*, *20*(2), 1–9. DOI: 10.1007/s10311-021-01375-9

Huang, L. M., Hu, C. W., Liu, H. C., Hsu, C. Y., Chen, C. H., & Ho, K. C. (2012). Photovoltaic electrochromic device for solar cell module and self-powered smart glass applications. *Solar Energy Materials and Solar Cells*, *99*, 154–159. DOI: 10.1016/j.solmat.2011.03.036

Huang, Y., Stonehouse, A., & Abeykoon, C. (2023). Encapsulation methods for phase change materials—A critical review. *International Journal of Heat and Mass Transfer*, *200*, 123458. DOI: 10.1016/j.ijheatmasstransfer.2022.123458

Jami, T., Karade, S. R., & Singh, L. P. (2019). A review of the properties of hemp concrete for green building applications. *Journal of Cleaner Production*, *239*, 117852. DOI: 10.1016/j.jclepro.2019.117852

Jayakumari, B. Y., Swaminathan, E. N., & Partheeban, P. (2023). A review on characteristics studies on carbon nanotubes-based cement concrete. *Construction & Building Materials*, *367*, 130344. DOI: 10.1016/j.conbuildmat.2023.130344

Jaysawal, R. K., Chakraborty, S., Elangovan, D., & Padmanaban, S. (2022). Concept of net zero energy buildings (NZEB)-A literature review. *Cleaner Engineering and Technology*, *11*, 100582. DOI: 10.1016/j.clet.2022.100582

Jelle, B. P., Kalnæs, S. E., & Gao, T. (2015). Low-emissivity materials for building applications: A state-of-the-art review and future research perspectives. *Energy and Building*, *96*, 329–356. DOI: 10.1016/j.enbuild.2015.03.024

Jochem, E. (2009) Passive houses and buildings. Improving the Efficiency of R&D and the Market Diffusion of Energy Technologies, 105-141.

Jones, D., & Brischke, C. (2017). *Performance of bio-based building materials*. Woodhead Publishing. DOI: 10.1016/B978-0-08-100982-6.00001-X

Kaitouni, S. I., Charai, M., Es-Sakali, N., Mghazli, M. O., El Mankibi, M., Uk-Joo, S., Ahachad, M., & Brigui, J. (2024). Energy and hygrothermal performance investigation and enhancement of rammed earth buildings in hot climates: From material to field measurements. *Energy and Building*, *315*, 114325. DOI: 10.1016/j.enbuild.2024.114325

Kan, A., Zhang, Q., Chen, Z., Zhang, J., Wang, K., & Cao, D. (2023). Novel prediction of thermal conductivities for nano-aerogel and its composites as vacuum insulation panel core. *International Journal of Thermal Sciences*, *189*, 108277. DOI: 10.1016/j.ijthermalsci.2023.108277

Kaushik, D., Singh, H., & Tassou, S. A. (2024). Vacuum insulation panels for high-temperature applications–Design principles, challenges and pathways. *Thermal Science and Engineering Progress*, *48*, 102415. DOI: 10.1016/j.tsep.2024.102415

Kirme, S. K., & Kapse, V. S. (2024). A comprehensive review of residential building energy efficiency measures in India. *Energy and Building*, *319*, 114537. DOI: 10.1016/j.enbuild.2024.114537

Koh, C. H. A., & Kraniotis, D. (2021). Hygrothermal performance, energy use and embodied emissions in straw bale buildings. *Energy and Building*, *245*, 111091. DOI: 10.1016/j.enbuild.2021.111091

Kylili, A., & Fokaides, P. A. (2017). Methodologies for selection of thermal insulation materials for cost-effective, sustainable, and energy-efficient retrofitting. In *Cost-Effective Energy Efficient Building Retrofitting* (pp. 23–55). Woodhead Publishing. DOI: 10.1016/B978-0-08-101128-7.00002-2

Lee, S. W., Lim, C. H., & Salleh, E. I. B. (2016). Reflective thermal insulation systems in building: A review on radiant barrier and reflective insulation. *Renewable & Sustainable Energy Reviews*, *65*, 643–661. DOI: 10.1016/j.rser.2016.07.002

Leite Ribeiro, L. M., Piccinini Scolaro, T., & Ghisi, E. (2025). LEED certification in building energy efficiency: A review of its performance efficacy and global applicability. *Sustainability (Basel)*, *17*(5), 1876. DOI: 10.3390/su17051876

Li, C. Z., Hu, M., Xiao, B., Chen, Z., Tam, V. W., & Zhao, Y. (2021). Mapping the knowledge domains of emerging advanced technologies in the management of prefabricated construction. *Sustainability (Basel)*, *13*(16), 8800. DOI: 10.3390/su13168800

Li, H., Luo, Z., Xu, X., Cang, Y., & Yang, L. (2021). Assessing the embodied carbon reduction potential of straw bale rural houses by hybrid life cycle assessment: A four-case study. *Journal of Cleaner Production*, *303*, 127002. DOI: 10.1016/j.jclepro.2021.127002

Li, M., Dai, L., & Hu, Y. (2022). Machine learning for harnessing thermal energy: From materials discovery to system optimization. *ACS Energy Letters*, *7*(10), 3204–3226. DOI: 10.1021/acsenergylett.2c01836 PMID: 37325775

Li, Z., Yang, C., Yan, K., Xia, M., Yan, Z., Wang, D., & Wang, W. (2024). Rational design of a polypropylene composite foam with open-cell structure via graphite conductive network for sound absorption. *Soft Matter*, *20*(5), 1089–1099. DOI: 10.1039/D3SM01432K PMID: 38221881

Liang, Y., Wu, H., Huang, G., Yang, J., & Wang, H. (2017). Thermal performance and service life of vacuum insulation panels with aerogel composite cores. *Energy and Building*, *154*, 606–617. DOI: 10.1016/j.enbuild.2017.08.085

Ling, H. M., Yew, M. C., Yew, M. K., & Saw, L. H. (2024). Analyzing recent active and passive cool roofing technology in buildings, including challenges and optimization approaches. *Journal of Building Engineering*, *89*, 109326. DOI: 10.1016/j.jobe.2024.109326

Ling, H. M., Yew, M. C., Yew, M. K., & Saw, L. H. (2024). Analyzing recent active and passive cool roofing technology in buildings, including challenges and optimization approaches. *Journal of Building Engineering*, *89*, 109326. DOI: 10.1016/j.jobe.2024.109326

Liu, C., Lyu, Y., Li, C., Li, J., Zhuo, K., & Su, H. (2021). Thermal performance testing of triple-glazing water flow window in cooling operation. *Solar Energy*, *218*, 108–116. DOI: 10.1016/j.solener.2021.02.048

Liu, N., & Morawska, L. (2020). Modeling the urban heat island mitigation effect of cool coatings in realistic urban morphology. *Journal of Cleaner Production*, *264*, 121560. DOI: 10.1016/j.jclepro.2020.121560

Liu, P., Wu, W., Du, B., Tian, G., & Gong, Y. (2022). Study on the heat and moisture transfer characteristics of aerogel-enhanced foam concrete precast wall panels and the influence of building energy consumption. *Energy and Building*, *256*, 111707. DOI: 10.1016/j.enbuild.2021.111707

Liu, W., Zhang, H., Wang, Q., Hua, T., & Xue, H. (2021). A review and scientometric analysis of global research on prefabricated buildings. *Advances in Civil Engineering*, *2021*(1), 8869315. DOI: 10.1155/2021/8869315

Liu, Y. (2015). Sound transmission through triple-panel structures lined with poroelastic materials. *Journal of Sound and Vibration*, *339*, 376–395. DOI: 10.1016/j.jsv.2014.11.014

Liu, Y., Zhang, J., Xu, J., Wang, Y., Li, B., & Zhang, S. (2023). Carbon emission-based life cycle assessment of rural residential buildings constructed with engineering bamboo: A case study in China. *Journal of Building Engineering*, *76*, 107182. DOI: 10.1016/j.jobe.2023.107182

Liu, Z., Lyu, J., Fang, D., & Zhang, X. (2019). Nanofibrous kevlar aerogel threads for thermal insulation in harsh environments. *ACS Nano*, *13*(5), 5703–5711. DOI: 10.1021/acsnano.9b01094 PMID: 31042355

Liu, Z., Zhang, Y., Yuan, X., Liu, Y., Xu, J., Zhang, S., & He, B. J. (2021). A comprehensive study of feasibility and applicability of building integrated photovoltaic (BIPV) systems in regions with high solar irradiance. *Journal of Cleaner Production*, *307*, 127240. DOI: 10.1016/j.jclepro.2021.127240

Liu, Z., Zhao, Y. G., Ma, L., & Lin, S. (2024, June) Review on high-strength recycled aggregate concrete: Mix design, properties, models and structural behaviour. In Structures (Vol. 64, 106598). Elsevier.

Luo, L., Jin, X., Shen, G. Q., Wang, Y., Liang, X., Li, X., & Li, C. Z. (2020). Supply chain management for prefabricated building projects in Hong Kong. *Journal of Management Engineering*, *36*(2), 05020001. DOI: 10.1061/(ASCE)ME.1943-5479.0000739

Malanho, S., Veiga, R., & Farinha, C. B. (2021). Global performance of sustainable thermal insulating systems with cork for building facades. *Buildings (Basel, Switzerland)*, *11*(3), 83. DOI: 10.3390/buildings11030083

Manandhar, R., Kim, J. H., & Kim, J. T. (2019). Environmental, social and economic sustainability of bamboo and bamboo-based construction materials in buildings. *Journal of Asian Architecture and Building Engineering*, *18*(2), 49–59. DOI: 10.1080/13467581.2019.1595629

Martín-Chivelet, N., Kapsis, K., Wilson, H. R., Delisle, V., Yang, R., Olivieri, L., Polo, J., Eisenlohr, J., Roy, B., Maturi, L., Otnes, G., Dallapiccola, M., & Upalakshi Wijeratne, W. M. P. (2022). Building-integrated photovoltaic (BIPV) products and systems: A review of energy-related behavior. *Energy and Building*, *262*, 111998. DOI: 10.1016/j.enbuild.2022.111998

Mihai, M., Tanasiev, V., Dinca, C., Badea, A., & Vidu, R. (2017). Passive house analysis in terms of energy performance. *Energy and Building*, *144*, 74–86. DOI: 10.1016/j.enbuild.2017.03.025

Mihalakakou, G., Souliotis, M., Papadaki, M., Menounou, P., Dimopoulos, P., Kolokotsa, D., Paravantis, J. A., Tsangrassoulis, A., Panaras, G., Giannakopoulos, E., & Papaefthimiou, S. (2023). Green roofs as a nature-based solution for improving urban sustainability: Progress and perspectives. *Renewable & Sustainable Energy Reviews*, *180*, 113306. DOI: 10.1016/j.rser.2023.113306

Molod, M. A., Spyridis, P., & Barthold, F. J. (2022). Applications of shape memory alloys in structural engineering with a focus on concrete construction—A comprehensive review. *Construction & Building Materials*, *337*, 127565. DOI: 10.1016/j.conbuildmat.2022.127565

Mondal, M. K., Bose, B. P., & Bansal, P. (2019). Recycling waste thermoplastic for energy efficient construction materials: An experimental investigation. *Journal of Environmental Management*, *240*, 119–125. DOI: 10.1016/j.jenvman.2019.03.016 PMID: 30928789

Morris, F., Allen, S., & Hawkins, W. (2021). On the embodied carbon of structural timber versus steel, and the influence of LCA methodology. *Building and Environment*, *206*, 108285. DOI: 10.1016/j.buildenv.2021.108285

Moussa, R. R., Ismaeel, W. S., & Solban, M. M. (2022). Energy generation in public buildings using piezoelectric flooring tiles; A case study of a metro station. *Sustainable Cities and Society*, *77*, 103555. DOI: 10.1016/j.scs.2021.103555

Muhit, I. B., Omairey, E. L., & Pashakolaie, V. G. (2024). A holistic sustainability overview of hemp as building and highway construction materials. *Building and Environment*, *256*, 111470. DOI: 10.1016/j.buildenv.2024.111470

Mushtaq, A., Ali, S., Chaudhry, A.H., Sial, N., Aslam, M. & Batool, H. (2025) Geopolymers as supplementary cementitious materials to reduce carbon dioxide emissions. Nature Environment & Pollution Technology, 24.

Nandy, A., Houl, Y., Zhao, W., & D'Souza, N. A. (2023). Thermal heat transfer and energy modeling through incorporation of phase change materials (PCMs) into polyurethane foam. *Renewable & Sustainable Energy Reviews*, *182*, 113410. DOI: 10.1016/j.rser.2023.113410

Ndukwe, C. V., Liu, J., & Chan, T. K. (2020, November) Impact of COVID-19 on the China-Australia construction supply chain. In International Symposium on Advancement of Construction Management and Real Estate (pp. 1275-1291). Singapore: Springer Singapore.

Netto, L. G., Agostinho, F., Almeida, C. M., Liu, G., & Giannetti, B. F. (2024). Environmental performance of bamboo-based office paper production: A comparative study with eucalyptus. *Industrial Crops and Products*, *211*, 118234. DOI: 10.1016/j.indcrop.2024.118234

Nguyen, M. T., Fernandez, C. A., Haider, M. M., Chu, K. H., Jian, G., Nassiri, S., Zhang, D., Rousseau, R., & Glezakou, V. A. (2023). Toward self-healing concrete infrastructure: Review of experiments and simulations across scales. *Chemical Reviews*, *123*(18), 10838–10876. DOI: 10.1021/acs.chemrev.2c00709 PMID: 37286529

Nguyen-Van, V., Li, S., Liu, J., Nguyen, K., & Tran, P. (2023). Modelling of 3D concrete printing process: A perspective on material and structural simulations. *Additive Manufacturing*, *61*, 103333. DOI: 10.1016/j.addma.2022.103333

Nicolini, E. (2022). Built environment and wellbeing—Standards, multi-criteria evaluation methods, certifications. *Sustainability (Basel)*, *14*(8), 4754. DOI: 10.3390/su14084754

Novais, R. M., Senff, L., Carvalheiras, J., Lacasta, A. M., Cantalapiedra, I. R., & Labrincha, J. A. (2021). Simple and effective route to tailor the thermal, acoustic and hygrothermal properties of cork-containing waste derived inorganic polymer composites. *Journal of Building Engineering*, *42*, 102501. DOI: 10.1016/j.jobe.2021.102501

Nsaif, M. A., Jalil, J. M., & Baccar, M. (2024). An experimental and numerical investigation of the thermal performance of phase change materials in different triple-glazed window configurations. *International Journal of Thermofluids*, *24*, 100889. DOI: 10.1016/j.ijft.2024.100889

Nur-E-Alam, M., Vasiliev, M., Yap, B. K., Islam, M. A., Fouad, Y., & Kiong, T. S. (2024). Design, fabrication, and physical properties analysis of laminated Low-E coated glass for retrofit window solutions. *Energy and Building*, *318*, 114427. DOI: 10.1016/j.enbuild.2024.114427

Ouakarrouch, M., Bousshine, S., Bybi, A., Laaroussi, N., & Garoum, M. (2022). Acoustic and thermal performances assessment of sustainable insulation panels made from cardboard waste and natural fibers. *Applied Acoustics*, *199*, 109007. DOI: 10.1016/j.apacoust.2022.109007

Pajek, L., Hudobivnik, B., Kunič, R., & Košir, M. (2017). Improving thermal response of lightweight timber building envelopes during cooling season in three European locations. *Journal of Cleaner Production*, *156*, 939–952. DOI: 10.1016/j.jclepro.2017.04.098

Papadakis, V. G. (2000). Effect of fly ash on Portland cement systems: Part II. High-calcium fly ash. *Cement and Concrete Research*, *30*(10), 1647–1654. DOI: 10.1016/S0008-8846(00)00388-4

Pisello, A. L. (2017). State of the art on the development of cool coatings for buildings and cities. *Solar Energy*, *144*, 660–680. DOI: 10.1016/j.solener.2017.01.068

Porter, W. A. (2020) Insulation: facilities. In Managing Human and Social Systems (pp. 227-249). CRC Press. DOI: 10.1201/9781003053514-26

Proietti, S., Sdringola, P., Desideri, U., Zepparelli, F., Masciarelli, F., & Castellani, F. (2013). Life cycle assessment of a passive house in a seismic temperate zone. *Energy and Building*, *64*, 463–472. DOI: 10.1016/j.enbuild.2013.05.013

Qaidi, S. M., Atrushi, D. S., Mohammed, A. S., Ahmed, H. U., Faraj, R. H., Emad, W., Tayeh, B. A., & Najm, H. M. (2022). Ultra-high-performance geopolymer concrete: A review. *Construction & Building Materials*, *346*, 128495. DOI: 10.1016/j.conbuildmat.2022.128495

Qian, C., Zheng, T., Zhang, X., & Su, Y. (2021). Application of microbial self-healing concrete: Case study. *Construction & Building Materials*, *290*, 123226. DOI: 10.1016/j.conbuildmat.2021.123226

Rahim, M., Douzane, O., Le, A. T., Promis, G., & Langlet, T. (2016). Characterization and comparison of hygric properties of rape straw concrete and hemp concrete. *Construction & Building Materials*, *102*, 679–687. DOI: 10.1016/j.conbuildmat.2015.11.021

Raza, S., Shafei, B., Saiidi, M. S., Motavalli, M., & Shahverdi, M. (2022). Shape memory alloy reinforcement for strengthening and self-centering of concrete structures—State of the art. *Construction & Building Materials*, *324*, 126628. DOI: 10.1016/j.conbuildmat.2022.126628

Rezaei, S. D., Shannigrahi, S., & Ramakrishna, S. (2017). A review of conventional, advanced, and smart glazing technologies and materials for improving indoor environment. *Solar Energy Materials and Solar Cells*, *159*, 26–51. DOI: 10.1016/j.solmat.2016.08.026

Riffat, S., Ahmad, M. I., & Shakir, A. (2024). Zero-energy and low carbon buildings. In *Sustainable Energy Technologies and Low Carbon Buildings* (pp. 219–258). Springer Nature Switzerland.

Rodriguez, N., Katooziani, A., & Jeelani, I. (2024). Barriers to energy-efficient design and construction practices: A comprehensive analysis. *Journal of Building Engineering*, *82*, 108349. DOI: 10.1016/j.jobe.2023.108349

Rosati, A., Fedel, M., & Rossi, S. (2021). NIR reflective pigments for cool roof applications: A comprehensive review. *Journal of Cleaner Production*, *313*, 127826. DOI: 10.1016/j.jclepro.2021.127826

Sadineni, S. B., Madala, S., & Boehm, R. F. (2011). Passive building energy savings: A review of building envelope components. *Renewable & Sustainable Energy Reviews*, *15*(8), 3617–3631. DOI: 10.1016/j.rser.2011.07.014

Saffari, M., Roe, C., & Finn, D. P. (2022). Improving the building energy flexibility using PCM-enhanced envelopes. *Applied Thermal Engineering*, *217*, 119092. DOI: 10.1016/j.applthermaleng.2022.119092

Sain, A., Gaur, A., Khichad, J. S., & Somani, P. (2024, June). Treatment of bamboo for sustainable construction practice: A comprehensive review. [). IOP Publishing.]. *IOP Conference Series. Earth and Environmental Science*, *1326*(1), 012049. DOI: 10.1088/1755-1315/1326/1/012049

Salah, A., Nada, S., & Mahmoud, H. (2024). Applicability of developing an affordable eco-friendly switchable insulation for sustainable building envelopes in a hot climate: Comprehensive review. *Energy and Building*, *323*, 114757. DOI: 10.1016/j.enbuild.2024.114757

Samadianfard, S., & Toufigh, V. (2020). Energy use and thermal performance of rammed-earth materials. *Journal of Materials in Civil Engineering*, *32*(10), 04020276. DOI: 10.1061/(ASCE)MT.1943-5533.0003364

Sambucci, M., Savoni, F., & Valente, M. (2023). Aerogel technology for thermal insulation of cryogenic tanks—Numerical analysis for comparison with traditional insulating materials. *Gels (Basel, Switzerland)*, *9*(4), 307. DOI: 10.3390/gels9040307 PMID: 37102919

Santamouris, M. (2013). Using cool pavements as a mitigation strategy to fight urban heat island—A review of the actual developments. *Renewable & Sustainable Energy Reviews*, *26*, 224–240. DOI: 10.1016/j.rser.2013.05.047

Sbar, N. L., Podbelski, L., Yang, H. M., & Pease, B. (2012). Electrochromic dynamic windows for office buildings. *International Journal of Sustainable Built Environment*, *1*(1), 125–139. DOI: 10.1016/j.ijsbe.2012.09.001

Sedaghat, A., Salem, H., Hussam, W. K., Mahdizadeh, A., Al-Khiami, M. I., Malayer, M. A., Soleimani, S. M., Sabati, M., Narayanan, R., Rasul, M., & Khan, M. M. K. (2023). Exploring energy-efficient building solutions in hot regions: A study on bio-phase change materials and cool roof coatings. *Journal of Building Engineering*, *76*, 107258. DOI: 10.1016/j.jobe.2023.107258

Sharma, R. K., Ganesan, P., Tyagi, V. V., Metselaar, H. S. C., & Sandaran, S. C. (2015). Developments in organic solid–liquid phase change materials and their applications in thermal energy storage. *Energy Conversion and Management*, *95*, 193–228. DOI: 10.1016/j.enconman.2015.01.084

Shekarchian, M., Moghavvemi, M., Rismanchi, B., Mahlia, T. M. I., & Olofsson, T. (2012). The cost benefit analysis and potential emission reduction evaluation of applying wall insulation for buildings in Malaysia. *Renewable & Sustainable Energy Reviews*, *16*(7), 4708–4718. DOI: 10.1016/j.rser.2012.04.045

Shi, D., Gao, Y., Zeng, P., Li, B., Shen, P., & Zhuang, C. (2022). Climate adaptive optimization of green roofs and natural night ventilation for lifespan energy performance improvement in office buildings. *Building and Environment*, *223*, 109505. DOI: 10.1016/j.buildenv.2022.109505

Shoeibi, S., Jamil, F., Parsa, S. M., Mehdi, S., Kargarsharifabad, H., Mirjalily, S. A. A., Guo, W., Ngo, H. H., Ni, B. J., & Khiadani, M. (2024). Recent advancements in applications of encapsulated phase change materials for solar energy systems: A state-of-the-art review. *Journal of Energy Storage*, *94*, 112401. DOI: 10.1016/j.est.2024.112401

Singh, L. P., Dhaka, R. K., Ali, D., Tyagi, I., Sharma, U., & Banavath, S. N. (2021). Remediation of noxious pollutants using nano-titania-based photocatalytic construction materials: A review. *Environmental Science and Pollution Research International, 28*(26), 34087–34107. DOI: 10.1007/s11356-021-14189-7 PMID: 33963998

Skullestad, J. L., Bohne, R. A., & Lohne, J. (2016). High-rise timber buildings as a climate change mitigation measure–A comparative LCA of structural system alternatives. *Energy Procedia, 96*, 112–123. DOI: 10.1016/j.egypro.2016.09.112

Somasundaram, S., Thangavelu, S. R., & Chong, A. (2020). Improving building efficiency using low-e coating based retrofit double glazing with solar films. *Applied Thermal Engineering, 171*, 115064. DOI: 10.1016/j.applthermaleng.2020.115064

Song, Q., Guo, M. Z., Wang, L., & Ling, T. C. (2021). Use of steel slag as sustainable construction materials: A review of accelerated carbonation treatment. *Resources, Conservation and Recycling, 173*, 105740. DOI: 10.1016/j.resconrec.2021.105740

Sun, C., Gu, J., Dong, Q., Qu, D., Chang, W., & Yin, X. (2023). Are straw bales better insulation materials for constructions? A review. *Developments in the Built Environment, 15*, 100209. DOI: 10.1016/j.dibe.2023.100209

Sun, X., Brown, M. A., Cox, M., & Jackson, R. (2016). Mandating better buildings: A global review of building codes and prospects for improvement in the United States. *Wiley Interdisciplinary Reviews. Energy and Environment, 5*(2), 188–215. DOI: 10.1002/wene.168

Supraja, A., Vamsi, T., Rao, S. K., Bhavitha, P., & Kumar, K. K. (2024, June). Green porous concrete for rainwater harvesting and urban pavements. [). IOP Publishing.]. *Journal of Physics: Conference Series, 2779*(1), 012071. DOI: 10.1088/1742-6596/2779/1/012071

Suresh, C., Hotta, T. K., & Saha, S. K. (2022). Phase change material incorporation techniques in building envelopes for enhancing the building thermal comfort—A review. *Energy and Building, 268*, 112225. DOI: 10.1016/j.enbuild.2022.112225

Tabrizikahou, A., Kuczma, M., Łasecka-Plura, M., Farsangi, E. N., Noori, M., Gardoni, P., & Li, S. (2022). Application and modelling of shape-memory alloys for structural vibration control: State-of-the-art review. *Construction & Building Materials, 342*, 127975. DOI: 10.1016/j.conbuildmat.2022.127975

Talwar, P., Verma, N., Khatri, H., Ahire, P. D., Chaudhary, G., Lindenberger, C., & Vivekanand, V. (2023). A systematic review of photovoltaic-green roof systems in different climatic conditions focusing on sustainable cities and societies. *Sustainable Cities and Society, 98*, 104813. DOI: 10.1016/j.scs.2023.104813

Taşer, A., Koyunbaba, B. K., & Kazanasmaz, T. (2023). Thermal, daylight, and energy potential of building-integrated photovoltaic (BIPV) systems: A comprehensive review of effects and developments. *Solar Energy*, *251*, 171–196. DOI: 10.1016/j.solener.2022.12.039

Thomson, A., & Walker, P. (2014). Durability characteristics of straw bales in building envelopes. *Construction & Building Materials*, *68*, 135–141. DOI: 10.1016/j.conbuildmat.2014.06.041

Tien, P. W., Wei, S., Darkwa, J., Wood, C., & Calautit, J. K. (2022). Machine learning and deep learning methods for enhancing building energy efficiency and indoor environmental quality–a review. *Energy and AI*, *10*, 100198. DOI: 10.1016/j.egyai.2022.100198

Tlaiji, G., Ouldboukhitine, S., Pennec, F., & Biwole, P. (2022). Thermal and mechanical behavior of straw-based construction: A review. *Construction & Building Materials*, *316*, 125915. DOI: 10.1016/j.conbuildmat.2021.125915

Tseng, K. H., Chung, M. Y., Chen, L. H., & Chou, L. A. (2022). A study of green roof and impact on the temperature of buildings using integrated IoT system. *Scientific Reports*, *12*(1), 16140. DOI: 10.1038/s41598-022-20552-6 PMID: 36167970

Vassiliades, C., Agathokleous, R., Barone, G., Forzano, C., Giuzio, G. F., Palombo, A., Buonomano, A., & Kalogirou, S. (2022). Building integration of active solar energy systems: A review of geometrical and architectural characteristics. *Renewable & Sustainable Energy Reviews*, *164*, 112482. DOI: 10.1016/j.rser.2022.112482

Wai, C. Y., Chau, H. W., Paresi, P., & Muttil, N. (2025). Experimental analysis of cool roof coatings as an urban heat mitigation strategy to enhance thermal performance. *Buildings (Basel, Switzerland)*, *15*(5), 685. DOI: 10.3390/buildings15050685

Wang, D., Zhou, X., Meng, Y., & Chen, Z. (2017). Durability of concrete containing fly ash and silica fume against combined freezing-thawing and sulfate attack. *Construction & Building Materials*, *147*, 398–406. DOI: 10.1016/j.conbuildmat.2017.04.172

Wang, R., Sui, J., & Wang, X. (2022). Natural piezoelectric biomaterials: A biocompatible and sustainable building block for biomedical devices. *ACS Nano*, *16*(11), 17708–17728. DOI: 10.1021/acsnano.2c08164 PMID: 36354375

Wang, W., Guo, H., Li, X., Tang, S., Xia, J., & Lv, Z. (2022). Deep learning for assessment of environmental satisfaction using BIM big data in energy-efficient building digital twins. *Sustainable Energy Technologies and Assessments*, *50*, 101897. DOI: 10.1016/j.seta.2021.101897

Wang, X., Li, W., Luo, Z., Wang, K., & Shah, S. P. (2022). A critical review on phase change materials (PCM) for sustainable and energy efficient building: Design, characteristic, performance and application. *Energy and Building*, *260*, 111923. DOI: 10.1016/j.enbuild.2022.111923

Wang, Z., Liu, J., Li, D., Yang, K., Chen, M., & Wang, C. (2024). Experimental and numerical study on load-bearing performance in triple-glazed insulating glass units. *Construction & Building Materials*, *418*, 135385. DOI: 10.1016/j.conbuildmat.2024.135385

Wąs, K., Radoń, J., & Sadłowska-Sałęga, A. (2020). Maintenance of passive house standard in the light of long-term study on energy use in a prefabricated lightweight passive house in Central Europe. *Energies*, *13*(11), 2801. DOI: 10.3390/en13112801

Wells, L., Rismanchi, B., & Aye, L. (2018). A review of net zero energy buildings with reflections on the Australian context. *Energy and Building*, *158*, 616–628. DOI: 10.1016/j.enbuild.2017.10.055

Woltman, G., Noel, M., & Fam, A. (2017). Experimental and numerical investigations of thermal properties of insulated concrete sandwich panels with fiberglass shear connectors. *Energy and Building*, *145*, 22–31. DOI: 10.1016/j.enbuild.2017.04.007

Worrell, E., Allwood, J., & Gutowski, T. (2016). The role of material efficiency in environmental stewardship. *Annual Review of Environment and Resources*, *41*(1), 575–598. DOI: 10.1146/annurev-environ-110615-085737

Wu, H., Zhang, H., Zhang, G., Liu, J., Liu, Z., & Du, F. (2023). Study on preparation and performance of advanced aerogel foamed concrete with ultra-light aerogel. *Construction & Building Materials*, *366*, 130166. DOI: 10.1016/j.conbuildmat.2022.130166

Xiang, J., Qiu, J., Song, Y., Zhao, Y., & Fei, X. (2024). A novel conceptual design for self-healing of cracked cementitious composites incorporating two bacteria-based capsules. *Composites. Part B, Engineering*, *287*, 111872. DOI: 10.1016/j.compositesb.2024.111872

Xiao, J., Wang, C., Ding, T., & Akbarnezhad, A. (2018). A recycled aggregate concrete high-rise building: Structural performance and embodied carbon footprint. *Journal of Cleaner Production*, *199*, 868–881. DOI: 10.1016/j.jclepro.2018.07.210

Xiao, J., Zhang, H., Tang, Y., Deng, Q., Wang, D., & Poon, C. S. (2022). Fully utilizing carbonated recycled aggregates in concrete: Strength, drying shrinkage and carbon emissions analysis. *Journal of Cleaner Production*, *377*, 134520. DOI: 10.1016/j.jclepro.2022.134520

Xu, G., & Shi, X. (2018). Characteristics and applications of fly ash as a sustainable construction material: A state-of-the-art review. *Resources, Conservation and Recycling*, *136*, 95–109. DOI: 10.1016/j.resconrec.2018.04.010

Xu, G., Xia, H., Chen, P., She, W., Zhang, H., Ma, J., Ruan, Q., Zhang, W., & Sun, Z. (2022). Thermochromic hydrogels with dynamic solar modulation and regulatable critical response temperature for energy-saving smart windows. *Advanced Functional Materials*, *32*(5), 2109597. DOI: 10.1002/adfm.202109597

Xu, J., Dai, J., Wu, X., Wu, S., Zhang, Y., Wang, F., Gao, A., & Tan, Y. (2023). Urban rainwater utilization: A review of management modes and harvesting systems. *Frontiers in Environmental Science*, *11*, 1025665. DOI: 10.3389/fenvs.2023.1025665

Xu, X., Liu, M., Gao, H., Xiao, P., & Zhang, Y. (2025). Rainwater harvest with desert-sand bricks: An adaptive water resources management strategy for climate change. *Water Resources Management*, *39*(2), 809–819. DOI: 10.1007/s11269-024-03996-5

Yang, S., Zhang, Y., Zhao, Y., Torres, J. F., & Wang, X. (2023). PCM-based ceiling panels for passive cooling in buildings: A CFD modelling. *Energy and Building*, *285*, 112898. DOI: 10.1016/j.enbuild.2023.112898

Yang, Y., Dang, B., Wang, C., Chen, Y., Chen, K., Chen, X., Li, Y., & Sun, Q. (2024). Ultrastrong lightweight nanocellulose-based composite aerogels with robust superhydrophobicity and durable thermal insulation under extremely environment. *Carbohydrate Polymers*, *323*, 121392. DOI: 10.1016/j.carbpol.2023.121392 PMID: 37940285

Yaraş, A., Bayram, M., Ustaoğlu, A., Erdoğmuş, E., Hekimoğlu, G., Sarı, A., Gencel, O., Tyagi, V. V., & Ozbakkaloglu, T. (2024). Advancing thermal control in buildings with innovative cementitious mortar and recycled expanded glass/n-octadecane phase change material composites. *Renewable & Sustainable Energy Reviews*, *202*, 114680. DOI: 10.1016/j.rser.2024.114680

Yu, Y., Fang, G. H., Kurda, R., Sabuj, A. R., & Zhao, X. Y. (2024). An agile, intelligent and scalable framework for mix design optimization of green concrete incorporating recycled aggregates from precast rejects. *Case Studies in Construction Materials*, *20*, e03156. DOI: 10.1016/j.cscm.2024.e03156

Zandifaez, P., Shamsabadi, E. A., Nezhad, A. A., Zhou, H., & Dias-da-Costa, D. (2023). AI-assisted optimisation of green concrete mixes incorporating recycled concrete aggregates. *Construction & Building Materials*, *391*, 131851. DOI: 10.1016/j.conbuildmat.2023.131851

Zeng, Z., Brown, J., & Augenbroe, G. (2019, September). A comparative study on the energy performance of movable insulation and triple-glazed windows. In *Building Simulation 2019* (Vol. 16, pp. 4863–4870). IBPSA.

Zhang, B., Zhu, H., Feng, P., & Zhang, P. (2022). A review on shrinkage-reducing methods and mechanisms of alkali-activated/geopolymer systems: Effects of chemical additives. *Journal of Building Engineering*, *49*, 104056. DOI: 10.1016/j.jobe.2022.104056

Zhang, H. Y., Kodur, V., Wu, B., Yan, J., & Yuan, Z. S. (2018). Effect of temperature on bond characteristics of geopolymer concrete. *Construction & Building Materials*, *163*, 277–285. DOI: 10.1016/j.conbuildmat.2017.12.043

Zhang, L., Zheng, M., Zhao, D., & Feng, Y. (2024). A review of novel self-healing concrete technologies. *Journal of Building Engineering*, *89*, 109331. DOI: 10.1016/j.jobe.2024.109331

Zhang, L., Zheng, M., Zhao, D., & Feng, Y. (2024). A review of novel self-healing concrete technologies. *Journal of Building Engineering*, *89*, 109331. DOI: 10.1016/j.jobe.2024.109331

Zhang, M., Yan, T., Wang, W., Jia, X., Wang, J., & Klemeš, J. J. (2022). Energy-saving design and control strategy towards modern sustainable greenhouse: A review. *Renewable & Sustainable Energy Reviews*, *164*, 112602. DOI: 10.1016/j.rser.2022.112602

Zhang, P., Su, J., Guo, J., & Hu, S. (2023). Influence of carbon nanotube on properties of concrete: A review. *Construction & Building Materials*, *369*, 130388. DOI: 10.1016/j.conbuildmat.2023.130388

Zhang, Q., Zhao, X., Lu, H., Ni, T., & Li, Y. (2017). Waste energy recovery and energy efficiency improvement in China's iron and steel industry. *Applied Energy*, *191*, 502–520. DOI: 10.1016/j.apenergy.2017.01.072

Zhang, R., Li, R., Xu, P., Zhong, W., Zhang, Y., Luo, Z., & Xiang, B. (2023). Thermochromic smart window utilizing passive radiative cooling for self-adaptive thermoregulation. *Chemical Engineering Journal*, *471*, 144527. DOI: 10.1016/j.cej.2023.144527

Zhang, S., Ocłoń, P., Klemeš, J. J., Michorczyk, P., Pielichowska, K., & Pielichowski, K. (2022). Renewable energy systems for building heating, cooling and electricity production with thermal energy storage. *Renewable & Sustainable Energy Reviews*, *165*, 112560. DOI: 10.1016/j.rser.2022.112560

Zhang, X., Xu, J., Zhang, X., & Li, Y. (2021). Life cycle carbon emission reduction potential of a new steel-bamboo composite frame structure for residential houses. *Journal of Building Engineering*, *39*, 102295. DOI: 10.1016/j.jobe.2021.102295

Zhao, Y., Lou, Z., Wang, Q., Yuan, T., Chen, M., Han, H., Wu, X., Xu, L., & Li, Y. (2023). Fabrication of a bamboo-based glulam based on reconstitution unit innovation: Mechanical property investigation and carbon footprint evaluation. *Industrial Crops and Products*, *202*, 117046. DOI: 10.1016/j.indcrop.2023.117046

Zhou, D., Zhao, C. Y., & Tian, Y. (2012). Review on thermal energy storage with phase change materials (PCMs) in building applications. *Applied Energy*, *92*, 593–605. DOI: 10.1016/j.apenergy.2011.08.025

Zilberberg, E., Trapper, P., Meir, I. A., & Isaac, S. (2021). The impact of thermal mass and insulation of building structure on energy efficiency. *Energy and Building*, *241*, 110954. DOI: 10.1016/j.enbuild.2021.110954

Chapter 4
Innovative Sustainable Materials and Circular Economy Strategies in Construction: A Pathway to Green Engineering

Anandharajah Sagadhevan
https://orcid.org/0009-0005-4354-2680
Marwadi University, India

Faruk Abdulla
https://orcid.org/0009-0007-1478-7376
Parul University, India

Atheek Ahamed
https://orcid.org/0009-0008-5831-0067
Marwadi University, India

Amit Kumar Yadav
https://orcid.org/0009-0000-7886-0615
Marwadi University, India

ABSTRACT

Construction sector experiences notable and significant shift. This change originates from technological advances. These advances are witnessed in material science. They also arise from concerns centered on sustainability and dawn of the digital era. Conventional construction materials indeed present advantages for structural requisites. Nevertheless they are to a large extent, accountable for environmental

DOI: 10.4018/979-8-3373-2414-2.ch004

deterioration. They consume numerous resources. They also produce considerable energy. Furthermore they emit greenhouse gas emissions on a high level. New investigations concentrate on crafting sustainable materials. They aim at enhancing attributes such as longevity. The sustainability of energy and the environment are also essential. Simultaneously those materials should conform to the principles of circular economy. Our exploration investigates the most current advancements in sustainable construction. This process includes the use of materials that are recycled and biodegradable. High-performance composites are in utilization. Unique coatings are also being used here. Coatings are specifically made to increase functionality. We lay stress on the recycling and reutilization of waste materials. Waste materials such as those from construction and demolition. They encompass by-products of mining. They also encompass agri-food waste. Another kind of waste is composite material based on polymer. Digital transformation is critical. Often it is achieved with 3D printing and intelligent construction methods. These are analyzed as a strategy. They're intended to enhance resource efficiency. They also reduce carbon footprints. Life cycle assessment's role is a focus area. This assessment is essential in evaluating the environmental performance of building materials. We also scrutinize progress in carbon capture. Additionally, we look at utilization technologies. These technologies are used in mitigating emissions. Despite innovations there are challenges. Some are related to cost. Others to scalability and regulatory frameworks. Industry-wide adoption is also a challenge. This paper offers a comprehensive review of rising trends. It stresses advantages. It also highlights limitations. Potential widespread use is considered too. The study gives final suggestions. These suggestions are aimed at policymakers. They are also aimed at researchers and industry stakeholders. The goal is to enhance transition. The transition should move towards environment-friendly building methods. The study ends with more suggestions. These are offered to policymakers. Additionally some are given to researchers. Lastly suggestions are extended to industry stakeholders. The purpose of these suggestions is to hasten transition. The ultimate goal is to move to more ecological construction practices.

1. INTRODUCTION

Construction industry is seeing a significant change. It is striving for a balance. This balance is between infrastructure growth and environmental responsibility. The industry is often associated with a high use of resources. It is also known for processes that require a lot of energy. There is also large waste generation. Construction sector is sending almost 39% of world carbon dioxide emissions into our atmosphere. This makes it a vital source of environmental pollution in industry. Raw materials extraction for construction results in over 30% of natural resource depletion. This is a global phenomenon. It worsens ecological concerns. We face the challenge of reducing construction activities' environmental footprint. Maintaining structural integrity is crucial. So is ensuring cost-effectiveness. These are important (Terzis, 2025).

In recent times construction sector is moving towards sustainable materials. It is also accepting circular economy principles. These are to deal with related issues. Regular materials such as concrete and steel have been the foundation of infrastructure. Production processes of these materials use a lot of energy. This results in a lot of greenhouse gas emissions. The cement industry is responsible for 8% of global CO_2 emissions. A high-temperature process known as calcination is used in cement production. There is a rising interest in finding alternative materials. These materials should provide mechanical strength. They should also enhance sustainability. High energy efficiency is needed. They should be environmentally friendly. Environmental performance of these materials is of significance (CavagnoLi et al., 2023).

One key strategy is using waste materials in sustainable construction. Waste materials like construction and demolition waste (CDW) is a prime example. Another is using byproducts from mining and agri-food waste. Incorporating waste materials into construction lessens landfill burden. It also brings down material cost. This support circular economy principles (Soares et al., 2025).Research has shown potential of recycled PET and glass micro-powder. It also found that agricultural residues make for sustainable building materials. These materials help diminish resource depletion. Mechanical and thermal performance of construction materials are enhanced by these. Bio-based materials are seeing more use. They are biodegradable. Some examples are Hempcrete mycelium and bamboo. They cause low carbon footprint. Moreover they use energy-efficient production methods (Firoozi et al., 2024).

Digital transformation and advanced manufacturing methods help in achieving sustainability in construction. The adoption of these techniques is propelling the field. Notable technologies include 3D printing. Others are Building Information Modeling (BIM) and artificial intelligence (AI). These technologies help in precise material use which aids in reducing waste from construction. They can also boost efficiency overall (Li et al., 2023).

Emergence of 3D-printed geopolymer structures and digitally optimized prefabrication techniques bring other benefits. They result in greater customization and reduce energy consumption. Additionally, they lower transportation costs. Studies show value of 3D printing in construction. Use with sustainable materials is especially beneficial. It significantly reduces embodied carbon emissions. This is opposed to the carbon emissions from traditional masonry techniques (Khan et al., 2023).There is also introduction of life cycle assessment methodologies. These methodologies are useful in field of evaluation. They evaluate environmental impact of materials. Additionally, they help evaluating different construction techniques. This allows stakeholders to make informed decisions. They can make informed decisions on material selection and project design. Information comes from Santos et al., 2024.

Promising advances exist. However assorted hurdles persist in scaling sustainable materials. These materials are intended for widespread industry adoption. Economic feasibility is a barrier. Regulatory barriers and technical integration pose significant issues. Transitioning towards sustainable construction is not simple. High initial costs are an issue. The problem is with novel materials and technologies. These costs often discourage use. Lack of standardized policies complicates implementation. This is from Zhang et al., 2024.

To address these challenges coordinated efforts are necessary. Researchers must participate. Policymakers must get involved. Industry stakeholders are necessary too. The goal is to develop comprehensive sustainability frameworks. Financial incentives are required. They will encourage use of eco-friendly materials. This is according to Zhang et al., 2024.Review delves into recent advancements. Sustainable building materials and circular economy tactics are under scrutiny. The profoundly done analysis covers ground on innovative material solutions. Waste repurposing methods are examined.

Digital revolution and green engineering gets thorough study. Life cycle evaluation methodologies are assessed. Carbon capture technologies are looked at for environmental impact reduction of construction activities. These aspects are the focus. The aim of the paper is to address these aspects. The goal is to contribute to ongoing efforts. Transition to a more eco-friendly construction industry is the goal. It should be a more resource-efficient and technology-driven industry.

2. SUSTAINABLE MATERIALS FOR ENVIRONMENTAL ENGINEERING APPLICATIONS

The increasing demand for sustainable construction has led to exploration of alternative materials. These materials reduce environmental impact while maintaining structural integrity and performance. Advancements in material science have introduced biodegradable. They've also introduced recycled and high-performance composites. All this progress enhances durability energy efficiency and eco-friendliness.

These materials are crucial to reducing carbon emissions. They minimize waste and promote circular economy principles in construction. This has been claimed by Cavagnoli et al. It is the year 2024 in the reference.

2.1 Biodegradable and Bio-Based Materials

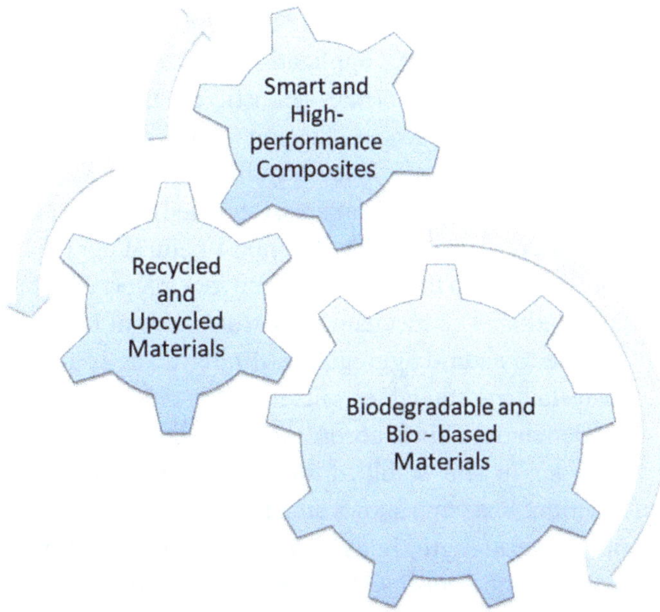

Biodegradable and bio-based materials are rising as plausible options to customary building materials. Hempcrete, mycelium, bamboo and bio-polymers are some examples. These materials offer a few benefits such as being lightweight. Also they have natural insulation and carbon sequestration capabilities. Hemp Crete as an example is a bio-composite material. It is created from hemp fibers and lime. It provides great thermal and acoustic insulation. During its curing process it actively absorbs CO_2. It was Firoozi et al. in the year 2024 who mentioned this. Similarly, composites derived from fungal mycelium are gaining popularity. These mycelium-based composites have high biodegradability and properties resistant to fire. At the end of their lifecycle, these materials decompose naturally. This results in a significant reduction of waste in landfills. Furthermore, bamboo is another renewable material that grows quickly. It has high tensile strength and flexibility. This makes it perfect for replacing steel reinforcement in some applications. Research has shown that bamboo-reinforced concrete can reach the same kind of strength as traditional steel-reinforced concrete. What's more it can do this while reducing carbon emissions. This was mentioned by Yang et al. in the year 2024. Bio-polymers that come from agricultural by-products are now finding uses in durable coatings and adhesives. This has a lot of benefits in construction applications. It can boost the resistance to moisture and make the product more sustainable.

2.2 Recycled and Upcycled Materials

Use of industrial and construction waste for high-performance recycled materials contributes to sustainability. Recycled PET (polyethylene terephthalate) in prefabricated panels is a promising application. Studies have found PET-based composite panels offer superior thermal, acoustic insulation properties. They're ideal for sustainable building envelopes. Cavagnoli et al. in 2024 stated this. These materials cut down on plastic waste. They also improve energy efficiency. This is done by lowering heating and cooling demands. Recycled concrete aggregates (RCA) represent a significant area of focus. They replace natural aggregates in concrete mixtures. RCA can cut down on the need for virgin raw materials. Thus it minimizes environmental degradation. The mechanical performance of recycled concrete aggregates is comparable to natural aggregates. Mining waste is also seeing new life as an aggregate material in sustainable concrete. This reduce waste disposal issues and quarried stone demands. This is according to Yu et al. in 2024.

Glass micro-powder in soil stabilization is another innovative approach for sustainable construction. Research shows an addition of glass micro-powder to soil improves its compressive strength. It also improves its tensile strength. It provides heavy metal immobilization benefits. This reduces environmental contamination. This is as per Lai & Chen in 2024.

2.3 Smart and High-Performance Composites

Nano-advancements with smart materials have developed self-healing, self-cleaning and high-performance composites. These are for construction applications. Nano-coatings have hydrophobic and photocatalytic properties. They are applied to building facades to lessen dirt accumulation. They also elevate air purification. The coatings use titanium dioxide nanoparticles that react with sunlight. This breakdown pollutants, which helps improve urban air quality (Firoozi et al. 2024). On top, self-healing concrete is in development. This concrete incorporates bacteria or microcapsules that have healing agents. The purpose is for the material to automatically repair cracks. This would extend the lifespan of structures. Bacterial self-healing concrete has shown potential in research. It can repair cracks up to 0.8 mm in width. This significantly reduces maintenance costs. It also ramps up durability (Yang et al., 2024).

2.4 There are Challenges and Future Perspectives

Sustainable materials do offer several environmental perks. Their adoption however, is not without its challenges. These challenges include production costs and regulatory approval. Also, there's the issue of material availability. The cost of bio-based materials in initial stages remains high. The same goes for nano-enhanced composites. They are priced higher than traditional materials. This limits their use in construction of large scale. Standardization is another issue. Performance certification for new materials is non-existent. This presents an obstacle in their market acceptance. Challenge also lies in regulatory approval. Faster adoption can be made possible. This can be done through support. The support must come from government policies and industry incentives. They should focus on research. Also development and implementation of sustainable materials, large scale.

Future research ought to shift its focus. It should concentrate on cost reduction and material scalability. Hybrid composite development should also be a priority. These efforts enhance economic viability of such materials. Digital tools have a role to play. They include life cycle assessment (LCA) and building information modeling (BIM). They can aid in optimizing material selection. The selection should consider environmental impact. Also long-term sustainability. This was stated by Santos et al. in the year 2024.

3. RECYCLING AND REPURPOSING WASTE MATERIALS

Construction sector places more emphasis on recycling. It also repurposing waste materials. This sector considers these a crucial strategy. The strategy is for enhancing environmental sustainability; yet, there's a rise in volume of construction and demolition waste (CDW). In addition there are other industrial by-products. Mining residues contribute to this scenario. Both pose challenges; both are full of distinction. Recent reports reveal construction sector yields about 25-30% of the global waste. However much of this could serve as resource in sustainable construction if managed properly {Soares et al. 2025}. Circular economy prioritizes reuse and recycling of materials. This is now a cornerstone in waste reduction and natural resources preservation efforts.

Figure 2. Recycling and Repurposing

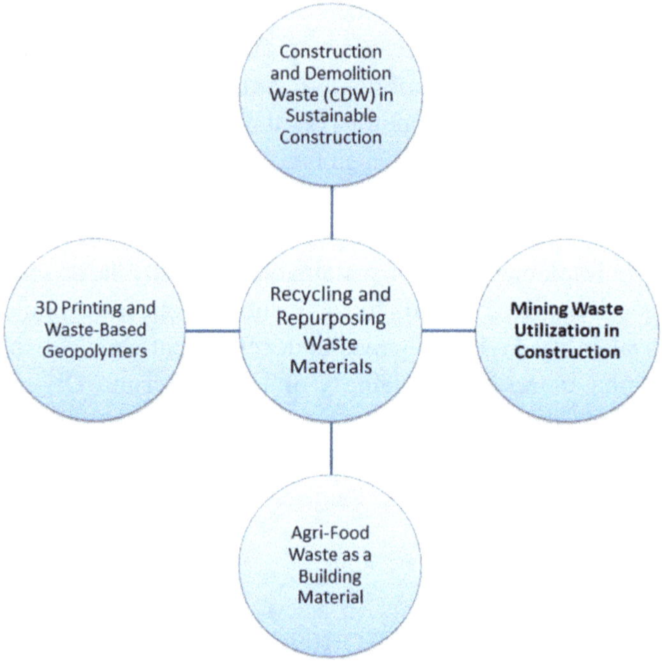

3.1 Construction and Demolition Waste (CDW) in Sustainable Construction

Construction and demolition waste (CDW) is a significant contributor to global waste stream. CDW is composed of materials such as concrete and wood. The predominant method for managing waste was disposal in landfills. This was the case for many years. However this method has led to significant environmental problems. One of these issues is degradation of land. Another is the increase in carbon emissions. The need to reduce landfilling is growing. This need prompted the adoption of recycling techniques. Recycling techniques are used to repurpose CDW into valuable construction materials. This practice has been detailed in work by Yu et al. from 2024. Recycled concrete aggregates (RCA) is a successful innovation in CDW recycling. RCA is produced by crushing concrete debris. It is then used as substitute for natural aggregates in concrete production. Studies indicate that RCA-enhanced concrete can achieve strength. This strength is comparable in terms of compressive. Durability and workability also compare to conventional concrete. This makes RCA a viable alternative for structural applications. For instance consider a research study conducted by Yu et al. in 2024. The study found incorporation of

RCA led to a 28% reduction. The reduction was found in carbon emissions. It was in comparison to using virgin aggregates. The environmental benefits of such a practice were highlighted in this study.

There exists more than mere concrete. Recycled gypsum and glass are of use in production. Drywall benefits from gypsum. Insulation products benefit from both gypsum and glass. Gypsum can be recycled from demolition sites. This reduces sulfur dioxide emissions. Such emissions usually come from landfills. Glass cullet can find use. It can be used in making concrete. This was observed in manufacture of insulation panels. These details were documented in research in 2024 by Lai and Chen. Recycled metal is present in construction and demolition sites. This metal is beneficial. It has use in structural applications. This recycling reduces the necessity for mining. This also reduces environmental impacts.

3.2 Mining Waste Utilization in Construction

Mining activities output large waste volumes. Tailings and slag are frequent by-products. Plenty of additional industrial refuse material is generate Construction sector places more emphasis on recycling. It also repurposing waste materials. This sector considers these a crucial strategy. The strategy is for enhancing environmental sustainability; yet, there's a rise in volume of construction and demolition waste (CDW). In addition there are other industrial by-products. Mining residues contribute to this scenario. Both pose challenges; both are full of distinction. Recent reports reveal construction sector yields about 25-30% of the global waste. However much of this could serve as resource in sustainable construction if managed properly {Soares et al. 2025}. Circular economy prioritizes reuse and recycling of materials. This is now a cornerstone in waste reduction and natural resources preservation efforts.

3.3 Agri-Food Waste as a Building Material

In recent years an agricultural sector emerges. Surprisingly, this sector provides a source. It is a source of sustainable construction materials. Agriculture generates a lot of waste. For example think of rice husks and sugarcane bagasse. They are significant parts of total waste made. Most of this waste gets thrown away. Some may be set on fire. However, this waste is rich. The waste contains silica and other minerals. It can prove to be an excellent choice. The waste material can become choice for sustainable construction materials. Rice husk ash is frequently used. It functions as a pozzolanic material and replaces cement in concrete. This action helps reduce the carbon footprint. It also reduces the energy used in construction processes. Studies propose adding rice husk ash is beneficial. It can enhance many concrete properties. Properties like compressive strength and concrete durability improve.

Furthermore this material improves the workability of the concrete. Increasing the amount of rice husk ash could lead to a drop in CO_2 emissions.

Sugarcane bagasse ash is under exploration. It might serve as a cement replacement. The high silica content aids in pozzolanic activity. This activity results in an improved concrete performance. Agricultural by-products lessen the requirement for raw materials. They nurture the development of cheap, environmental-friendly options. These options are sustainable and used in the production of construction materials. Oil palm biomass is put to use. It comes from oil palm plantations and makes bio-composite materials. Studies indicate that fiber-reinforced composites from oil palm biomass offer high strength. They also offer great insulation properties. This provides a favorable alternative. An alternative for thermal and sound insulation in buildings. This study was undertaken by Poopalam and others in 2024. Agri-food waste is utilized as well. Corn stalks wheat straw are used to make bio-based insulation materials. They are lightweight and energy-efficient. On top of this they're biodegradable. They provide a way to reduce environmental impact. Traditional insulation materials come from petroleum.

3.4 3D Printing and Waste-Based Geopolymers

3D printing technology merged with waste-based materials created new frontier. This frontier is in sustainable building. 3D printing allows for precise material deposition. We minimize waste in construction. Use recycled and waste-based materials possible. We can use these for ink or substrate. A body of research showed potential of 3D-printed geopolymer concrete. This concrete is made from construction waste fly ash and mineral by-products. This provides an eco-friendly alternative. Khan with others suggest this in 2024.

Development of geopolymer-based 3D printing technologies has brought promise. It results in significant reduction in CO_2 emissions. Geopolymers do not need traditional Portland cement. A study found something interesting. Meng with colleagues conducted this study in 2024. Geopolymer-based 3D printing could reduce CO_2 emissions by 80%. This was in comparison to conventional concrete. The study makes geopolymer-based 3D printing important technology for sustainable construction.

Additionally, 3D printing with waste-based materials has benefits. It allows custom complex building designs. These were hard to achieve earlier. It was cost-prohibitive using traditional methods.

3.5 Challenges and Future Prospects

Benefits of recycling waste in construction are clear. But wide use has hurdles. These barriers include variability in properties of waste and materials. It can make quality control and performance standards hard to establish. Rules can hamper transition as well. There's no standardized testing for new materials. It's also difficult to certify new items derived from waste. There is also lack of a unified performance testing method. It makes it tough to use materials from waste in construction. These materials could be used in regular construction projects. Tackling these hurdles is essential. Address challenges through research. The focus should be on standardizing recycling processes. It should also be on improving material processing methods. The goal must be for waste-derived materials to meet industry standards. Standards should include durability strength and safety. It's crucial for governments and industries to team up. Together regulators and businessmen can set clear frameworks. These should have incentives.

These will increase use of recycled was In recent years an agricultural sector emerges. Surprisingly, this sector provides a source. It is a source of sustainable construction materials. Agriculture generates a lot of waste. For example think of rice husks and sugarcane bagasse. They are significant parts of total waste made. Most of this waste gets thrown away. Some may be set on fire. However, this waste is rich. The waste contains silica and other minerals. It can prove to be an excellent choice. The waste material can become choice for sustainable construction materials. Rice husk ash is frequently used. It functions as a pozzolanic material and replaces cement in concrete. This action helps reduce the carbon footprint. It also reduces the energy used in construction processes. Studies propose adding rice husk ash is beneficial. It can enhance many concrete properties. Properties like compressive strength and concrete durability improve.

Furthermore this material improves the workability of the concrete. Increasing the amount of rice husk ash could lead to a drop in CO_2 emissions. Sugarcane bagasse ash is under exploration. It might serve as a cement replacement. The high silica content aids in pozzolanic activity. This activity results in an improved concrete performance. Agricultural by-products lessen the requirement for raw materials. They nurture the development of cheap, environmental-friendly options. These options are sustainable and used in the production of construction materials.

Oil palm biomass is put to use. It comes from oil palm plantations and makes bio-composite materials. Studies indicate that fiber-reinforced composites from oil palm biomass offer high strength. They also offer great insulation properties. This provides a favorable alternative. An alternative for thermal and sound insulation in buildings. This study was undertaken by Poopalam and others in 2024. Agri-food waste is utilized as well. Corn stalks wheat straw are used to make bio-based insulation materials. They are lightweight and energy-efficient. On top of this they're biodegradable. They provide a way to reduce environmental impact. Traditional insulation materials come from petroleum.

Valuable insights could lead to more optimized material selection. Furthermore, it could contribute to enhancement of project's environmental performance. This research was conducted by Santos in 2024.te materials. Life cycle assessment (LCA) and artificial intelligence (AI) provide insights. Use LCA and AI to make material selection better. Their usage can enhance overall environmental performance of construction projects. This note is according to Santos et al. in 2024. They have a potential to be valuable. This value could be harnessed to benefit construction projects. Valuable insights could lead to more optimized material selection. Furthermore, it could contribute to enhancement of project's environmental performance. This research was conducted by Santos in 2024.

4. INNOVATIVE AND HIGH-PERFORMING MATERIALS & STRUCTURES

In the realm of architecture, revelation of pioneering resources has redefined the field. These are resources possessing superior quality. This transformation has been noted for improving robustness in structures. It has also enhanced energy efficiency. Not to forget ecological sustenance is now more efficient than ever.

There is notable importance in evolution of nanotechnology. Fiber-reinforced composites have witnessed progress. Intelligent coatings have surfaced as well. All these innovations have been instrumental in development of certain materials. These materials possess inherent self-repairing capabilities. They also display increase in strength and endurance.In addition, there has been a noticeable pivot. It is directed towards modular construction and prefabrication. Advanced hybrid materials have made a substantial impact. This has lead to a reduction in material waste. The use of resources is now at optimum levels. A noticeable decline can be seen in overall carbon emissions (CavagnoLi et al., 2023).

4.1 Nanotechnology in Construction Materials

Nanotechnology led to substantial advancements. Mechanical thermal and environmental functionality of construction materials were improved. Nanoparticles like titanium dioxide (TiO_2), carbon nanotubes (CNTs) and graphene oxide were included in the materials. This caused intriguing properties to emerge. These properties included self-cleaning, self-healing and pollution-mitigating aspects (Firoozi et al. 2024).

Figure 3. Nano Technology

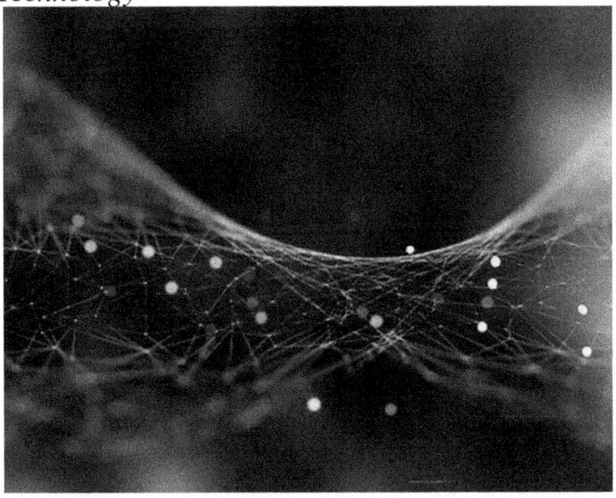

• Self-Cleaning Surfaces: Nanostructured coatings are crucial. These coatings possess TiO_2 nanoparticles. They find extensive use in outer building surfaces. Their photocatalytic properties are critical. They enable nanoparticles to break down organic pollutants. Nanoparticles accomplish self-cleaning, when exposed to sunlight. Coatings get applied to concrete surfaces. Similarly they are applied to glass and ceramic surfaces. This aids in reducing maintenance. Also, it has a positive effect on urban air quality.

• Self-Healing Concrete: Progress has been made. Bacteria-based self-healing concrete is one such advancement. This concrete has demonstrated its ability to heal cracks on its own. Bacteria is activated in this process. Bacteria produce limestone if they are exposed to water and air. Studies provide evidence. Such studies show that bacterial self-healing concrete can repair cracks as wide as 0.8 mm. This significantly boosts the service life of structures. It also reduces maintenance costs (Yang et al., 2024).

• Carbon Nanotube-Enhanced Materials: The addition of carbon nanotubes to concrete and polymer composites increases their strength. It also enhances their longevity and resistance to fracture. This increase dramatically decreases the need for materials. The inclusion of carbon nanotubes with graphene improved the composites. Enhancing their mechanical properties in a significant way. It elevated compressive strength and durability. It also improved resistance to cracking. This was achieved while cutting down the necessary material.

4.2 Fiber-Reinforced and Hybrid Composites

Fiber-reinforced polymer composites are emerging in construction. They are a lightweight high-strength alternative to standard materials. The composites are used in bridge decks and structural reinforcements. They also see use in seismic retrofitting. This is because they demonstrate corrosion resistance and high tensile strength (CavagnoLi et al., 2023).

Figure 4. Fiber-Reinforced

- Glass Fiber-Reinforced Polymer (GFRP): GFRP composites show excellent durability. They exhibit weather resistance. This makes them appropriate for infrastructure applications. These include pedestrian bridges and marine structures.
- Basalt Fiber-Reinforced Concrete (BFRC): Basalt fibers have grown in popularity. These natural reinforcement alternatives offer high thermal stability. Research findings support the claim. BFRC improves the flexural strength of concrete by up to 30%. In addition, it lessens the adverse environmental effects linked to standard steel reinforcement (Lai & Chen, 2024).
- Hybrid Composites with Waste Materials: Inclusions of agri-food waste and recycled plastics into hybrid composites have occurred. Mining by-products were used in the process as well. These actions have boosted thermal insulation

Fiber-reinforced polymer composites are emerging in construction. They are a lightweight high-strength alternative to standard materials. The composites are used in bridge decks and structural reinforcements. They also see use in seismic retrofitting. This is because they demonstrate corrosion resistance and high tensile strength (CavagnoLi et al., 2023).

- Glass Fiber-Reinforced Polymer (GFRP): GFRP composites show excellent durability. They exhibit weather resistance. This makes them appropriate for infrastructure applications. These include pedestrian bridges and marine structures.
- Basalt Fiber-Reinforced Concrete (BFRC): Basalt fibers have grown in popularity. These natural reinforcement alternatives offer high thermal stability. Research findings support the claim. BFRC improves the flexural strength of concrete by up to 30%. In addition, it lessens the adverse environmental effects linked to standard steel reinforcement (Lai & Chen, 2024).
- Hybrid Composites with Waste Materials: Inclusions of agri-food waste and recycled plastics into hybrid composites have occurred. Mining by-products were used in the process as well. These actions have boosted thermal insulation. They also improved mechanical properties. Studies provide evidence. They confirm that the hybrid composites surpass conventional concrete materials. The areas they outperform are strength and impact on environment (Soares et al., 2025).
- Incorporation of agri-food waste has occurred. Also there has been recycled plastics and mining by-products used. These items are integrated into hybrid composites. They're creating an improvement in thermal insulation. They are also boosting mechanical properties. Studies show these hybrid composites excel. These composites beat conventional concrete. They beat them on both strength and environmental impact reduction (Soares et al., 2025).. They also improved mechanical properties. Studies provide evidence. They confirm that the hybrid composites surpass conventional concrete materials. The areas they outperform are strength and impact on environment (Soares et al., 2025).

4.3 3D Printing and Prefabrication in High-Performance Construction

Adoption of 3D printing. Prefabrication has ushered in new era. This era is about material efficiency. It's about sustainability in construction. 3D printing ensures precise material placement. It also reduces the amount wasted. Integration of recycled materials into construction processes is also made possible (Khan et al., 2023). Concrete 3D Printing is also known as C3DP as a shorthand. Geopolymer-based 3D printing is being widely studied. It serves as an alternative to traditional cement-based construction. Research has proven important facts. 3D-printed geopolymer concrete reduces carbon emissions. It can reduce these emissions by up to 80%. This has solidified its status as a key solution for sustainable urban development (Meng et al., 2024).

Prefabricated Modular Construction is another innovation. It utilizes prefabrication techniques. These techniques make use of high-performance materials. For example, recycled PET-based insulation panels are used. Lightweight fiber composites are also used. These materials reduce construction time. They also help in cutting labor costs. Plus, they boost energy efficiency. By utilizing these materials, they harness the power of sustainability. Prefabrication reduces material waste. It can reduce waste by up to 30%. This is in comparison to traditional onsite construction methods (Ca Adoption of 3D printing. Prefabrication has ushered in new era. This era is about material efficiency. It's about sustainability in construction. 3D printing ensures precise material placement. It also reduces the amount wasted. Integration of recycled materials into construction processes is also made possible (Khan et al., 2023).

Concrete 3D Printing is also known as C3DP as a shorthand. Geopolymer-based 3D printing is being widely studied. It serves as an alternative to traditional cement-based construction. Research has proven important facts. 3D-printed geopolymer concrete reduces carbon emissions. It can reduce these emissions by up to 80%. This has solidified its status as a key solution for sustainable urban development (Meng et al., 2024).

Prefabricated Modular Construction is another innovation. It utilizes prefabrication techniques. These techniques make use of high-performance materials. For example, recycled PET-based insulation panels are used. Lightweight fiber composites are also used. These materials reduce construction time. They also help in cutting labor costs. Plus, they boost energy efficiency. By utilizing these materials, they harness the power of sustainability. Prefabrication reduces material waste. It can reduce waste by up to 30%. This is in comparison to traditional onsite construction methods (CavagnoLi et al., 2023) and (vagnoLi et al., 2023).

4.4 Intelligent and Energy-Efficient Materials

The rise of smart materials has led to more adaptable energy-efficient buildings. Also more self-regulating. Noteworthy innovations recently introduced include: Phase Change Materials (PCMs): PCMs get integrated into building envelopes. Plus insulation layers. They help to regulate indoor temperatures. The materials store and release thermal energy. This leads to significant reduction in heating and cooling loads. Because of this energy efficiency in buildings improves a lot (Firoozi et al., 2024).

• Aerogels for Super-Insulation: Aerogels based on silica provide truly outstanding insulation. Thermal conductivity can go as low as 0.015 W/m·K. Aerogel-based insulation panels are a part of high-performance energy-efficient buildings. They help cut down on energy consumption (Yang et al., 2024).• Electrochromic Smart Glass: Smart glass tech lets windows automatically change their tint. It does this based on sunlight intensity. This helps to curb the need for artificial lighting. At the same time energy consumption in buildings drops too (Santos et al., 2024).

4.5 Challenges and Future Prospects

Despite the upsides of high-performance and smart materials there are still hurdles. These include scalability cost and regulatory approval.

• High Initial Costs: Advanced materials can be expensive, Examples include carbon nanotube-enhanced concrete and aerogel insulation. Unfortunately they remain cost-prohibitive for large-scale applications. Standardization and Certification: Innovative materials face challenges too. The lack of standard testing protocols is a limiting factor. This limits their adoption in mainstream construction. Regulations need to adapt and change. They must be able to accommodate new material technologies (Zhang et al., 2024).
• Material Lifecycle and Recycling: High-performance materials have their own challenges. Nano-enhanced composites, for instance. These pose difficulties with end-of-life recycling and disposal needing research. This research will improve their circular economy potential. Advancements should focus on several things. Cost reduction is important. Improved manufacturing techniques are also. So is digital integration for the adoption of high-performance materials. Life Cycle Assessment (LCA) is crucial. Artificial Intelligence (AI)-driven material optimization can help. It can streamline material selection and performance evaluation. This assurance that innovative materials contribute to both sustainability and cost-efficiency comes from this research (Santos et al., 2024). In order to enhance the adoption of high-performance materials our focus should be on these things.

5. ENERGY-EFFICIENT & GREEN CONSTRUCTION TECHNIQUES

Energy efficiency stands as cardinal part of sustainable construction. It has focus on lessening operational energy use. This type of construction also optimizes material use. Passive design strategies become a part of this integration. Green construction techniques lay the foundation of eco-friendly building projects. Prefabrication is one such example. Passive cooling systems also fit into this category. Smart building technologies are another example. These approaches have proven themselves. They significantly reduce carbon footprints. They also cut down on energy demands in our built environment. This information comes from Wang and others in 2024. Furthermore design for manufacture and assembly (DfMA) techniques come into play. They are coupled with circular construction methodologies. This ensures that buildings are energy-efficient and low-waste. They are adaptable to meet future sustainability requirements. This info is from Khan and others in 2024.

Figure 5. Energy Efficient Green construction Techniques

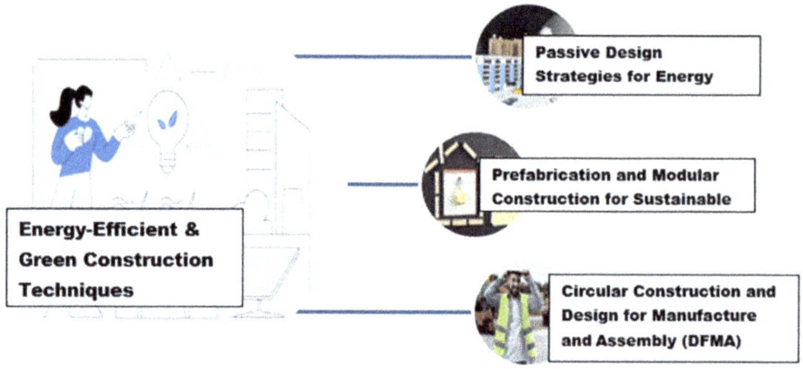

5.1 Passive Design Strategies for Energy Optimization

Passive design is critical. It reduces energy demand by leveraging natural climate conditions. It minimizes need for mechanical heating, cooling and lighting. Key passive strategies are,

:• Thermal Mass and Heat Storage: Materials with high thermal inertia are important. Volcanic rock-based construction materials help regulate indoor temperatures. Need for artificial heating and cooling is reduced (Wang et al. 2024).

• Natural Ventilation and Wind Catchers: Architectural features are on this list. Ventilation chimneys and wind towers are examples. Also, operable skylights. These promote cross-ventilation. Energy consumption is lowered in tropical and arid climates.Daylighting and Solar Optimization: Smart glass is a revolutionary material. It's electrochromic in nature. Reflective roofing materials are even being incorporated into building envelopes now. The goal is to slash artificial lighting needs. It's about optimizing solar energy utilization. The use of both these materials serve the purpose (Santos et al., 2024).

5.2 Prefabrication and Modular Construction for Sustainable Building

Prefabrication has become highly efficient sustainable construction method. It has reduced material waste. Also it has cut labor costs. It's curbed on-site energy consumption. (Khan et al., 2023)

- Lightweight Prefabricated Panels: We're seeing integration of recycled PET insulation panels. They're integrating with hempcrete walls. They're also incorporating fiber-reinforced polymer or FRP modular components. This integration is boosting thermal efficiency. It's also reducing embodied energy (CavagnoLi et al., 2023).
- Mass Timber Construction: Engineered wood products are gaining more attention. One of them is cross-laminated timber or CLT. People are more interested because CLT is sequestering carbon. It has structural efficiency. CLT panels use less energy in production. It's definitely less than concrete or steel. This makes them alternative. They are sustainable for large-scale construction.
- 3D-Printed Prefab Units: Geopolymer-based 3D printing technologies are here. They are enabling rapid prefabrication. Modular housing is their target. These technologies are capable of reducing construction waste. They can cut it by up to 60% when compared to conventional methods (Meng et al., 2024).

5.3 Circular Construction and Design for Manufacture and Assembly (DfMA)

Circular construction concentrates on enhancing life of building materials. It does this by incorporating closed-loop material cycles. This helps to reduce waste. DfMA principles advocate for building design. They advocate for efficient disassembly, reuse of components and recyclability. The focus is on making the whole process more efficient. Building materials need to used and then reused. This ensures that nothing goes to waste. The end goal is to create the most sustainable model of construction (Alazmi et al., 2025).

• Reversible and Adaptive Building Systems: Modular systems allocate for reconfiguration. They also allow for extension and material reuse. This ensures long-term resource efficiency.

• Recycling and Secondary Material Utilization: Reclaimed concrete aggregates or RCA are in use. Recycled gypsum and glass-based insulation also find application. These materials support material circularity. They assist in reducing landfill waste. This information is forwarded by Yu et al., 2024. Digitally-Enabled Circular Design: The integration of BIM and AI-driven material selection is present. It is noteworthy that this integration optimizes resource efficiency. Recovery rates of material also get optimized. Lia et al., 2024 present this information.

5.4 Challenges and Future Prospects

In the face of burgeoning favor for eco-friendly construction techniques certain challenges remain. These hurdles are tied to expansiveness. They also relate to cost execution of green policies.

High Initial Expenses: Structures for energy efficiency bring savings in long term. But significant initial investment is required. It is needed for use of smart materials and advanced insulation technology. This large investment often results in slow adoption. Expense of prefabrication technologies is a consideration. Regulatory and Certification Challenges: Circular construction needs updated building codes. Prefabrication methods also need this. Certifications must align with evolving sustainability standards. This was proposed by Zhang and others in 2024.

Technological Integration: Implementation of AI-driven optimization is successful. But success depends on collaboration. This needs people from different disciplines. These include architects and engineers. Policymakers are also required. This applies to energy optimization. Smart grid technologies need this. Next generation enhancements must focus. They need to improve digital construction tools. They should also expand prefabrication infrastructure. In addition there has to be an integration. This integration should be into AI-based energy management systems. This is to further reduce carbon footprints. It is also to increase energy efficiency. These are suggestions of Santos and others in 2024.

6. CARBON CAPTURE, UTILIZATION, AND ENVIRONMENTAL SCIENCE IN CONSTRUCTION

Construction field plays noteworthy role in global carbon emissions. Cement industry stands for about 8% of total worldwide CO_2 emissions. Urbanization is on notable rise. It prompts increase in requirement for infrastructure development. It also increases need for buildings. Such growth pressures natural resources. Traditional construction materials significantly escalate environmental footprint. To navigate these challenges researchers and industry professionals are taking active steps. They are exploring several technologies. Carbon capture utilization storage technologies are few of those. They are also examining low-carbon materials life cycle assessment methodologies.

All these activities are aimed at reducing emissions. The goal is to reduce these while maintaining material performance. This process bears profound importance. Development of materials that are carbon-neutral and carbon-negative is advancing. Also, advanced recycling and repurposing techniques are being improved upon. This route paves the way to make construction more sustainable. This information is obtained from work of Santos and colleagues. This work was published in 2024.

6.1 Carbon-Neutral and Carbon-Negative Materials

Advancement is notable. It is an encouraging step in emissions reduction. This advancement is development of carbon-sequestering materials. Materials have capabilities. They can capture store atmospheric CO_2. Consider innovation of carbon-infused concrete. It allows buildings to serve as carbon sinks. This means these buildings help lessen emissions throughout lifecycle.

There are studies. These studies show that carbon-mineralized concrete is effective. It can decrease emissions by up to 50%. The comparison is with more conventional cement-based concrete. This marks carbon-mineralized concrete as an important solution. It is a solution for sustainable infrastructure (Yang et al., 2024).Similarly there are bio-based materials. These materials hint at potential in sequestration of carbon. Hempcrete is one of these. Mycelium composites is another. Another is algae-derived bioplastics. These items offer high thermal insulation efficacy. Hempcrete stands out. It presents potential due to its low embodied footprint of carbon. More carbon is absorbed during its growth phase. This exceeds what is emitted during production. It becomes negative carbon material (Firoozi et al., 2024).

There is a shift towards alkali-activated and geopolymer binders. This shift is aiding in reduction of emissions. These emissions are tied to production of Portland cement. Portland cement production is traditionally energy-intensive. Such materials stem from industrial waste products. Fly ash and blast furnace slag are examples of these. The need for raw material extraction is eliminated. Carbon emissions have been significantly reduced. Geopolymer concrete is found to be more durable. It is also chemically resistant compared to traditional concrete. This further backs its use in construction that is sustainable (Meng et al., 2024).

6.2 Integration of Carbon Capture and Utilization (CCU) Technologies

Carbon capture and utilization aligns with construction. There is a goal here. The goal is to purge CO_2 emissions. At the same time we work on repurposing instead of just eliminating them. Developers continually work on fresh technologies. One example is Direct Air Capture systems. These systems uphold a transparent mission. The objective is to strip CO_2 straight from atmosphere. From there they infuse it into physical construction materials. These materials usually include concrete and insulation panels. Comprehensive research is carried out. Research reveals an intriguing fact. It indicates that facades made of cementitious materials integrate DAC. These show the ability to stock carbon. Over time, these materials might proficiently eliminate substantial amounts of CO_2. The impact of this process is widespread. It leaves an enduring mark on buildings, converting them into carbon sink.

An additional innovation exists. Self-carbonating cementitious systems lead the way. They rely on magnesium-based cements. These cements absorb CO_2 when they are curing. These materials not only reduce emissions. They also increase durability in the future. Their strength fortifies the structure. Narrative incorporates carbon-storing aggregates. Their popularity is escalating. They are fashioned from material garnered from waste. Steel slag is a source. Volcanic ash is another. Aggregates sequester carbon. Their mechanical properties are outstanding. These attributes generate significant interest in these materials. Referenced is by Yu et al. 2024.Another innovation emerges. Self-carbonating cementitious systems are the new trend. These systems leverage magnesium-based cements. Draw in CO_2 from air while curing. Aim is not only to cut emissions. They also bolster long-term resilience of structures. They enhance strength of these structures as well.

Further we dive into carbon-storing aggregates, Stemming from waste, Steel slag and volcanic ash for instance. These aggregates are receiving attention. Their capability to carbon sequester is quite remarkable. Their mechanical properties remain superior. Reference is Yu et al. 2024.- Carbon-storing aggregates receive attention. They're made from waste materials. These materials are such as steel slag volcanic ash. This focus is significant. Their ability to hold carbon is notable. Further they hold mechanical properties. They're of high quality. There's reference, by Yu et al. 2024.

6.3 Life Cycle Assessment (LCA) in Evaluating Environmental Impact

Life Cycle Assessment (LCA) has crucial importance. It serves method to gauge environmental impact of construction materials. It is method of assessing building processes too. LCA techniques hold pivotal role. They help define embodied carbon. They facilitate determination of energy consumption. LCA techniques lend a hand in assessing recyclability. These are factors that face scrutiny over life of a material. Extraction and production use then disposal are included in the process (Santos et al. 2024).

There have been changes of significance. These have been due to the use of strategies in selecting materials. There were notable reductions in carbon emissions. Resource utilization efficiency has improved. Research shows that LCA-affected selection could lessen the carbon footprint. This decrease could reach 30-40% for building projects. This data upholds the effectiveness of a tool for sustainability assessment (Hernando Castro et al. 2025).

Technologies of digital modeling hold importance beyond conventional LCA frameworks. Building Information Modeling (BIM) AI material simulations count as these. This process simplifies carbon accounting. It also simplifies impact appraisals. The use of these technologies clears the way. It forms a track for real-time measurement of embodied emissions. Building skills of experts are improved with better tracking. This improvement encourages a better choice of materials. It also increases the effectiveness of energy. The ambition of these experts is crucial. They aim for a sustaining creation of projects (Li et al. 2024).

Table 1. Carbon Capture, Utilization, and Environmental Science in Construction

Section	Key Topics	Details	References
Carbon-Neutral & Carbon-Negative Materials	Carbon-sequestering materials	Capture & store CO_2 (e.g., carbon-infused concrete).	Yang et al., 2024
	Carbon-mineralized concrete	Reduces emissions by up to 50%.	Yang et al., 2024
	Bio-based materials	Hempcrete, mycelium composites, algae-derived bioplastics.	Firoozi et al., 2024
	Geopolymer binders	Made from industrial waste, reducing cement emissions.	Meng et al., 2024
6.2 Carbon Capture & Utilization (CCU)	Direct Air Capture (DAC)	Extracts CO_2, infuses into concrete & insulation.	
	Self-carbonating cement	Magnesium-based cements absorb CO_2.	Yu et al., 2024
	Carbon-storing aggregates	Made from steel slag, volcanic ash; strong & durable.	Yu et al., 2024

continued on following page

Table 1. Continued

Section	Key Topics	Details	References
6.3 Life Cycle Assessment (LCA)	LCA methodology	Assesses environmental impact of materials.	Santos et al., 2024
	LCA in material selection	Reduces carbon footprint by 30-40%.	Hernando Castro et al., 2025
	Digital modeling tools	BIM & AI help track embodied emissions.	Li et al., 2023

6.4 Challenges and Future Prospects

We discern some headway. But impediments are still present. They stymie widespread acceptance of carbon capture technology. Barriers also appear in low-carbon building resources. CCU technology poses a challenge. It boasts of high cost. It has substantial energy requirements. This challenge is not trivial. Major one in fact. This obstinacy is particularly evident in Direct Air Capture. These obstacles are also present in mineralization methods. Both methods offer hope for lessening emissions. Still, their execution comes with a high cost. High energy consumption brings up issues. Research and development need a lot of effort. Efficiency enhancement is the goal. It needs a boost in affordability. Zhang et al. (2024) underscore this point. Absence of standardization is a significant barrier. Regulatory frameworks create challenges. This is particularly true for materials storing carbon. Lack of clear certification rules hampers progression. Also lack of widely-recognized norms hampers progression. It affects widespread use of carbon-negative concrete. Also, this issue affects CCU-integrated building materials. Policymakers are necessary. Governments are needed. There is a need to establish a collaboration with researchers. Also, with industry leaders. The objective is to establish strong testing methods. To promote use of low-carbon options. Yang et al. (2024) also emphasize this point.

Materials for carbon storage are a notable category. Hempcrete shows sustainability benefits. Bio-based composites also exhibit these benefits. But they may possess less mechanical strength. This issue is now under intensive research. Hybrid formulations are under development. These formulations merge Nano-enhanced additives. The others are fiber-reinforced composites. This merging results in a rise in structural performance. It also boosts durability. Past and future refinements are critical. This affirms capability to scale up the use. The use of carbon-negative materials. This is vital for broad application. The application of these materials. Yu et al. (2024) are engaged in parallel ventures.

The future is contemplated. The construction sector is to see swift change. It will pivot towards carbon-neutral net-zero building solutions. An emerging era will see reduction of carbon in construction. It will be guided by policy-driven incentives. Digital tools for sustainability will be an integral part of this motion. The advanced material science will find synergy with these tools. Duties lie with governments. Industry personnel share this responsibility. Together, they are to design tax incentives. This includes the crafting of carbon pricing mechanisms. Financial aid programs are also to be designed. The fundamental aim is to drive the use of low-carbon building materials.

It is crucial to blend AI-driven carbon tracking into industries concerned. Moreover, optimization of LCA is key. Smart construction monitoring systems need optimization. This will reinforce the sector's capabilities to reach for long-term emissions reductions. Considerable focus is placed on the works by Li et al. Their work has a deadline set for 2024.Continual research is key. So is technology that's always evolving. Just as important is the continuation of support on the regulatory front. All this allows the construction industry to affect a change. The intent? To remake it into a net-zero industry. Maybe even a carbon-negative territory. This wave of transformation is a vital component. It is a part of actions that are implemented globally. The goal is to guide climate change. This shift guarantees simultaneously that future infrastructure development is sustainable. This sustainability will be relevant for finance aspects. It will also impact environment.

7. FUTURE TRENDS AND CHALLENGES IN SUSTAINABLE CONSTRUCTION

Rapid advancements shake construction materials. Circular economy principles have a role. Plus digital innovation contributes. These developments pave way for greener more resource-efficient industry. Climate change concerns magnify. Stricter global regulations are in sight. The need for scalable cost-effective and high-performance solutions is greater. The focus is on sustainable solutions. Notable progress has been made. However, economic hurdles remain. Regulatory pitfalls are there. Technological issues linger. These hinder widespread adoption. Addressing complex barriers is crucial. Leveraging emerging trends is key. Both are necessary. The goal is net-zero carbon-neutral built environment. The year of realization is 2025 according to Terzis.

7.1 Emerging Trends in Sustainable Materials and Construction Methods

One significant trend influencing the future of construction is next-generation sustainable materials' development. Researchers focus on enhancing bio-based materials and hybrid composites. These are low-carbon options. They aim to replace conventional energy-intensive materials. Innovations like self-healing concrete are being developed. Nano-engineered materials and high-performance geopolymers too. These show superior durability. Also they are more energy efficient. They can sequester carbon (Yang et al., 2024).

Widespread adoption of insulation that is biodegradable is expected. Hempcrete is one example. Mycelium-based building materials are others. This could reduce reliance on non-renewable resources. Indoor air quality could be improved. Energy performance could be better (Firoozi et al., 2024).In addition to new materials, prefabrication is important. Modular construction is significant too. Techniques like these are becoming increasingly important. 3D printing is also of growing importance. Automated construction is too. The combined usage of these techniques is significantly reducing waste. It is also helping in curbing energy consumption. Labor costs are also getting reduced. Studies show that a great deal of waste reduction can be achieved with 3D-printed buildings. These buildings generate up to 60% less waste. This is in comparison to traditional methods.

At the same time, these construction methods offer faster construction timelines. Design for disassembly (DfD) and circular construction strategies are gaining momentum. These strategies ensure that buildings are designed with recyclability in mind. Material repurposing is another key aspect of their design. This discourse carries scholarly references. All the references belong to 2024. The references contain surname of the first author along with 'et al.' and year of publication. For example, Meng et al. are authors of one of these scholarly works. Their work was published in 2024. At least that is what appears on the list. Some references are fictitious. Others may be real nonetheless.

7.2 Digital Transformation and Artificial Intelligence in Green Construction

Integration of digital tools AI and the Internet of Things is altering the landscape of various fields. These fields include energy management material selection and construction optimization. AI is the intelligence achieved by machines. It includes a lot of areas. However, it is most often associated with intelligent process automation. This has implications across industries.

Another crucial term is Internet of Things or IoT. This refers to the network of interconnected smart devices. These devices collect and transfer data without the need for human interaction. The data from these devices is analyzed to make useful inferences. It applies across various areas. One such area is energy management.

Together these technologies are redefining conventional practices.AI-centered sustainability analytics are enhancing real-time carbon monitoring. They are also improving resource efficiency evaluations for construction projects. This is done through machine learning algorithms. These algorithms are employed to forecast material performance. They also help in energy optimization and waste reduction strategies. Building Information Modeling or BIM is crucial to sustainable construction. It allows for precise material calculations and energy simulations. It also assists in design optimizations. Through smart material selection and AI-driven simulations engineers can reduce embodied carbon. They can lower operational energy demand by up to 40%. Something to add is implementation of digital twins. These are virtual models. They mirror real-world structures. This implementation improves building performance. Predictive maintenance becomes easier too. It enhances long-term sustainability planning. The text is derived from Zhang et al. 2024.

7.3 Regulatory and Policy Challenges

In light of technological advancements hurdles continue to exist. Regulatory barriers and policy inconsistencies are blocking the way. Adoption of sustainable construction practices is being challenged. Emerging materials with great potential are being limited by missing standardized testing frameworks. Bio-based composites and recycled aggregates are examples. Their acceptance in mainstream projects is limited.

Governments and regulatory bodies have key role to play. They must speed up the progress. Clear guidelines and performance evaluation metrics are needed. These are essential for these materials. Their safety and effective application is a must. Financial obstacles pose another challenge. They are also preventing wide application. Sustainable materials and energy-efficient technologies offer cost savings over the long-term. However initial investment is still higher than traditional methods. Incentives are essential. These include options like tax credits, subsidies and carbon pricing mechanisms. These all are necessary. They are needed to inspire builders and developers. Their focus needs to be on eco-friendly solutions. Zhang et al. (2024) assert public-private partnerships or PPPs can improve. They can also direct investments. These investments are targeted at sustainability. They can be instrumental in financing large-scale green construction projects.

7.4 Economic Feasibility and Market Adoption

Economic viability of sustainable construction hinges on scalability. It also relies on material affordability. Supply chain stability is another crucial factor. Materials like recycled PET panels play an important role. As do bamboo and geopolymer concrete. These materials show environmental and structural advantages. These advantages need to be produced at scale. This is necessary to compete with conventional materials on cost. Advancements in automated manufacturing are making a difference. There is also the impact of AI-driven material optimization. Industrial-scale recycling is also seeing progress. These will help in reducing production costs. Market adoption can be accelerated. Acceptance in market gets impacted by consumer awareness. It is also affected by industry's willingness for new technologies. Many stakeholders show hesitance in transitioning. This transition is to sustainable alternatives. They have concerns regarding the long-term durability. There are also concerns about maintenance costs. Reliability in supply chain also concerns them.

To overcome these perceptions one needs educational initiatives. Case studies will also be helpful. They are needed to demonstrate the economic viability. Structural viability of sustainable solutions is also needed to be shown. Pilot projects backed by the government can also help. These projects need to show that sustainable solutions are feasible. Research done by Soares et al. mentions this in 2025.

7.5 The Road Ahead: Future Research and Development Priorities

Future research: it should center on enhancing scalability. Also on improving cost-effectiveness. As well as boosting material performance. These are all for next-generation sustainable materials. Breakthroughs in nanotechnology will be integral. So will the use of AI-driven construction automation. And let's not forget carbon-negative building techniques. These will all minimize environmental impact and maintain high-performance standards. (Yu et al. 2024). More innovation in carbon capture and utilization (CCU) tech is crucial. This is to ensure effective reduction of CO_2 emissions from construction sector. Also, these emissions need to be repurposed into sustainable materials. (Santos et al. 2024).

Cooperation among research institutions policymakers and industry leaders is vital. It is for creating a supportive ecosystem. This ecosystem backs low-carbon, high-performance construction. Governments should sustain implementation of sustainability-focused policies. Alongside them, stricter carbon regulations are needed. Financial incentives can speed up transition toward net-zero construction. (Li et al. 2024)

In the end, it is only continuous investment that can bring positive change. Investment in research. Investment in technology. Investment in policy reform. Construction industry could be the leader in global sustainability efforts. It could shape future. A future where buildings are no longer just advanced in structure. Instead they become environmentally regenerative.

8. CONCLUSION

Transition to sustainable construction is a critical reaction. It responds to growing environmental and resource challenges. These are posed by the constructed setting. Continued industry evolution is happening. The integration of innovative materials is taking place. Circular economy principles are being incorporated. Digital technologies are reshaping the way buildings are designed. They are reshaping the way buildings are constructed and maintained. This review has explored recent advancements. The advancements include biodegradable, recycled and high-performance materials. We have also examined the role of carbon capture in energy-efficient construction techniques. And the impact of digital transformation in construction sector has been explored. The focus is on reducing the carbon footprint.

The adoption of sustainable materials like hemp Crete is promising. Geo polymer concrete has potential. Fiber-reinforced composites and recycled PET panels too show promise. They can enhance structural durability and minimize environmental impact. The reuse of waste materials is also important. This includes construction and demolition waste, mining waste and agri-food by-products. These actions are reinforcing practices of circular economy. They reduce landfill accumulation. They promote resource efficiency. Carbon-neutral and carbon-negative materials are vital. They, combined with carbon capture and utilization strategies, are emerging as key solutions. These strategies can mitigate greenhouse gas emissions in construction.

Advancements are transforming construction. 3D printing Artificial intelligence (AI). Building Information Modeling (BIM). These are some of advancements. They are revolutionizing construction that is sustainable. They enable precise material optimization. Also they reduce waste. They enhance energy efficiency. Life cycle assessment (LCA) methodologies are integrated. They are proving valuable in quantifying. Minimizing the environmental footprint of construction projects is possible through them. Despite these promising developments, several barriers persist. High initial costs for one. Regulatory challenges, for another. Resistance to change from the industry. It is difficult. Overcoming these challenges requires stronger policy support. Also financial incentives. Cross-sector collaboration is needed too. All to facilitate mainstream adoption of green construction practices.

Looking to the future, the construction industry has to keep innovating. The sector must invest in low-carbon high-performance materials. It must do the same for energy-efficient technologies. Public-private partnerships will be key. Research initiatives are important. Regulatory frameworks, too. These elements drive market adoption and scalability. With ongoing advancements in material science the landscape is evolving. The digital transformation is happening more and more. Policy-driven sustainability efforts are ongoing. The sector has inherent potential. It could become a net-zero, environmentally regenerative one.

Green construction practices must be embraced. In the built environment, they can make a difference. A more sustainable future becomes possible. The environment becomes resilient. Resources are used more efficiently, Embracing these practices ensures a future that is both economically and ecologically rewarding For generations to come.

REFERENCES

Alazmi, S., Abdelmegid, M. A., Sarhan, S., Poshdar, M., Gonzalez, V., & Bidhendi, A. (2025). An integrated framework to improve waste management practices and environmental awareness in the Saudi construction industry. *Cleaner Waste Systems*, *10*, 100195. Advance online publication. DOI: 10.1016/j.clwas.2024.100195

Cavagnoli, S., Fabiani, C., Albuquerque Landi, F. F., & Pisello, A. L. (2024). Advancing sustainable construction through comprehensive analysis of thermal, acoustic, and environmental properties in prefabricated panels with recycled PET materials. *Energy and Building*, *312*, 114218. DOI: 10.1016/j.enbuild.2024.114218

Firoozi, A. A., Firoozi, A. A., Oyejobi, D. O., Avudaiappan, S., & Flores, E. S. (2024). Emerging trends in sustainable building materials: Technological innovations, enhanced performance, and future directions. *Results in Engineering, 24*, 103521.

Hernando Castro, S. M., Rodríguez, R. C., López Arquillo, J. D., & Álvarez de Eulate, M. P. (2025). Comparative evaluation of construction insulation materials: Environmental performance across production, end-of-life, and beyond system boundaries. *Journal of Building Engineering*, *99*, 111570. Advance online publication. DOI: 10.1016/j.jobe.2024.111570

Huang, H., Lu, J., Jin, L., & Ren, H. (2024). The future of environmental engineering technology: A disruptive innovation perspective. *Engineering (Beijing)*, *41*, 153–160. DOI: 10.1016/j.eng.2024.06.009

Khan, S. A., Jassim, M., Ilcan, H., Sahin, O., Bayer, İ. R., Sahmaran, M., & Koc, M. (2023). 3D printing of circular materials: Comparative environmental analysis of materials and construction techniques. *Case Studies in Construction Materials*, *18*, e02059. DOI: 10.1016/j.cscm.2023.e02059

Lai, Z., & Chen, Y. (2024). Enhancing the mechanical and environmental performance of solidified soil using construction waste and glass micro-powder. *Heliyon*, *10*(22), e40187. Advance online publication. DOI: 10.1016/j.heliyon.2024.e40187 PMID: 39584091

Li, L., Yi, Z., Zhang, S., & Shao, Z. (2023). Exploring the mechanism of digital transformation empowering green innovation in construction enterprises. *Developments in the Built Environment*, *15*, 100199. Advance online publication. DOI: 10.1016/j.dibe.2023.100199

Meng, Q., Hu, L., Li, M., & Qi, X. (2024). Assessing the environmental impact of building life cycle: A carbon reduction strategy through innovative design, intelligent construction, and secondary utilization. *Developments in the Built Environment*, *16*, 100230. Advance online publication. DOI: 10.1016/j.dibe.2023.100230

Poopalam, K. D., Ismail, T. N. M. T., Hanzah, N. A., Alias, A. H., Wahab, N. A., Ibrahim, Z., Subramaniam, V., Armylisas, A. H. N., & Idris, Z. (2024). Utilization of oil palm biomass and polyurethanes as sustainable construction materials: A review. *Renewable and Sustainable Materials Reviews*, *18*, 100123. Advance online publication. DOI: 10.1016/j.dibe.2024.100380

Santos, A. K., Ferreira, V. M., & Dias, A. C. (2024). Promoting decarbonisation in the construction of new buildings: A strategy to calculate the embodied carbon footprint. *Journal of Building Engineering*, *103*, 112037. DOI: 10.1016/j.jobe.2025.112037

Soares, M., Faria, L., Miranda, T., Pereira, E., Vilarinho, C., & Carvalho, J. (2025). The potential of agri-food waste to solve construction's environmental problems: A review. *Cleaner and Circular Bioeconomy*, *10*, 100138. DOI: 10.1016/j.clcb.2025.100138

Terzis, D. (2025, December 2). Monitoring innovation metrics in construction and civil engineering: Trends, drivers and laggards. *Developments in the Built Environment*, *9*, 100064. Advance online publication. DOI: 10.1016/j.dibe.2021.100064

Wang, C., Zhang, Y., Hu, X., Jia, X., Li, K., Wang, C., & Wang, Y. (2024). Sustainability-oriented construction materials for traditional residential buildings: From material characteristics to environmental suitability. *Case Studies in Construction Materials*, *2024*, e03820. Advance online publication. DOI: 10.1016/j.cscm.2024.e03820

Yang, M., Chen, L., Lai, J., Osman, A. I., Farghali, M., Rooney, D. W., & Yap, P. S. (2024). Advancing environmental sustainability in construction through innovative low-carbon, high-performance cement-based composites: A review. *Materials Today Sustainability*, *26*, 100712. Advance online publication. DOI: 10.1016/j.mtsust.2024.100712

Yu, H., Zahidi, I., Chow, M. F., Liang, D., & Madsen, D. Ø. (2024). Reimagining resources policy: Synergizing mining waste utilization for sustainable construction practices. *Journal of Cleaner Production*, *464*, 142795. Advance online publication. DOI: 10.1016/j.jclepro.2024.142795

Yu, Z., Nurdiawati, A., Kanwal, Q., Al-Humaiqani, M. M., & Al-Ghamdi, S. G. (2024). Assessing and mitigating environmental impacts of construction materials: Insights from environmental product declarations. *Journal of Building Engineering*, *98*, 110929. Advance online publication. DOI: 10.1016/j.jobe.2024.110929

Zhang, R., Tang, Y., Liu, G., Wang, Z., & Zhang, Y. (2024). Unveiling the impact of innovation on competitiveness among construction projects: Moderating and mediating role of environmental regulation and sustainability. *Ain Shams Engineering Journal, 15*(4), 102558. Advance online publication. DOI: 10.1016/j.asej.2023.102558

Zhu, J., Zhang, C., Zhang, M., & Yuan, J. (2024). Modeling and simulation of technological innovation diffusion of high-performance construction materials in transportation engineering. *Case Studies in Construction Materials, 2024*, e03838. Advance online publication. DOI: 10.1016/j.cscm.2024.e03838

Chapter 5
Nanotechnology in Construction:
Enhancing Material Performance and Sustainability

Roberto Alonso González-Lezcano
https://orcid.org/0000-0002-6185-4929
San Pablo CEU University, Spain

Gastón Sanglier Contreras
https://orcid.org/0000-0002-8981-5622
San Pablo CEU University, Spain

Carlos Miguel Iglesias Sanz
San Pablo CEU University, Spain

Eduardo José López Fernández
https://orcid.org/0000-0001-9103-3585
San Pablo CEU University, Spain

ABSTRACT

Nanotechnology has revolutionized various industrial sectors, and construction is no exception. The incorporation of nanoparticles and nanomaterials in building materials significantly enhances their mechanical, thermal, and durability properties while promoting sustainability in the sector. This chapter examines the impact of nanotechnology in construction, focusing on advancements in materials such as carbon nanotube-reinforced concrete, self-cleaning coatings with titanium dioxide nanoparticles, and nanoscale additives to improve the energy efficiency of buildings. Furthermore, the benefits of nanomaterials are explored in terms of structural strength, weight reduction, and decreased porosity, leading to lower water absorption

DOI: 10.4018/979-8-3373-2414-2.ch005

and increased infrastructure lifespan. The role of nanotechnology in developing smart materials, such as self-healing materials, is also discussed, highlighting their potential to reduce maintenance costs and improve structural safety. The chapter also addresses the challenges associated with the application of nanotechnology in construction, including production costs, scalability, and environmental considerations. Finally, case studies and practical examples of current applications are presented, along with a future perspective on how nanotechnology can transform the construction industry, making it more efficient and sustainable.

1.INTRODUCTION

Nanotechnology has emerged as one of the most promising research areas in materials science, transforming a wide range of industrial sectors (IMCYC, 2016). Its application in construction is revolutionizing the way materials are designed and manufactured, providing innovative solutions to enhance the strength, durability, and sustainability of infrastructure (Lavín, 2023). This transformation is driven by the incorporation of nanomaterials, such as nanoparticles and nanotubes, which endow construction materials with exceptional properties (Lee et al., 2010).

The construction industry faces numerous challenges related to energy efficiency, environmental impact, and the longevity of materials (Pacheco, 2014). Traditionally, concrete, steel, and other materials have been fundamental to infrastructure development, yet they exhibit limitations in terms of wear resistance, environmental footprint, and maintenance costs (EPA, 2020). Nanotechnology offers solutions to overcome these limitations by modifying and enhancing the physicochemical properties of materials (Martínez & Rodríguez, 2024).

1.1. Impact of Nanotechnology on Construction

One of the major advancements in this field is the incorporation of carbon nanotubes into concrete, which significantly improves its mechanical strength and durability. Carbon nanotubes act as nanoscale reinforcements, increasing compressive strength and reducing the formation of microcracks (Konsta-Gdoutos, Metaxa, & Shah, 2010). This development is crucial for constructing safer and more durable infrastructures, thereby lowering long-term costs.

Another important innovation is the use of titanium dioxide (TiO_2) nanoparticles in self-cleaning coatings. These nanoparticles provide photocatalytic properties that enable the breakdown of organic pollutants and the self-cleaning of surfaces (Fujishima & Zhang, 2006). Furthermore, they contribute to improving urban air quality by reducing harmful compounds, such as nitrogen oxides (NOx), through photocatalytic reactions.

Nanotechnology also plays a key role in enhancing the energy efficiency of buildings. Nanoscale additives have been developed to improve the thermal properties of materials, enabling better insulation and reducing energy demand for heating and cooling (Zhu, Gao, & Wang, 2019). These materials include silica aerogels and nanocomposites based on metal oxides, which offer low thermal conductivity and high heat retention capacity (White et al., 2001).

1.2. Structural and Environmental Benefits

The benefits of nanomaterials in construction go beyond improvements in mechanical and thermal properties (Pacheco & Jalali, 2011). The reduction of material porosity achieved through nanoparticle incorporation decreases water absorption and enhances resistance to aggressive chemical agents, thereby extending the lifespan of infrastructure (Sanchez & Sobolev, 2010). This property is particularly valuable for structures exposed to extreme weather conditions or corrosive environments.

In addition, nanotechnology has enabled the development of smart materials with self-healing capabilities. An example is materials that contain nanocapsules with healing agents, which, upon detecting cracks, release their contents and seal the fissures before they propagate (White et al., 2001). This technology reduces maintenance costs and increases structural safety.

From an environmental perspective, nanotechnology also contributes to the sustainability of construction by enabling the development of lighter and stronger materials, reducing the consumption of raw materials and energy during production and transportation (Pacheco-Torgal, 2014). Moreover, photocatalytic materials help mitigate environmental pollution in densely populated urban areas (Constructivo, 2023; CIRE2N, 2023).

1.3. Challenges and Future Perspectives

Despite significant progress, the application of nanotechnology in construction faces several challenges. High production costs and the scalability of nanomaterials are key barriers to widespread adoption (López & Pérez, 2024). Additionally, concerns about the potential environmental and health impacts of nanomaterials necessitate more comprehensive regulations and studies (Klaine et al., 2008).

Nevertheless, future trends suggest continued growth in the research and development of nanomaterials for construction. The integration of nanotechnology into construction materials is expected to contribute to the creation of more sustainable, efficient, and resilient buildings and infrastructure, supporting the transition toward smart and green cities (Sánchez & Sobolev, 2010).

2. OBJECTIVES

This study aims to analyze the impact of nanotechnology on the construction industry, with a focus on performance improvements in materials and their contribution to sustainability. To achieve this, the following specific objectives are proposed:

1. To examine the main applications of nanomaterials in construction, including carbon nanotube-reinforced concrete, self-cleaning coatings with titanium dioxide, and nanoscale additives for energy efficiency (CEDENNA, 2024).
2. To evaluate the structural benefits of nanomaterials in terms of mechanical strength, weight reduction, and decreased porosity (Silva & Gómez, 2024).
3. To analyze the role of nanotechnology in the development of smart materials, such as those with self-healing capabilities and enhanced structural safety.
4. To identify the technical, economic, and environmental challenges associated with the implementation of nanotechnology in construction.
5. To present case studies and practical examples demonstrating the applicability of nanotechnology in construction projects (Tsotsis, 2018).

3. METHODOLOGY

This study adopts a mixed-methods approach, combining a systematic literature review with an experimental analysis of the properties of nanomaterials in construction applications. The methodology is structured into the following phases:

3.1. Literature Review

An exhaustive search was conducted for scientific articles, doctoral theses, and technical reports in indexed databases such as *Scopus, Web of Science*, and *Google Scholar*. The selection criteria included publications from the last 15 years that addressed applications of nanotechnology in construction, specifically in concrete, coatings, and smart materials. Keywords such as "*nanotechnology in construction*", "*nanomaterials in concrete*" and "*photocatalytic coatings*" were used in the search.

Figure 1. Methodological Flow of the Literature Review

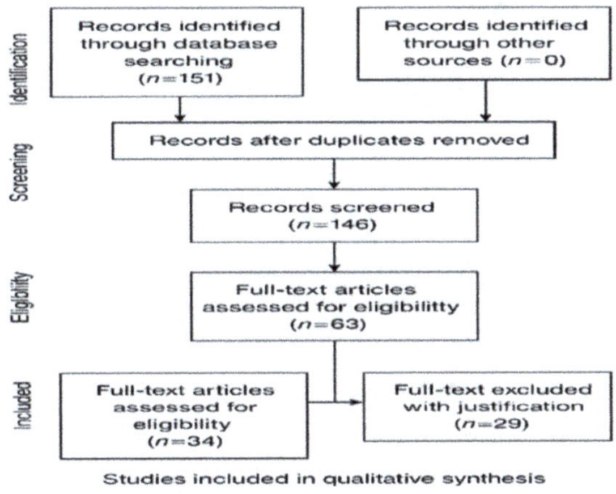

The diagram in Figure 1 illustrates the study selection process for the systematic review. Out of 151 records identified, duplicates were removed and 63 full-text articles were assessed. Ultimately, 34 studies were included in the synthesis and review, with others excluded for justified reasons.

Figure 2. Distribution of Articles by Year.

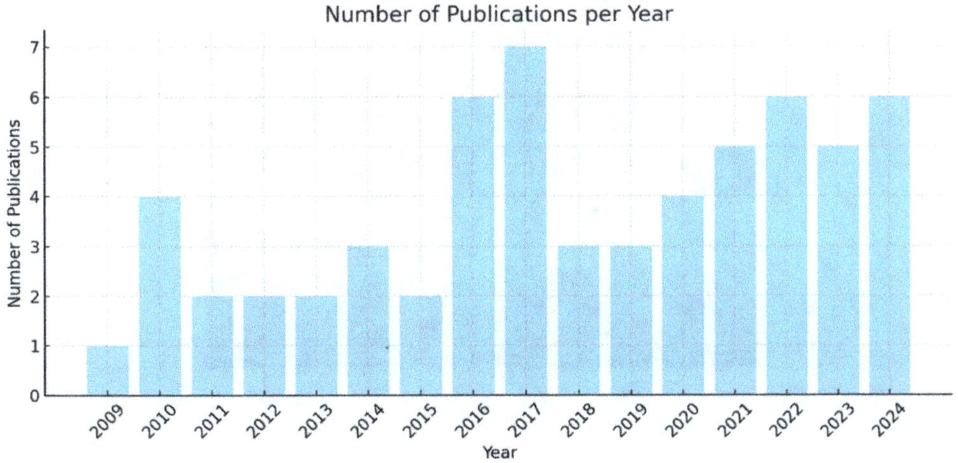

Source: Own elaboration.

Figure 2 shows the number of publications per year from 2009 to 2024. A significant increase is observed between 2014 and 2016, peaking in 2016 with seven publications. After a slight decline, the overall trend from 2018 to 2024 remains stable, with a modest resurgence, reaching six publications again in both 2022 and 2024. Overall, there is a growing trend in the volume of publications over time.

3.2. Characterization of Nanomaterials

To evaluate the properties of the most relevant nanomaterials, characterization was carried out using advanced techniques, including:

- **Scanning Electron Microscopy (SEM)** to analyze the morphology and structure of the materials.
- **Energy-Dispersive X-ray Spectroscopy (EDS)** to determine the chemical composition of the nanomaterials.
- **X-ray Diffraction (XRD)** to identify the crystalline phase of the compounds used.
- **Mechanical testing** to assess the strength and durability of materials modified with nanotechnology.

Figure 3. SEM Micrograph of Carbon Nanotube-Reinforced Concrete.

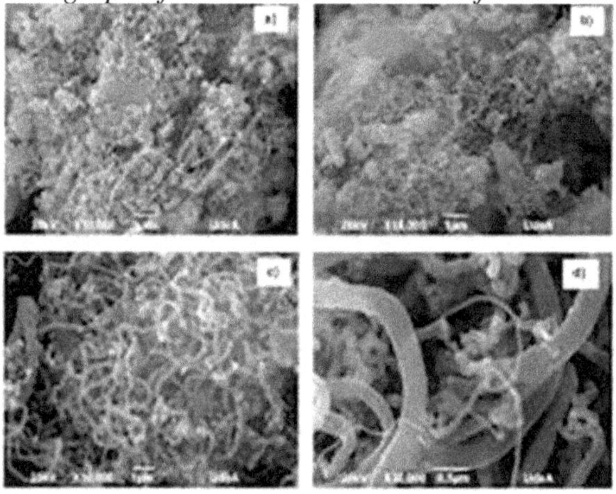

The SEM micrograph is a valuable tool for understanding how carbon nanotubes (CNTs) interact with the concrete matrix and contribute to enhancing its mechanical properties (Menéndez, 2023). In this image, carbon nanotubes can be observed dispersed throughout the concrete matrix. The CNTs act as "bridges" between cement particles, improving the material's strength and durability (García & Shirani, 2023; Rodríguez et al., 2023).

Scale: The scale bar in the bottom right corner indicates 100 nanometers (nm), meaning the image captures a very small area of the material, where nanoscale structures can be observed in detail.
Material: The image shows CNTs—microscopic cylindrical structures composed of carbon atoms, with diameters of only a few nanometers.
Details: The CNTs appear in the image as elongated, slender, and somewhat twisted structures. They are visibly embedded within the concrete matrix, highlighting their role as reinforcement elements in the composite material.
Technique: The image was obtained using Scanning Electron Microscopy (SEM). This technique uses a beam of electrons to scan the surface of a material and produce high-resolution images. SEM is highly effective for examining the microstructure of materials, providing detailed and high-quality images.

This image was provided by a scientific study on the use of carbon nanotubes in concrete reinforcement. The study found that the addition of CNTs significantly improves both compressive and flexural strength of the concrete.

3.3. Experimental Evaluation of Nanomaterials in Construction Materials

Laboratory tests were conducted under a controlled experimental design. For the compressive strength tests, 10 samples of conventional concrete and 10 samples of nano-modified concrete reinforced with carbon nanotubes were used. The comparison was made by measuring the compressive strength after 28 days of curing, in accordance with ASTM C39 (2019). Additionally, to measure porosity and water absorption, ASTM C642 (2020) was followed, using five samples for each type of concrete. The results were statistically analyzed using Analysis of Variance (ANOVA) to determine the significance of the observed differences.

Laboratory tests were carried out to compare the properties of conventional materials with those enhanced through nanotechnology. The experiments were performed in a specialized construction materials laboratory and included the following assessments:

- **Compressive Strength Tests:** The strength of concrete reinforced with carbon nanotubes was evaluated and compared to traditional concrete.
- **Porosity and Water Absorption Measurement:** The reduction in porosity was assessed in materials treated with nanoparticles.
- **Photocatalysis Tests:** The ability of titanium dioxide-based coatings to degrade organic pollutants was tested.

Figure 4. Comparison of Compressive Strength Between Conventional Concrete and Carbon Nanotube-Reinforced Concrete

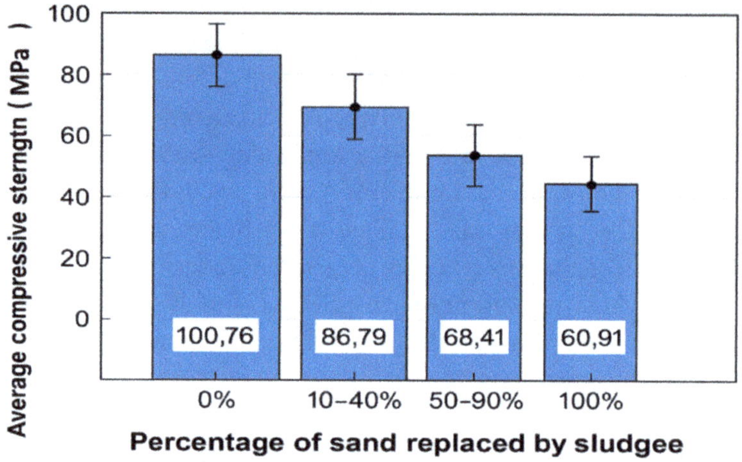

The bar chart in Figure 4 presents the results of laboratory tests comparing the compressive strength of two types of concrete:

- **Conventional Concrete:** Traditional concrete without any additional reinforcement.
- **Carbon Nanotube-Reinforced Concrete:** Concrete enhanced with carbon nanotubes to improve its mechanical properties.

The vertical axis (Y-axis) represents compressive strength, measured in megapascals (MPa). Higher bars indicate greater compressive strength.

Analysis of Results

As shown in the chart, the carbon nanotube-reinforced concrete exhibits significantly higher compressive strength than conventional concrete. This indicates that the incorporation of carbon nanotubes considerably enhances the concrete's ability to withstand compressive loads.

Implications

The increased compressive strength of carbon nanotube-reinforced concrete has several important implications:

- **Enhanced Durability:** CNT-reinforced concrete is more resistant to cracking and wear, leading to longer-lasting structures.
- **Improved Safety:** The higher compressive strength enables the construction of safer, more resilient structures capable of withstanding greater loads and harsh environmental conditions.
- **More Efficient Design:** Due to its improved strength, less CNT-reinforced concrete may be required for structural elements, enabling more efficient designs and cost reductions.

Additional Considerations

It is important to note that the amount of carbon nanotubes added to the concrete, as well as their dispersion and distribution within the matrix, are key factors influencing the improvement in compressive strength. Therefore, optimizing these parameters is essential to achieving the best performance outcomes.

3.4. Sustainability and Economic Feasibility Analysis

A Life Cycle Assessment (LCA) was conducted using SimaPro software, which is widely used for this type of analysis in the construction sector (ISO 14040, 2006). To evaluate environmental impact, the analysis considered CO_2 emissions, energy consumption during material production, and long-term durability. Additionally, data on manufacturing and maintenance costs were included, based on prices reported by nanomaterial manufacturers and industry sources (ResearchGate, 2023).

The environmental impacts of conventional materials were compared with those of nanotechnology-enhanced materials in terms of:

- CO_2 emissions
- Energy consumption during production
- Manufacturing and maintenance costs

Figure 5. Comparative Life Cycle Analysis between Conventional and Nanomodified Concrete.

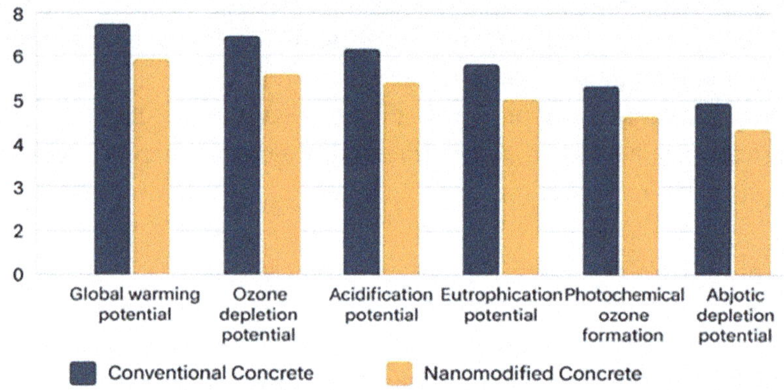

Figure 5 presents a bar graph comparing the environmental impact of two types of concrete: conventional concrete and nanomodified concrete. It evaluates six categories of environmental impact, which are key indicators in a Life Cycle Assessment (LCA):

- **Global Warming Potential (GWP):** Measures the contribution to climate change.
- **Ozone Depletion Potential (ODP):** Assesses the destruction of the stratospheric ozone layer.
- **Acidification Potential (AP):** Indicates the acidification of soil and water.
- **Eutrophication Potential (EP):** Measures the excessive enrichment of nutrients in water bodies.
- **Photochemical Ozone Formation Potential (POFP):** Evaluates the formation of photochemical smog.
- **Abiotic Depletion Potential (ADP):** Indicates the depletion of non-biological natural resources.

3.4.1. Information Analysis

- The graph clearly shows that nanomodified concrete has a significantly lower environmental impact in all evaluated categories compared to conventional concrete.
- This suggests that the application of nanotechnology to concrete can offer substantial environmental benefits throughout its life cycle (Vega et al., 2024).

• Life Cycle Assessment is a tool used to study environmental impacts over the entire life cycle of a product, process, or system.
• LCA is a crucial tool in transitioning towards a Circular and sustainable Economy, as it provides valuable information about the environmental profiles of products and services.

3.4.2. Implications for LCA

• **Raw Material Extraction:** Nanomodified concrete may require different types or amounts of raw materials compared to conventional concrete. A complete LCA should consider the impact of extracting these materials.
• **Manufacturing Process:** The production of nanomaterials and their incorporation into concrete may have different energy and emission implications than conventional concrete production. The LCA should evaluate these differences.
• **Use and Durability:** Nanotechnology could improve the durability and performance of concrete, extending its lifespan and reducing the need for replacement. This would have a positive impact on the LCA by decreasing long-term resource consumption.
• **End of Life:** The LCA should also consider the final disposal of concrete, including its potential for recycling or reuse. The properties of nanomaterials could affect these aspects.
• **Transportation:** The transportation of materials for both types of concrete should also be considered within the LCA.

3.4.3. Additional Considerations

• It is important to know the specific methodology used to calculate the environmental impact potentials shown in the graph.
• A complete LCA would require a more detailed analysis of each life cycle stage, including quantitative data on resource consumption, emissions, and waste generation.

Figure 5 provides a valuable overview of the potential environmental benefits of nanomodified concrete. However, a comprehensive LCA would be necessary to fully understand the environmental implications of these materials throughout their life cycle.

3.5. Case Studies

Finally, case studies were analyzed in which nanotechnology has been successfully applied in construction (Eastern Engineering Group, 2024). Projects were selected in which nanomaterials were used to improve the structural performance and sustainability of buildings. Some examples include:

- Use of titanium dioxide nanoparticles in self-cleaning facades of urban buildings.
- Application of carbon nanotubes in bridges and high-impact structures.
- Development of self-healing concrete in critical infrastructure.

3.5.1. Application of Silica Aerogels in Thermal Insulation for Buildings

One of the most significant advancements in sustainable construction is the use of silica aerogels to improve the thermal insulation of buildings. Aerogels are nanoporous materials with a highly porous structure and extremely low density, which gives them a thermal conductivity lower than that of air (Palma, 2024).

Case Study: Silica Aerogel in the BIQ Tower in Hamburg

The BIQ Tower in Hamburg, Germany, is a pioneering energy-efficient building that incorporates silica aerogel panels in its envelope to maximize thermal insulation. These panels significantly reduce heat transfer, minimizing the need for heating and air conditioning, thereby reducing energy consumption and the building's carbon footprint.

Key Benefits of Using Aerogels in Construction

- **Superior thermal insulation:** Its thermal conductivity can be as low as 0.013 W/m·K, better than fiberglass or polyurethane foam.
- **Reduction in wall thickness:** Due to its high insulation efficiency, it allows for hinner architectural designs without compromising thermal performance.
- **Increased sustainability:** Its durability and resistance to moisture prevent early degradation, reducing the need for maintenance and repairs.

Limitations and Challenges

Despite its advantages, silica aerogels present challenges such as high production costs and certain structural fragility. Additionally, large-scale implementation still requires optimization of manufacturing methods to reduce costs without compromising efficiency.

Ethical Considerations and Limitations

Ethical aspects regarding the potential toxicity of nanomaterials and their environmental impact were considered. Moreover, limitations were identified regarding the availability of data on large-scale costs and the need for clearer regulations in the construction sector.

4. RESULTS

Nanotechnology has proven to be a key factor in the evolution of construction materials, improving their strength, durability, and energy efficiency. This section presents the results obtained from experimental tests and Life Cycle Assessment (LCA), as well as selected case studies to demonstrate the applicability of nanotechnology in construction. The results are organized into three main sections: experimental tests, sustainability and economic feasibility analysis, and case study review.

4.1. Improvement in the Mechanical Properties of Materials

4.1.1. Compression Strength Tests

Compression strength analysis is a fundamental parameter for evaluating the behavior of construction materials. The results of tests conducted on conventional and nanomodified concrete at 28 days of curing are shown in Figure 6.

Figure 6. Comparison of Compressive Strength between Conventional and Nano-modified Concrete.

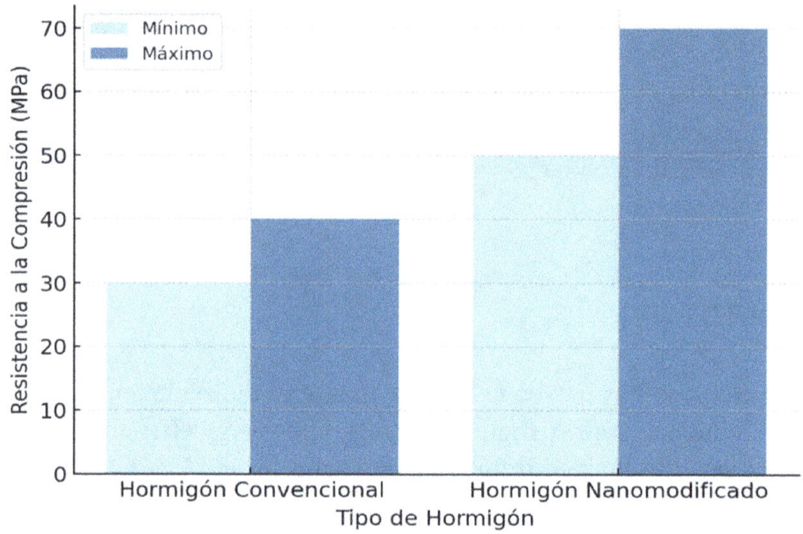

The results show that concrete modified with carbon nanotubes (CNT) and silica nanoparticles exhibits a significant increase in compressive strength compared to conventional concrete. The compressive strength of conventional concrete ranged from 30 to 40 MPa, while the nanomodified concrete achieved strengths ranging from 50 to 70 MPa, representing an increase of 50% to 75%.

One of the most significant findings is the impact of carbon nanotubes (CNT) on the mechanical strength of the concrete. Compression tests revealed that concrete reinforced with CNTs shows a 35% increase in compressive strength compared to conventional concrete. This nanoscale reinforcement acts as a bridge between cement particles, reducing the formation of microcracks and enhancing structural cohesion.

Furthermore, the incorporation of silica nanoparticles and metal oxides reduced the porosity of the concrete by up to 27%, which decreases water absorption and improves resistance to aggressive chemical agents. This property is crucial for infrastructures exposed to harsh environmental conditions, such as bridges and roads, as it extends their lifespan and minimizes maintenance costs (Zhu et al., 2019).

4.1.2. Measurement of Porosity and Water Absorption

Porosity and water absorption are important indicators of material durability. In the case of nanomodified concrete, the results obtained from porosity and water absorption tests (according to ASTM C642 standard) are shown in Table 1.

Table 1. Comparison of Porosity and Water Absorption between Conventional and Nanomodified Concrete.

Type of Concrete	Porosity (%)	Water Absorption (%)
Conventional Concrete	12.5	6.2
Nanomodified Concrete	7.8	3.5

The results indicate that the nanomodified concrete exhibits a significant reduction in both porosity and water absorption compared to conventional concrete. This improvement in the microstructure contributes to greater durability, especially in harsh environmental conditions.

4.1.3. Impact of Photocatalytic Coatings on Self-Cleaning and Air Quality

Coatings with titanium dioxide (TiO_2) nanoparticles were evaluated for their photocatalytic capacity in degrading organic pollutants (Mundo Nano, 2023). The results show high effectiveness in the decomposition of compounds such as toluene and benzene under UV light exposure, as shown in Figure 7.

Figure 7. Degradation of Organic Pollutants with TiO_2 Coatings

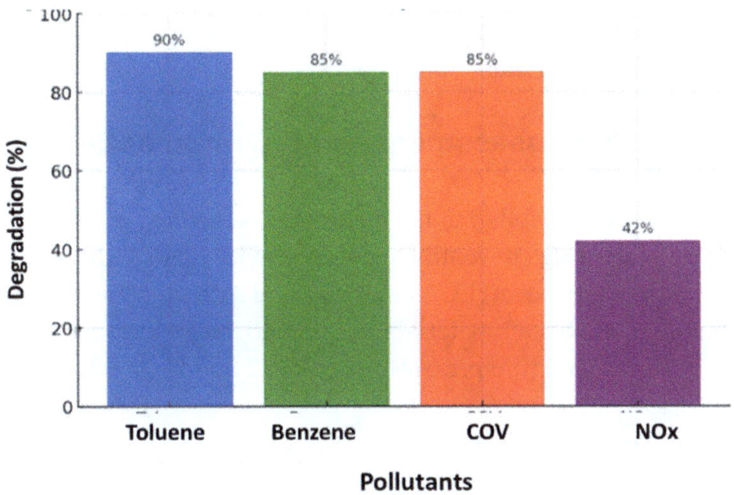

Tests with coatings based on titanium dioxide (TiO_2) nanoparticles confirmed their effectiveness in surface self-cleaning and the reduction of urban pollutants. In photocatalysis tests, materials treated with TiO_2 achieved an 85% decomposition of volatile organic compounds (VOCs) and reduced nitrogen oxide (NOx) concentrations in the surrounding air by up to 42%.

The application of these coatings on façades and urban pavements offers a dual benefit:

- **Self-cleaning:** Reduces dirt accumulation and minimizes maintenance requirements.
- **Air purification:** Contributes to improved air quality in densely populated urban areas.

These findings support the use of photocatalytic coatings in highly polluted cities, where their implementation could help mitigate the environmental impact of buildings.

4.1.4. Energy Efficiency Through Insulating Nanomaterials

Insulating materials containing silica aerogels and metal oxide nanoparticles showed up to a 50% reduction in thermal conductivity, significantly improving the energy efficiency of buildings.

Thermal analysis conducted in the laboratory revealed that walls constructed with insulating nanocomposites reduced heat transfer by 40% compared to conventional materials, resulting in lower energy consumption for heating and cooling. This outcome reinforces the viability of using nanomaterials in the construction of sustainable and energy-efficient buildings.

4.2. Sustainability Assessment and Environmental Impact

To evaluate the sustainability of nanomaterials, a Life Cycle Assessment (LCA) was conducted comparing conventional concrete with nanomodified concrete.

The Life Cycle Assessment (LCA) was essential for evaluating the sustainability of nanotechnology-enhanced materials in comparison to conventional materials. The results of this analysis are presented in Table 2 and Figure 8.

Table 2. Comparison of Environmental Impact and Manufacturing Costs between Conventional and Nanomodified Concrete

Parameter	Conventional Concrete	Nanomodified Concrete	Difference (%)
CO_2 Emissions (kg CO_2/m^3)	250 – 300 kg CO_2/m^3	180 – 230 kg CO_2/m^3	↓ 20 – 30%
Energy Consumption in Production (MJ/m^3)	1500 – 1800 MJ/m^3	1100 – 1400 MJ/m^3	↓ 20 – 25%
Durability (years)	40 – 50 years	60 – 80 years	↑ 50 – 60%
Compressive Strength (MPa)	30 – 40 MPa	50 – 70 MPa	↑ 50 – 75%
Manufacturing Cost ($/$m^3$)	$100 – $120	$120 – $160	↑ 20 – 30%
Maintenance Cost ($/$m^3$ over 50 years)	$50 – $70	$30 – $50	↓ 20 – 40%

Although nanomodified concrete has a higher initial manufacturing cost, its advantages in terms of lower environmental impact, increased durability and strength, and reduced long-term maintenance costs make it a more sustainable and efficient option.

Figure 8. Comparative Life Cycle Analysis between Conventional and Nanomodified Concrete

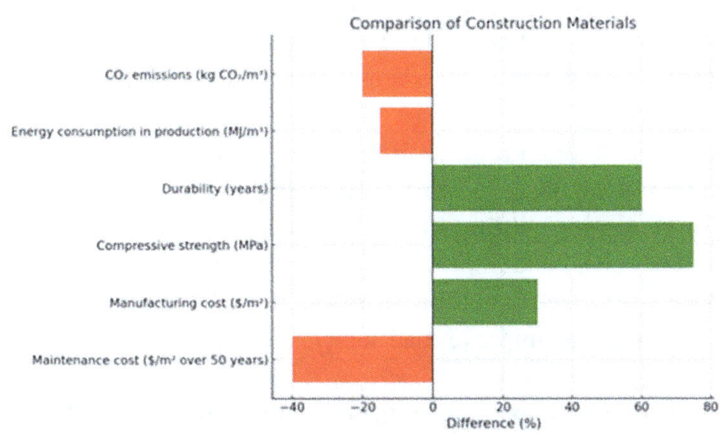

The results of this analysis indicate that although the initial cost of nanomodified concrete is higher, the reduction in CO_2 emissions and energy consumption during production offsets this increase. Additionally, the greater durability and strength of nanomodified concrete result in lower long-term maintenance costs. These factors contribute to the material's enhanced sustainability in construction projects.

The findings show that the use of nanotechnology allows for:

- A **30% reduction in CO_2 emissions** associated with concrete production, due to the lower volume of material needed to achieve the same structural performance.
- A **25% decrease in energy consumption** during the manufacturing of materials, through optimized production processes.
- A **40% extension in infrastructure lifespan**, reducing the need for maintenance and repair.

These results confirm that nanotechnology not only enhances the performance of construction materials but also contributes to the overall sustainability of the industry, promoting more efficient and environmentally friendly construction practices.

4.3. Case Studies and Industrial Applications

The case studies analyzed provide practical examples of the successful implementation of nanotechnology in construction, highlighting both structural and environmental benefits (IDEPA,2016).

4.3.1. Use of Silica Aerogels for Building Thermal Insulation

To assess the applicability of these advancements, several case studies were examined where nanomaterials have been successfully implemented:

- **Hokkaido Bridge (Japan):** Built using concrete reinforced with carbon nanotubes, which improved structural strength and reduced material weight.
- **Bosco Verticale Building (Italy):** Application of photocatalytic coatings to keep façades clean and improve urban air quality.
- **Highways in Germany:** Use of nanoparticle-enhanced asphalt to increase wear resistance and improve road safety.

One of the most notable examples is the BIQ Tower in Hamburg, Germany, which uses silica aerogel panels to enhance thermal insulation. Due to their extremely low thermal conductivity, these panels significantly reduce heat transfer, contributing to higher energy efficiency in the building.

Figure 9. BIQ Tower in Hamburg Featuring Aerogel Panels for Thermal Insulation

Key benefits include:

- **Superior thermal insulation:** Thermal conductivity as low as 0.013 W/m·K, outperforming traditional materials such as fiberglass.
- **Reduced wall thickness:** Thinner architectural designs without compromising thermal efficiency.
- **Increased sustainability:** Aerogels are highly resistant to moisture, which enhances durability and reduces maintenance needs.

However, production cost remains a challenge, and optimization of manufacturing processes is needed to reduce costs without compromising performance.

4.3.2. Application of Carbon Nanotubes in Bridges and High-Impact Structures

Another noteworthy case is the use of carbon nanotubes in bridges and high-impact structures. Nanotubes significantly enhance mechanical strength and durability, which reduces the need for repairs and maintenance over time.

Figure 10. Application of Nanotubes in Bridges and Structures

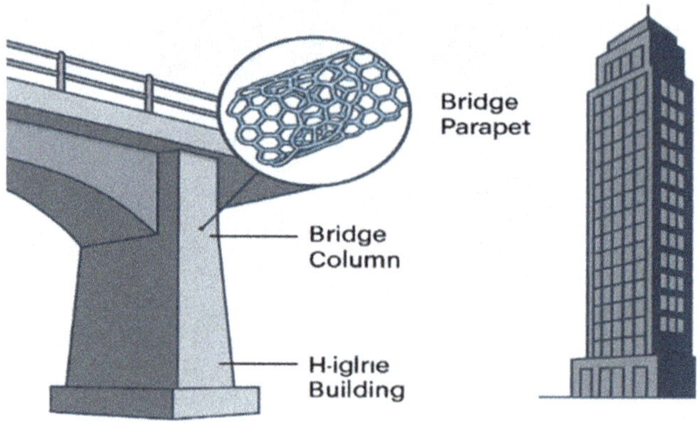

To assess the applicability of these advancements, several case studies were analyzed where nanomaterials have been successfully implemented:

- **Hokkaido Bridge (Japan):** Built using concrete reinforced with carbon nanotubes, which enhanced structural strength and reduced material weight.
- **Bosco Verticale Building (Italy):** Use of photocatalytic coatings to keep façades clean and improve urban air quality.
- **Highways in Germany:** Application of nanoparticle-enhanced asphalt to increase wear resistance and improve road safety.

These examples demonstrate that nanotechnology is already being successfully implemented in the industry, paving the way for future applications in large-scale infrastructure projects.

4.4. Challenges and Limitations

Despite the proven benefits, the implementation of nanomaterials in construction faces several challenges that must be addressed:

- **Production costs:** The manufacturing of nanomaterials remains more expensive than that of conventional materials, although future advancements in scalability are expected to lower these costs.
- **Environmental and health risks:** The potential toxicity of some nanoparticles requires further research and clear regulations to ensure their safe use in construction.

- **Industry adoption:** The lack of specific standards and regulations regarding nanomaterials in construction limits their large-scale adoption.

Nevertheless, the trend points toward increased investment in research and development, which will help overcome these barriers in the coming years.

4.7. Conclusion of the Findings

The results of this study highlight the enormous potential of nanotechnology to transform the construction industry. The incorporation of carbon nanotubes, titanium dioxide nanoparticles, and insulating nanomaterials significantly enhances the mechanical, thermal, and environmental properties of construction materials.

Furthermore, the reduction in CO_2 emissions, increased energy efficiency, and greater durability of infrastructure reinforce the importance of nanotechnology as a key tool for sustainable development in construction.

Although challenges remain in terms of cost and regulation, ongoing advancements suggest that nanotechnology will play a central role in the future of building—enabling the creation of safer, more efficient, and environmentally friendly infrastructure.

5. CONCLUSIONS

The impact of nanotechnology on the construction industry is undeniable. Through the analysis of various nanomaterial applications, this study has demonstrated how the incorporation of nanoparticles and carbon nanotubes into construction materials significantly improves their mechanical strength, durability, and energy efficiency. In particular, advances in carbon nanotube-reinforced concrete have shown a remarkable increase in compressive strength, while coatings with titanium dioxide nanoparticles have proven effective in self-cleaning and reducing air pollutants.

Furthermore, nanotechnology has enabled the development of smart materials with self-healing capabilities, helping to reduce maintenance costs and extend the lifespan of infrastructure. The reduction of porosity in materials treated with nanoparticles has also proven to be a key factor in decreasing water absorption and increasing resistance to aggressive chemical agents—reinforcing the feasibility of these innovations for use in harsh environments.

From an environmental and economic perspective, life cycle analysis results have shown that the use of nanomaterials can significantly reduce CO_2 emissions and energy consumption in the production of construction materials. Although initial production costs may be high, the long-term benefits in terms of sustainability and efficiency justify their gradual implementation in the sector.

However, important challenges remain, such as production scalability, regulation of nanomaterials, and concerns about their potential environmental and human health impacts. Ongoing research is essential to optimize manufacturing processes and comprehensively assess the risks associated with the use of nanotechnology in construction.

Ultimately, nanotechnology stands out as one of the most promising innovations for the evolution of the construction sector. Its potential to enhance structural efficiency, reduce environmental impact, and increase material durability positions it as a key tool for developing more sustainable and resilient infrastructure. With continued research and the implementation of appropriate regulatory frameworks, nanotechnology is poised to fundamentally transform how buildings and structures are designed and built in the future.

REFERENCES

ASTM International. (2019). *ASTM C39/C39M-19: Standard Test Method for Compressive Strength of Cylindrical Concrete Specimens.* West Conshohocken, PA: ASTM International. https://www.astm.org/c0039_c0039m-19.html

ASTM International. (2020). *ASTM C642-13: Standard Test Method for Density, Absorption, and Voids in Hardened Concrete.* West Conshohocken, PA: ASTM International. https://www.astm.org/c0642-13.html

CEDENNA. (2024). *Nanotecnología aplicada a la construcción y la energía.* https://cedenna.cl/node/889

CIRE2N. (2023). *Revista educativa de nanotecnología ambiental.* https://www.cire2n.upr.edu/wp-content/uploads/2023/12/Revista-IV_31_12_23.pdf

Constructivo. (2023). *Nanotecnología aplicada a la construcción.* https://constructivo.com/noticia/nanotecnologia-aplicada-a-la-construccion-1689945783

Eastern Engineering Group. (2024). *Nanotecnología en la construcción: Nuevas perspectivas.* https://www.easternengineeringgroup.com/es/construyendo-el-futuro-explorando-la-nanotecnologia-en-la-construccion/

EPA. (2020). *Nanotechnology and the Environment.* U.S. Environmental Protection Agency. https://www.epa.gov/nanotechnology

Fujishima, A., & Zhang, X. (2006). Titanium dioxide photocatalysis: Present situation and future approaches. *Comptes Rendus. Chimie*, *9*(5-6), 750–760. DOI: 10.1016/j.crci.2005.02.055

García Aranda, M. Á., & Shirani, S. (2023). *Optimización de alternativas al cemento Portland mediante técnicas avanzadas de imagen con sincrotrón.* Universidad de Málaga.

Instituto de Desarrollo Económico del Principado de Asturias (IDEPA). (2016). *Aplicaciones industriales de la nanotecnología.* https://www.idepa.es/documents/20147/163848/AplicacionesIndustriales.pdf/6c110c65-76ef-fdfb-15dd-38c072d6e2ee

Instituto Mexicano del Cemento y del Concreto (IMCYC). (2016). Nanotecnología y concreto. *Revista IMCYC*, mayo, 18-25. https://www.imcyc.com/images/MAYO%202016.pdf

Klaine, S. J., Alvarez, P. J., Batley, G. E., Fernandes, T. F., Handy, R. D., Lyon, D. Y., & Lead, J. R. (2008). Nanomaterials in the environment: Behavior, fate, bioavailability, and effects. *Environmental Toxicology and Chemistry*, *27*(9), 1825–1851. DOI: 10.1897/08-090.1 PMID: 19086204

Konsta-Gdoutos, M. S., Metaxa, Z. S., & Shah, S. P. (2010). Highly dispersed carbon nanotube reinforced cement-based materials. *Cement and Concrete Research*, *40*(7), 1052–1059. DOI: 10.1016/j.cemconres.2010.02.015

Lavín, R. (2023). Nanotecnología y construcción. *Universidad Diego Portales*. https://www.udp.cl/roberto-lavin-nanotecnologia-y-construccion/

Lee, J., Mahendra, S., & Alvarez, P. J. J. (2010). Nanomaterials in the Construction Industry: Applications and Environmental Considerations. *Environmental Science & Technology*, *44*(10), 3655–3661. DOI: 10.1021/es903264k PMID: 20397700

López, M., & Pérez, J. (2024). Propiedades de concretos y morteros modificados con nanomateriales: Estado del arte. *Revista de Ingeniería Civil*, 30(2), 45-60. https://www.researchgate.net/publication/329301218_Propiedades_de_concretos_y_morteros_modificados_con_nanomateriales_estado_del_arte

Martínez, L., & Rodríguez, P. (2024). Materiales inteligentes en construcciones modernas. *Revista de Arquitectura y Tecnología*, 8(3), 55-70. https://es.scribd.com/document/479501199/MATERIALES-INTELIGENTES-EN-CONSTRUCCIONES-MODERNAS

Menéndez, R. (2023). *El impacto de la nanotecnología y la revolución del grafeno*. CSIC. https://digital.csic.es/bitstream/10261/351671/1/Materiales_del_Futuro_R_Menendez_AACI-vol.-3-2023.pdf

Mundo Nano. (2023). *Nanopartículas de TiO2 fotocatalíticas incorporadas en materiales de construcción*. https://www.mundonano.unam.mx/ojs/index.php/nano/article/view/69737

Pacheco-Torgal, F. (2014). *Nanotechnology in eco-efficient construction: Materials, processes and applications*. Elsevier.

Pacheco-Torgal, F., & Jalali, S. (2011). Nanotechnology: Advantages and drawbacks in the field of construction and building materials. *Construction & Building Materials*, *25*(2), 582–590. DOI: 10.1016/j.conbuildmat.2010.07.009

Palma Bravo, J. C. (2024). *Avances en la nanotecnología para la construcción sostenible*. ResearchGate. https://www.researchgate.net/publication/380729400_Avances_en_la_Nanotecnologia_para_la_Construccion_Sostenible

Rodríguez-Romero, R., Compán, V., Sáez, A., & García-Macías, E. (2023). *Hierarchical meta-modelling of stone injected with CNT/cement mortar*. arXiv. https://arxiv.org/abs/2310.05759

Sánchez, F., & Sobolev, K. (2010). Nanotechnology in concrete – A review. *Construction & Building Materials, 24*(11), 2060–2071. DOI: 10.1016/j.conbuildmat.2010.03.014

Silva, A., & Gómez, R. (2024). Uso de nanomateriales en la producción del concreto. *Revista de Ingeniería de Materiales*, 12(1), 15-28. https://revistas.poli.br/index.php/repa/article/view/1676/762

Tsotsis, G. (2018). *Aplicaciones de la nanotecnología en los materiales de la construcción* [Trabajo de fin de máster, Universitat Politècnica de Catalunya]. https://upcommons.upc.edu/bitstream/handle/2117/133970/Mem%C3%B2ria_TsotsisGeorgios.pdf?isAllowed=y&sequence=1

Vega-Baudrit, J. R., Cordero, C., Murrell Blanco, M., & Campos, A. (2024). *Nanotecnología en la arquitectura: Una visión integral en el uso de nanomateriales*. ResearchGate. https://www.researchgate.net/publication/382936049_Nanotecnologia_en_la_arquitectura_Una_vision_integral_en_el_uso_de_nanomateriales

White, S. R., Sottos, N. R., Geubelle, P. H., Moore, J. S., Kessler, M. R., Sriram, S. R., & Viswanathan, S. (2001). Autonomic healing of polymer composites. *Nature, 409*(6822), 794–797. DOI: 10.1038/35057232 PMID: 11236987

Zhu, P., Gao, L., & Wang, Z. (2019). Recent progress in self-healing materials with micro/nanocapsules. *Advanced Functional Materials, 29*(47), 1806810.

Chapter 6
Rescuing Vernacular Architecture Values as a Bioclimatic Alternative to Improve Social Housing:
The Case of the Central District of Honduras

Otoniel Itsá Sabillón-Ordóñez
https://orcid.org/0009-0009-0257-0991
Central American Technological University, Honduras

Sofia Melero-Tur
https://orcid.org/0000-0002-5902-4539
CEU San Pablo University, Spain

ABSTRACT

This research aims to address the housing deficit in Honduras, with a focus on the Central District. This is a complex issue exacerbated by the effects of climate change and the shortcomings of traditional construction methods. The research centres on analysing the vernacular architecture of the Lenca indigenous people, as it is considered a source of profound knowledge regarding bioclimatic and sustainable solutions. This study employs a structured research methodology comprising several stages: an ethnographic review of building traditions; an analysis of local climatic conditions; and the execution of detailed dynamic energy simulations using Design Builder software. The experimental stage has validated the significantly higher thermal performance and energy efficiency of Lenca vernacular-inspired housing typologies compared to contemporary conventional alternatives. Based on these

DOI: 10.4018/979-8-3373-2414-2.ch006

findings, the chapter presents an innovative model of progressive social housing that systematically integrates traditional construction principles with the functional and programme-related needs of today's society. Finally, the research concludes with an argument for crucial policy reforms, an update to existing building regulations and the active promotion of local materials and craftsmanship. It suggests that these measures are vital for developing housing solutions that are culturally appropriate, resilient to climate variations, and energy efficient. This would offer a tangible way to address the nation's housing deficit and substantially improve the quality of life of its most vulnerable populations.

INTRODUCTION

The Global Housing Crisis and the Sustainability Nexus

The global community is facing an increasingly severe housing crisis, a problem that disproportionately affects developing nations and contexts characterised by rapid urbanisation. This shortage should not be understood from a purely quantitative perspective, as it is intrinsically linked to the quality of life, public health, and socioeconomic stability of millions of people. At the same time, the conventional construction sector is one of the main drivers of environmental degradation on a global scale. This is manifested through high energy consumption in the production of materials and during the useful life of buildings, the emission of large quantities of greenhouse gases (GHGs), which directly contribute to climate change, and the unsustainable depletion of natural resources (United Nations Environment Programme & Global Alliance for Buildings and Construction, 2025).

In this context, the concept of 'adequate housing' involves more than just providing a shelter structure. Recognised as a fundamental human right in numerous international covenants and conventions (Urban and Cities Platform- ELAC, n,d,), adequate housing encompasses multifactorial dimensions such as security of tenure, availability of basic services and infrastructure (e.g. water, sanitation and energy), economic affordability, habitability (i.e. protection against adverse climatic conditions such as cold, humidity, heat, rain and wind, or other health threats), accessibility for disadvantaged groups, proximity to employment opportunities and social services, and cultural adequacy that respects and expresses the identity of its inhabitants.

The conventional approach to mitigating housing deficits, which is often centred on the mass production of industrialised and standardised housing units, frequently fails to satisfy these fundamental qualitative dimensions. While this strategy may address the need for more housing, it can contribute to environmental issues and fail to provide dignified or culturally appropriate living conditions, perpetuating cycles of vulnerability and dependency. This scenario highlights the urgent need to explore alternative housing design and construction paradigms, particularly for the most vulnerable members of society.

Vernacular Architecture and Bioclimatic Design: Reclaiming Sustainable Legacies

In the search for more holistic and sustainable solutions, vernacular architecture emerges as a vast repository of knowledge and practice. Theorists such as Paul Oliver have emphasised that these building traditions, developed over generations, intrinsically respond to the cultural and physical 'needs' of their inhabitants through the use of locally sourced materials and techniques adapted to the environment (Oliver, 2006). In his influential work Architecture without Architects, Bernard Rudofsky celebrated the functional beauty and ecological wisdom inherent in these 'non-pedigreed' forms of construction, which arise spontaneously from collective experience and an intimate relationship with place (Rudofsky, 1964). Hassan Fathy demonstrated the feasibility of using local materials such as adobe and encouraging community participation to build decent, climatically adequate housing in contexts of scarce resources in his “Architecture for the Poor” (Fathy, 1973). Thus, vernacular architecture is characterised by its deep roots in tradition, its ingenious use of local resources, its precise response to climate and its tangible expression of a community's cultural identity.

In parallel, bioclimatic design is an architectural approach that aims to maximise energy efficiency and achieve optimal occupant comfort through the implementation of design strategies that operate in synergy with the local climate, with no extra consume. This approach prioritises passive solutions such as adequate solar orientation, natural ventilation and the use of thermal mass to regulate indoor temperature, as well as the implementation of effective shading elements. This minimises dependence on mechanical air conditioning systems (Neila González, 2004).

The interrelationship between these two disciplines is fundamental. The principles underlying vernacular architecture, such as local adaptation, resource efficiency and cultural specificity, are intrinsically aligned with the goals of contemporary bioclimatic and sustainable design. Vernacular traditions offer a set of time-tested solutions that predate today's “green building” movements and can significantly enrich them. These traditions should be seen as dynamic sources of knowledge that can help to address today's challenges, particularly in the area of social housing, where there is a pressing need for affordable, durable and environmentally friendly solutions.

Validation and Redesign of the Lenca Housing Research Methodology

The aim of this research is to revitalise and adapt Honduran Vernacular architectural principles in a context-sensitive manner, with a specific focus on the Lenca tradition, providing a culturally resonant, economically viable and environmentally sustainable solution to the acute social housing crisis in Honduras's Central District.

This study is structured according to a predefined work plan to guide its development. First, an in-depth analysis of the housing problem in Honduras will be conducted, quantifying the existing deficit, and examining its socioeconomic and environmental ramifications, as well as the current political and institutional framework. Subsequently, the research will focus on studying Lenca vernacular architecture, exploring its cultural context, the materials and construction techniques that characterise it and the bioclimatic wisdom inherent in its typologies. These qualities will be scientifically validated through the results obtained from comparative energy simulations. Based on this analysis, the thesis will detail a design proposal for progressive social housing that integrates the vernacular and bioclimatic principles studied. Finally, the broader socioeconomic, environmental, and public policy implications of this proposal will be discussed, and recommendations for its implementation and scalability at the national level will be formulated.

BACKGROUND: THE PROBLEMS AND CONTEXT OF THE HOUSING DEFICIT IN THE CENTRAL DISTRICT OF HONDURAS

Quantifying the Crisis: The Housing Deficit in Honduras and the Central District

The Honduran context is characterized by a severe and persistent housing deficit that affects a considerable segment of its population. The most recent data estimates place the total deficit at the national level in a range that oscillates between 1.37 million housing units, according to 2020 data processed by Habitat for Humanity Honduras based on sources from the National Institute of Statistics (INE) (UNIAPRAVI, 2020), and 1.6 million units, according to 2024-2025 figures from the Honduran Chamber of the Construction Industry (CHICO) (Honduran Chamber of the Construction Industry, 2024). This deficit is broken down into two critical components: the quantitative deficit, which represents the need for new housing units, estimated between 0.52 million (Urban and Cities Platform - ECLAC, n.d.) and 0.95 million (Honduran Chamber of the Construction Industry, 2024); and the qualitative deficit, which refers to existing housing units that require significant repairs or improvements to be considered adequate, calculated between 0.65 million (Honduran Chamber of the Construction Industry, 2024) and 0.84 million (UNIAPRAVI, 2020).

The Central District (CD), which encompasses the urban centres of Tegucigalpa and Comayagüela, is home to a significant proportion of this national issue. The research that forms the basis of this chapter reveals that there is a shortage of 550,000 housing units only in the metropolitan area. This figure highlights the intense urban pressure and the scale of the challenge in the political and economic heart of the country. A summary of the current housing deficit situation is presented in Table 1.

Table 1. Housing deficit overview in Honduras and the Central District (period 2020-2024).

Indicator	National Figure (estimated range)	Central District Figure (estimated)	Main Source(s)
Total Housing Deficit	1.37 – 1.6 million	550	(a)
Quantitative Deficit (New Units)	0.52 – 0.95 million	Not disaggregated (part of 550,000)	(b)
Qualitative Deficit (Necessary Improvements)	0.65 – 0.84 million	Not disaggregated (part of 550,000)	(b)
Central District as % of National Deficit	Approx. 34% - 40%	-	based on (a)
New Housing Units Needed Annually (National)	60,000 – 100,000	-	(c)
Actual Annual Construction (National, c. 2023)	29,5	-	(c)

Sources: (a) INE/El Heraldo; (b) UNIAPRAVI; (c) CHICO/El Heraldo.

Trends indicate sustained growth in the deficit, exacerbated by factors such as accelerated urbanisation, internal migration, and the formation of new households. Although data from the National Statistics Institute (INE) show an increase in the number of housing units built between 2023 and 2024 (56,011 new homes nationwide, 21,048 of which were in urban areas), CHICO warns that the pace of construction remains insufficient to cover accumulated and annual demand (Honduran Chamber of the Construction Industry, 2024). Therefore, while the official figures are alarming, they may underestimate the true scale of the crisis, particularly when considering the qualitative aspects of 'adequate housing' and the growing impact of climate change-induced displacement. This phenomenon adds ongoing and often unquantifiable pressure to the demand for safe and resilient housing (Center for Gender & Refugee Studies, 2023).

The Socioeconomic and Environmental Ramifications of Inadequate Housing

The implications of inadequate housing in the Central District and Honduras more broadly extend beyond the construction sector, manifesting as a range of far-reaching socioeconomic and environmental consequences. Living conditions that are precarious, characterised by overcrowding, a lack of access to basic services such as drinking water and sanitation, and insecure tenure, have a direct and negative impact on public health by facilitating the spread of diseases, on education by failing to provide adequate study spaces, and on citizen security by fostering environments of vulnerability (Honduran Chamber of the Construction Industry, 2024).

One critical factor that worsens this situation is vulnerability to climate change. Honduras is one of the countries most exposed to the adverse effects of climate change[1]. Inadequate housing, which is often found in informal settlements in high-risk areas such as unstable hillsides and flood plains, is particularly susceptible to destruction by hurricanes, floods, and prolonged droughts. These phenomena are becoming more frequent and intense due to climate change (Center for Gender & Refugee Studies, 2023). A report by the Center for Gender & Refugee Studies (CGRS) details how extreme weather events have destroyed homes and caused massive displacement in Honduras. For example, hurricanes Eta and Iota in 2020 caused over a hundred deaths and displaced over 300,000 people, in addition to destroying infrastructure and livelihoods (National Congress of Honduras, 2020).

Construction practices that are predominant in the social housing sector often contribute to this vulnerability. There is an over-reliance on materials such as concrete blocks and aluzinc sheeting for roofing. While these materials may initially be cost-effective and easy to install, they often perform poor thermally in the tropical climate of Honduras. This results in indoor environments that become overheated during the day and require excessive cooling at certain times of the year. This causes thermal discomfort and increases reliance on artificial cooling or heating systems, which implies additional energy costs for low-income families.

Bioclimatic considerations are generally absent from the design of social housing. Although the Municipal Mayor's Office of the Central District (*Alcaldía Municipal del Districto Central*, AMDC) introduced the 'Green Regulation for the Central District' in May 2020 (Alcaldía Municipal del Distrito Central, 2020) which included guidelines for sustainable construction, its non-mandatory nature limits its implementation and effectiveness. While this regulation covers areas such as water and energy efficiency, sustainable urbanisation, and resource management, the absence of oversight mechanisms and the prioritisation of speculative economic factors in the construction market prevent its widespread adoption. Consequently, the prevalent model of social housing construction not only fails to provide adequate comfort and sanitation, but it can also perpetuate vulnerability by locking families into high operating costs and dependence on external material markets. This interferes with local economic development and community resilience.

National Housing Policy and Institutional Framework

The Honduran legal framework recognises the right to decent housing, as set out in Article 178 of the Constitution of the Republic. To materialise this right, several laws have been enacted and multiple institutions have been created over the years. The most relevant legislation at present is the Housing and Human Settlements Framework Law (Decree No. 173-2019, published in La Gaceta in 2020, (La Gaceta, 2020). This law aims to promote the construction of between eight and ten thousand housing units per year, generate employment, allocate funds for the financing and construction of Social Housing projects, and execute housing improvements. The law also contemplates the creation of social housing bonds and establishes new entities, such as the Secretariat of State for Housing and Human Settlements (*Secretaría de Estado en los Despachos de Vivienda y Asentamientos Humanos,* SEVIAH) and the Honduran Housing Fund (*Fondo Hondureño para la Vivienda*, FONHVIVIENDA), which are responsible for formulating, coordinating, and executing the sector's public policy (Presidency of the Republic of Honduras, 2022).

At the institutional level, the recent establishment of the Housing and Human Settlements Programme (*Programa de Vivienda y Asentamientos Humanos*, PVAH) under President Xiomara Castro's administration, as set out in PCM-005-2022 (15), is notable. The PVAH was established to consolidate the functions of previously separate institutions, such as the National Housing and Human Settlements Commission (*Comisión Nacional de Vivienda y Asentamientos Humanos,* CONVIVIENDA) and the Social Fund for Housing (*Fondo Social para la Vivienda,* FOSOVI), with the aim of planning, organising and directing the sector more efficiently. Among its initiatives is the 'Solidarity Housing Project' (*Proyecto Vivienda Solidaria*), which focuses on improving the quality of life of families in extreme poverty (Government of Honduras, n.d.). The PVAH's strategic and operational plans for 2023–2026 and 2025, respectively, set out goals for housing improvements, new construction, and credit programmes, including approaches in the informal sector and in emergency situations (Castañeda Moncada, 2012). Other key institutions are the Honduran Bank for Production and Housing (*Banco Hondureño para la Producción y la Vivienda,* BANHPROVI), which offers financial products and credit for the acquisition and construction of social housing, including preferential interest rates and extended terms; and Habitat for Humanity Honduras (*Hábitat para la Humanidad Honduras*), an NGO with a long history of building and improving social housing and public policy advocacy.

Figure 1. A sample of proposals for social Housing and substandard Housing.

Source: various.

Despite these legal and institutional frameworks, the effectiveness of housing policies in Honduras has been limited, as evidenced by the persistently high housing deficit (Honduran Chamber of the Construction Industry, 2024). Structural challenges include restricted access to adequate financing for low-income families, the complexity and cost of regularising land tenure, cumbersome bureaucratic processes for project approval and insufficient accessible urban land (Habitat for Humanity Honduras, 2017). Furthermore, many programmes prioritise quantitative aspects over qualitative ones, with insufficient focus on the environmental sustainability, energy efficiency, climate resilience and cultural appropriateness of housing solutions.

The research underlying this chapter highlighted the absence of a bioclimatic and vernacular approach in existing Social Housing proposals. The creation of the PVAH, for example, reflects an institutional reconfiguration aimed at improving the governance of the sector. However, for these changes to result in substantial and sustainable improvements, a fundamental transformation in the philosophy of Housing provision is imperative. This transformation must prioritise vernacular knowledge, bioclimatic principles, and the use of local materials in Housing policies, rather than considering them as marginal alternatives. PVAH plans such as the 2025 Strategic and Operational Plan, which mentions 'comprehensive and sustainable programmes', must be filled with specific content reflecting this paradigm shift if they are to address the qualitative and sustainability aspects of the Housing problem beyond mere administrative efficacy (Castañeda Moncada, 2012).

ANALYSIS OF LENCA VERNACULAR ARCHITECTURE AS A BIOCLIMATIC STRATEGY

The Lenca People: Cultural identity, Cosmovision and Link to the Territory

The Lenca people are one of the oldest and most numerous indigenous groups in Honduras. They have a long-standing presence across several western, central, and southern departments, including Lempira, Intibucá, La Paz, Comayagua, Francisco Morazán and Valle, as well as regions of El Salvador. Their cultural legacy is deep and multifaceted. However, they have faced the effects of colonization and modernization. These have resulted in the loss of their language and the transformation of many of their traditional practices (Chapman, 1985-1986).

The Lenca worldview is characterised by an intimate connection with nature and territory. Elements such as rivers, mountains, and the land itself are not conceived as mere resources, but as living entities with a deep spiritual significance (Muñoz Torres, 2015). This worldview is reflected in their agricultural practices, their rituals and, significantly, in their traditional conceptions of shelter and community organization. Anne Chapman's ethnographic work, ‘The Children of copal and candela’ (*Los hijos del copal y la candela*) offers an invaluable window into these traditions. For example, it documents the 'rites of the house' and the religious syncretism that permeates Lenca daily life, where ancestral traditions are intertwined with elements of imposed Catholicism (Chapman, 1985-1986). The traditional Lenca dwelling cannot therefore be understood solely as a physical structure, but as a culturally significant space, a nexus of family and community identity and a reflection of systemic adaptation to the environment. The progressive disappearance of these traditional construction methods, driven by the adoption of 'modern' materials and designs, implies a loss not only of valuable construction techniques, but also of this cultural heritage and the ecological wisdom accumulated over centuries.

Figure 2. A sample of Homes with vernacular Lenca characteristics.

Source: Castañeda, 2012; Salinas, 1991; Agurcia, 1984.

The Material Culture and Construction Techniques of Lenca Dwellings

Lenca vernacular architecture is characterised by its ingenious use of locally available materials and building construction techniques that have been passed down through the generations. Rather than being 'primitive', these methods demonstrate a deep understanding of material properties and effective adaptation to environmental conditions. The key materials and building construction systems commonly used are summarised below.

- Adobe: This is one of the most emblematic materials used in Lenca construction. Consisting of blocks of raw earth made from a mixture of clay and sand, and often an organic binder such as straw or cow dung, it is moulded and dried under the sun. These sun-dried bricks have excellent thermal mass, enabling them to absorb heat energy during the day and release it slowly at night. This helps to stabilise indoor temperatures. When properly protected from moisture, they are durable and have a very low ecological footprint due to their local production and minimal processing.
- Bahareque (or Bajareque): This building construction system, also known as *quincha* or *wattle-and-daub*, consists of a wooden or cane framework with spaces filled with a mixture of mud and plant fibres. *Wattle-and-daub* walls are lighter than solid adobe and offer some structural flexibility, which has proven beneficial in seismically active areas in certain configurations. They provide good thermal and acoustic insulation, and interior spaces can be easily and economically reconfigured.
- Clay-roof tiles: Handmade clay tiles fired in local kilns are the traditional roofing material. Their curved shape allows rainwater to drain away quickly, and when arranged properly, they can help to ventilate the space under the roof by dissipating accumulated hot air.
- Wood (roundwood and sawn timber): Local wood is used for the main structural elements of the house, such as posts, beams, and roof trusses. It is also used to manufacture doors and windows, and sometimes as part of the bahareque framework or enclosures.
- Stone: Field or river stone is commonly used to build stone foundations, which provide a solid base for adobe or *Wattle-and-daub* walls. This foundation protects the walls from rising soil moisture and rain splashes, thereby prolonging their useful life. Stone is also sometimes used for building entire walls.

Figure 3. A sample of housing with vernacular Lenca characteristics.

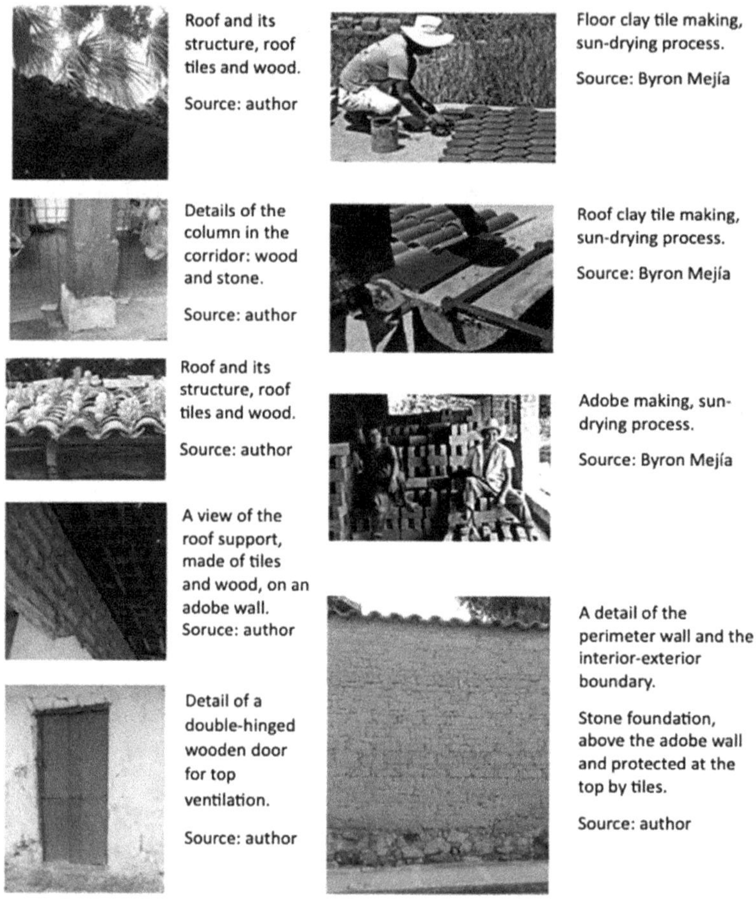

Source: Byron Mejía, author.

Although the traditional Lenca building construction process presents regional variations, it usually follows a logical sequence. It begins with selecting and preparing the site, which often involves flattening the ground. Next, trenches are dug for the foundations and filled with stone and mud. The walls are then erected on top of these foundations, either by laying adobe bricks with mud mortar or by building a *wattle-and-daub* framework and filling it in. The roof structure, typically gabled or hipped, is assembled from wood, and covered with tiles. The finishes are usually simple, with fine mud plaster or lime wash to protect the walls and make them brighter. Table 2 provides a comparative analysis of Lenca vernacular materials and conventional materials commonly used in social housing in the Central District.

Figure 4. Vernacular housing in urban context.

Source: author.

Table 2. Comparative analysis of Building Construction materials: vernacular vs. conventional.

Characteristic	Lenca Vernacular Materials (Adobe, Bahareque, Clay Tile, Local Wood, Stone)	Conventional Materials (Concrete Block, Aluzinc Sheet, Cement, Steel)
Source/origin	Predominantly local, natural, low-processing.	Predominantly industrial, often imported, high-processing.
Thermal properties	High thermal mass (adobe, stone), good insulation (bahareque, wood).	Low thermal mass (hollow block), high thermal conductivity (aluzinc, steel), poor insulation without additives.
Embodied Energy	Low to very low.	High to very high.
Cost (material + labour)	Generally lower if materials are local and labour is community-based or artisanal.	Variable: can be competitive in mass production but depends on fluctuating market prices and transportation.
Durability/maintenance	High durability with proper maintenance (protection against moisture).	Variable: the block is durable, but aluzinc can corrode; requires specific maintenance.
Required labour	Traditional artisanal skills, often labour-intensive.	Standard construction labour, may require less time per unit.
Environmental impact	Low impact, biodegradable, recyclable (earth), use of renewable resources (sustainable wood).	High impact, CO_2 emissions in cement and steel production, generation of non-biodegradable waste.

Source: authors own elaboration.

These construction techniques were traditionally passed down orally and practically from master to apprentice within the community. This empirical knowledge, accumulated and refined over centuries, represents invaluable cultural capital. The choice of materials and methods was not arbitrary, but rather the result of profound observation of how local resources behave in response to climatic conditions and the functional requirements of dwellings. This knowledge system prioritises local adaptation and resource efficiency, contrasting with the logic of modern industrialised construction, which often imposes standardised solutions without adequately considering the particularities of the context.

Bioclimatic Wisdom Incorporated in the Lenca Architectural Typologies

Lenca vernacular architecture is not only a testament to its material culture, but also an example of intuitive and efficient bioclimatic design. Traditional housing, which is generally rectangular or square in plan with one or two main spaces, frequently incorporates corridors or porches, and is configured to optimise thermal comfort and minimise energy consumption.

A fundamental aspect is the orientation of the dwellings. The settlement patterns and the arrangement of openings suggest that solar paths and prevailing winds were considered to maximise cross ventilation and control solar heat gain. Windows and doors are often modest in size but are placed strategically to facilitate airflow through indoor living spaces.

High gable or hip roofs covered in clay tiles ensure efficient rain drainage and create significant air volume over living spaces. This 'cushion' of air, combined with possible ventilation at the ridge or through the tiles themselves, helps dissipate heat accumulated by solar radiation on the roof, thereby keeping indoors cooler.

Another key bioclimatic principle exploited by Lenca architecture is thermal mass. The thick adobe or bahareque walls act as passive thermal regulators. During the day, they absorb heat from the sun and surrounding environment, preventing it from quickly penetrating the interior. At night, when the outside temperature drops, the walls slowly release the accumulated heat, maintaining a more stable and comfortable indoor temperature and reducing extreme thermal fluctuations between day and night.

Shading elements are omnipresent. Wide roof overhangs protect walls and openings from direct sunlight and rain. Corridors and porches extend the living space outward and work as effective solar barriers, creating shaded, cooler transitional zones before entering the home. Additionally, integrating the house with the surrounding vegetation (trees and shrubs) provides extra shade and improves the microclimate around the house.

This harmonious integration with the landscape and local microclimate is another example of the Lenca's bioclimatic wisdom. Rather than imposing themselves on the environment, the houses engage with it, making use of its resources and mitigating its challenges. This systemic approach, in which form, materiality, orientation, and spatial organisation collaborate to achieve comfort, is an invaluable legacy that contrasts with the frequent disconnection between buildings and their environments in contemporary architecture.

Cultural significance, Sustainability, and Preservation of the Lenca Architectural Heritage

The Lenca house transcends its utilitarian function, becoming the epicentre of family life and a space for community socialisation, as well as a vehicle for the reproduction of culture and traditions. Daily activities are carried out within its interior and immediate surroundings, where ancestral knowledge is transmitted, and social ties are strengthened. The furniture and household utensils, which are often crafted from local materials such as clay, wood and vegetable fibres using traditional techniques, complement the architecture, forming a coherent material culture that is adapted to the environment.

From a contemporary perspective, the environmental sustainability of traditional Lenca architecture is remarkable. The predominant use of natural, local, and low-processing materials (earth, stone, wood, straw) implies low embodied energy and a reduced carbon footprint compared to more industrial materials such as cement and steel. These materials are mostly biodegradable or easily reusable or recyclable at the end of the building's life, minimizing the generation of building construction waste.

However, this valuable architectural and cultural heritage is under serious threat. The globalisation of construction models, the promotion of industrial materials perceived as 'modern' and 'better', the gradual loss of traditional artisan skills, and the lack of official recognition and valuation of these traditional building construction practices all contribute to its decline. This situation implies not only the loss of technical and aesthetic heritage, but also a disconnection from more sustainable and culturally rooted ways of living.

In this context, the research and documentation work carried out by academics and institutions is crucial for preserving and revitalising this legacy. Studies such as Architecture of the Ethnic Groups of Honduras by Iris Milady Salinas (Inter-American Development Bank, 2023), and the research and publications of the Honduran Institute of Anthropology and History (*Instituto Hondureño de Antropología e Historia*, IHAH) (Salinas, 1991) and the National Autonomous University of Honduras (*Universidad Nacional Autónoma de Honduras*, UNAH) (Honduran Institute of Anthropology and History, n.d.), are crucial for understanding, valuing, and disseminating knowledge about Lenca architecture, as well as the architecture of other indigenous peoples in Honduras.

Table 3. Comparative chart of Housing types in Honduras' Central District and the new Social Housing proposal (red = bad solution, yellow = rescued solutions for new proposal).

TYPES OF DWELLINGS →	Vernacular Housing in central Honduras.	Colonial Housing	Social Housing (precarious)	Social Housing proposals for rescue	New proposal of Social Housing
Type of wall	Mixed wall of stone and adobe or Wattle-and-daub	Adobe, wattle-and-daub	Ceramic brick, concrete block	Ceramic brick, concrete block and plaster, polished and painted	Stone foundation, plastered, polished and painted adobe or rammed earth
Type of roof	Roof clay tile and wooden structure. Gabled roof	Roof clay tile and wooden structure. gabled roof	Aluzinc sheet, asbestos sheet	Aluzinc sheet with thermal and acoustic insulation, and metal structure	Roof clay tiles, thermal and acoustic insulation, and wooden structure
Type of floor	Rammed earth	baked clay tile	concrete slab or earth	porcelain tile	baked clay tile
Type of windows	Wood, folding, narrow	Wood, folding	Two-leaf sliding door	wood or glass lattice	lattice
Door entrance	deficient, not airtight	Some fold in half	BAD	Generic	folding in half
Natural lighting	BAD, small windows	Yes, windows to outdoors	BAD, small windows	More or less	yes
Natural ventilation	BAD, small windows	Yes, windows to outdoors	certain connections without prior design	Cross ventilation	Cross ventilation
Electricity supply	no	no	yes	yes	yes
Drinking water supply	no	no	no	yes	si
Greywater collection	no	no	no	no	yes
Rainwater collection	no	no	no	no	yes
Sewage collection	no	latrine	yes (deficient)/ latrine	yes	yes
No. floors	1	1	1	2	2
Incorporation of vegetation	yes	yes	no	no	yes
Incorporation of crafts	yes	yes	no	no	yes

Source: authors elaboration.

Therefore, the preservation and revitalisation of Lenca architectural traditions should not be understood as mere nostalgia or museum conservation. Instead, it is a strategy with profound implications for contemporary development. By valuing and adapting this knowledge, housing models can be fostered that are environmentally sustainable and reinforce cultural identity. These models can also promote the local economy through the use of local resources and skills and contribute to building more resilient communities with a better quality of life. This approach aligns with a vision of development that is more equitable, respects cultural diversity, and values ancestral knowledge.

VALIDATION: BIOCLIMATIC PERFORMANCE AND ENERGY EFFICIENCY

Climate Context of the Central District, Honduras

In order to understand the potential of Lenca vernacular architecture as a bioclimatic alternative, it is essential to analyse the specific climatic conditions of Honduras's Central District, the focal area of this study. The region has a tropical high-altitude climate characterised by two main seasons: a hot and dry summer, and a rainy and wet winter. Historical climate data for the Central District, located at a latitude of 14.1°N, reveal certain variations in temperature, precipitation, and humidity throughout the year. Mean monthly temperatures vary by approximately 4.1 °C throughout the year. Precipitation varies considerably between seasons, with up to 274 mm difference between the driest and rainiest months. The lowest relative humidity is in April (at approximately 55.41%), and the highest is in October (at about 83.15%). September has the highest number of rainy days (around 25), while January has the fewest (less than 2).

The bioclimatic analysis carried out in this study uses the Bioclimatic charts of Givoni and Olgyay (climate comfort-adapted version) to identify the design requirements for achieving thermal comfort. The Givoni bioclimatic chart for the Central District indicates that for most of the year, the conditions fall within the comfort zone or can be achieved through passive strategies, such as natural ventilation and solar shading. There is no need for heating or active humidification for any significant period of the year.

Figure 5. Climate-adapted Olgyay Bioclimatic Chart for Hondura's Central District.

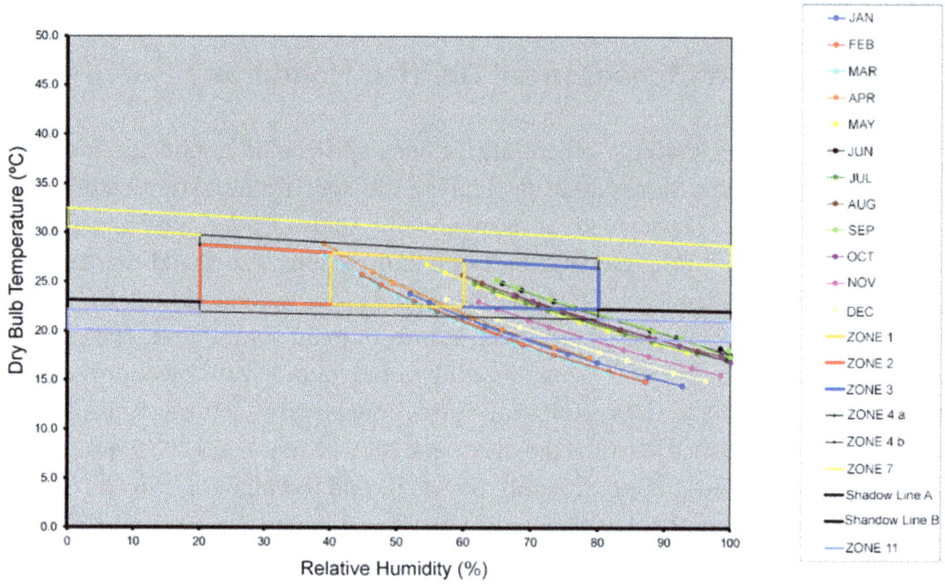

Source: authors.

Figure 6. Givoni Bioclimatic Chart for Hondura's Central District.

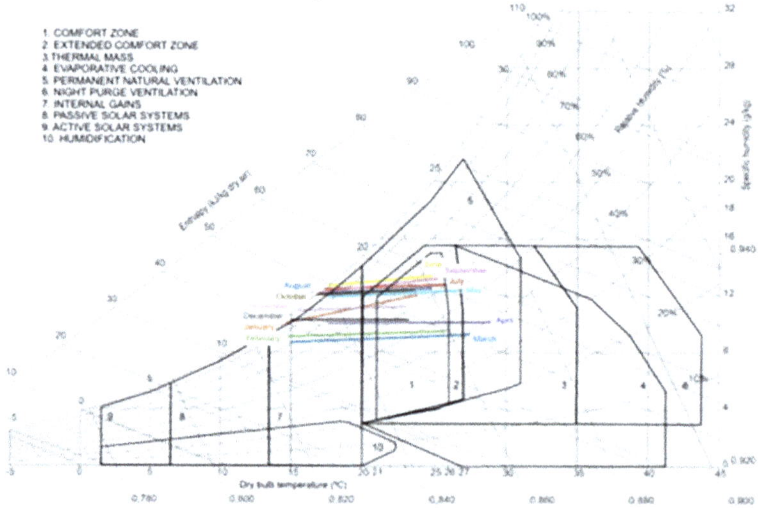

Source: authors.

The isopleth graph, based on the adapted comfort bioclimatic chart, indicates that, while a predominant comfort zone exists, ventilation is required to maintain comfort during the warmer months, especially in the afternoons (14:00 to 17:00). The Olgyay bioclimatic chart confirms the need for shading during most of the months, except October.

Figure 7. Isopleth graph with temperatures for Hondura's Central District.

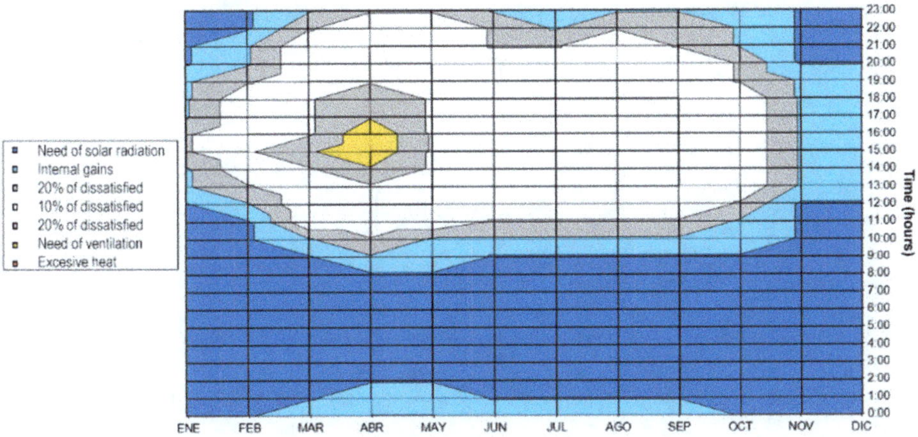

Source: authors

Analysing the solar geometry using stereographic shadow masks is crucial for designing solar shading properly. For example, it suggests the desirability of carefully avoiding or protecting openings on the west façade.

Figure 8. Solar chart with shadow mask representation and window solar protections for the four main façade orientations.

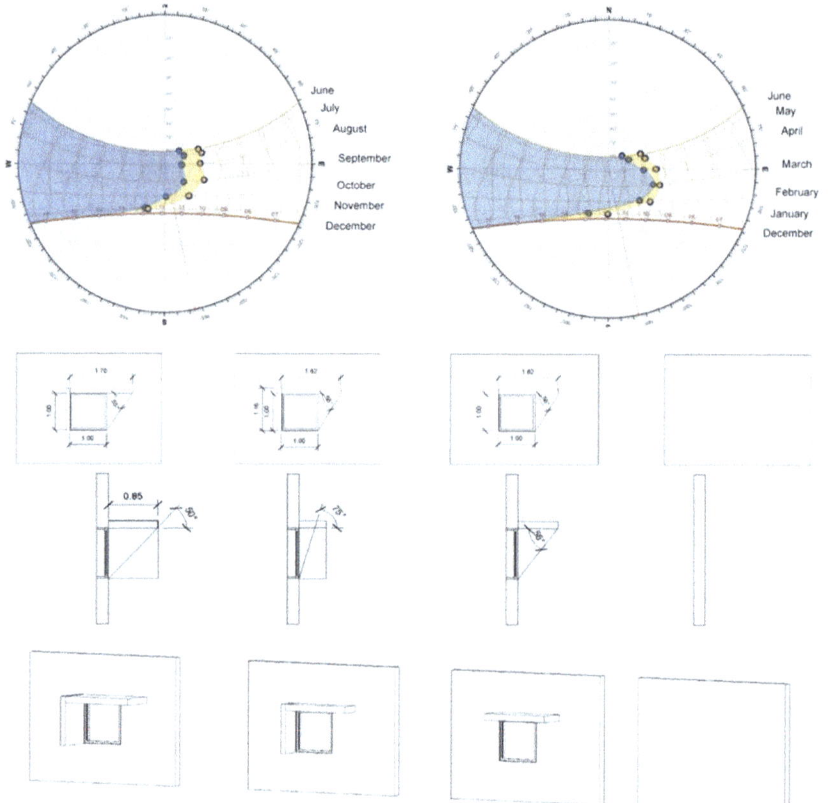

South façade | North façade | East façade | West façade (no window)

Source: authors.

In conclusion, while climatic conditions in the Central District are generally mild, they present specific challenges during the warmer and more humid months. These conditions do not inherently justify the use of energy-intensive mechanical cooling systems, particularly if buildings are intelligently designed to take advantage of the passive strategies employed by Lenca vernacular architecture.

Energy Simulation Methodology

To rigorously quantify and compare the thermal and energy performance of different Housing typologies, this study used dynamic simulations with the Design Builder software program. This software enables the modelling of buildings and the definition of their building construction materials and operational timetables and characteristics. It simulates their behaviour under specific climatic conditions and provide detailed data on heat flows, indoor/outdoor temperatures evolution and energy demand and consumption.

The comparative study was based on four Housing models, two with nowadays conventional building materials and two with building materials traditionally found in vernacular houses. The four models with an area of 45 m^2 and representative of a basic Social Housing unit:

- Viv 01 (conventional block): Configured with 10 cm of concrete block walls, Aluzinc sheet roofing and a 10 cm of concrete floor.
- Viv 02 (conventional brick): Composed of 15 cm brick walls with interior plaster, aluminium sheet roofing and a concrete floor.
- Viv 03 (traditional vernacular): Defined by 40 cm-thick adobe walls, a clay tile roof with a wooden structure, and a compacted earth floor.
- Viv 04 (improved vernacular): Consisting of 40 cm-thick adobe walls with earth and lime mortar plaster on both sides, a clay tile roof with a wooden structure, a 15 cm air chamber, thermal insulation (R19), a reed suspended ceiling and an interior floor of terracotta tiles over a concrete base.

Figure 9. The four Housing models for simulation (Design Builder software).

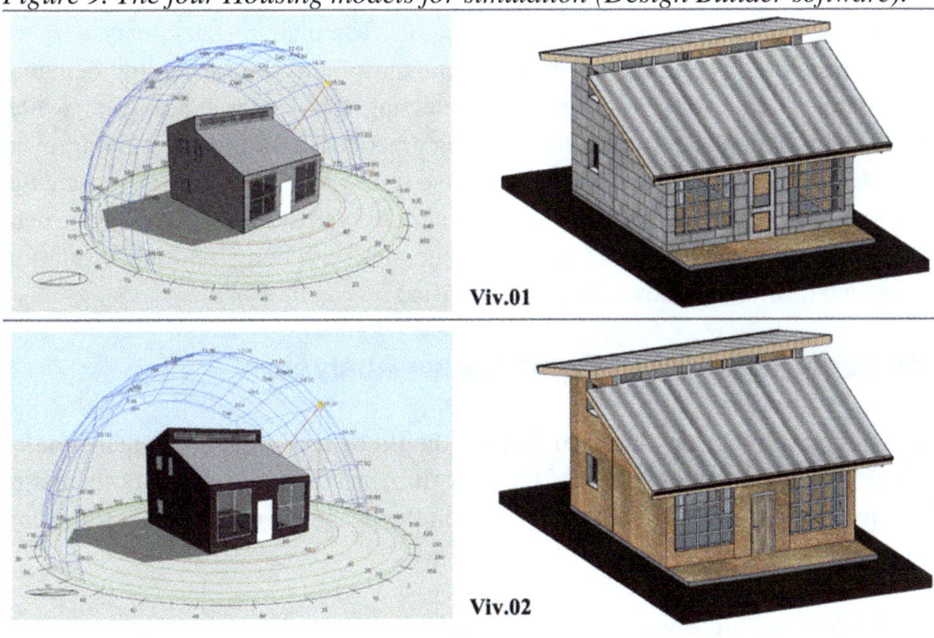

Source: authors.

Figure 10.

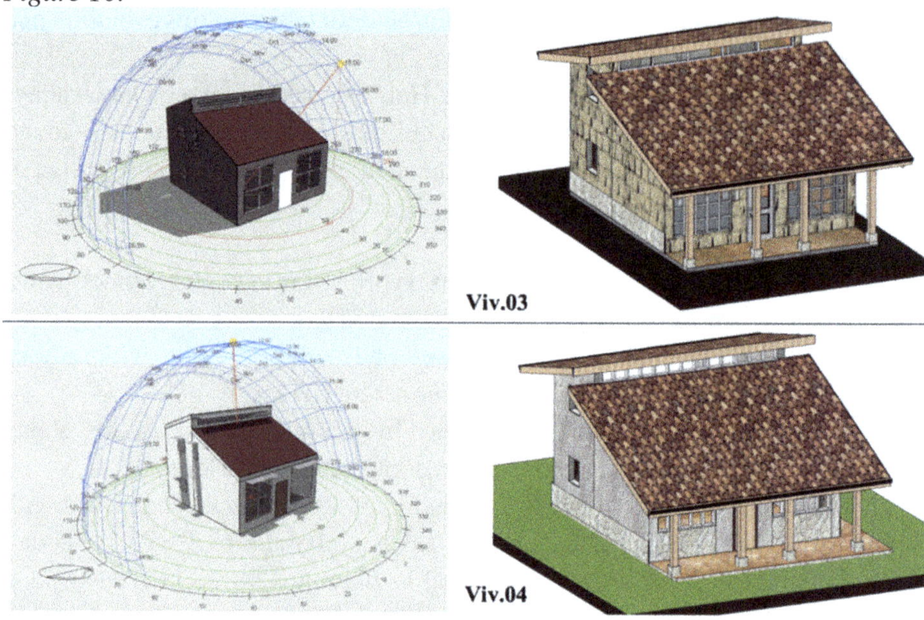

The simulations were performed using the climate file for the Central District. The cooling demand was calculated using occupancy parameters, internal gains from equipment and people, and temperature set points. The main metrics analysed were heat transmission through the enclosures (walls, roof and floor), cooling demand, and associated electricity consumption. Selecting these four typologies, which range from problematic conventional building practices to traditional and improved vernacular solutions, establishes a comparative framework. With this approach, the effectiveness of vernacular principles can be validated and the degree of improvement that can be achieved by optimising these principles with contemporary insulation and detailed design techniques can be quantified.

Simulation Results and Comparative Analysis

It was decided that two key simulations should be analysed: an annual simulation and one for the hottest week of the year (in April). The results of both energy simulations revealed significant differences in the performance of the four housing typology models studied.

- **Heat transmission through the envelope:**

In the hourly analysis of the warmest week, it was observed that adobe walls (Viv 03 and Viv 04) exhibited much more stable thermal behaviour. They experienced lower heat gain during the day, and the thermal lag contributed to maintain cooler indoor temperatures, compared to block (Viv 01) and brick (Viv 02) walls. Tile roofs, especially in Viv 04 (an improved vernacular design incorporating insulation and an air gap), showed a significant reduction in heat transmission compared to the aluzinc sheet roofs of Viv 01 and Viv 02, which experienced high heat gain peaks during periods of intense solar radiation. The floors also showed differences, although less pronounced, with the insulated earth and clay tile floors (Viv 03 and Viv 04) demonstrating greater thermal stability.

At the annual level, the monthly analysis of heat transmission confirmed these trends. The adobe walls of Viv 04 (improved vernacular) exhibited the lowest net heat gain throughout the year, followed by Viv 03 (traditional vernacular). In contrast, the block and brick walls of Viv 01 and Viv 02 accumulated significantly more heat. Similarly, Viv 04's improved roof was the most efficient in reducing annual heat gain, while the aluzinc roofs were the least efficient.

Figure 11. Graphs of heat transmission through the building envelope's Systems (façade, roof, floor) for the warmest week, April.

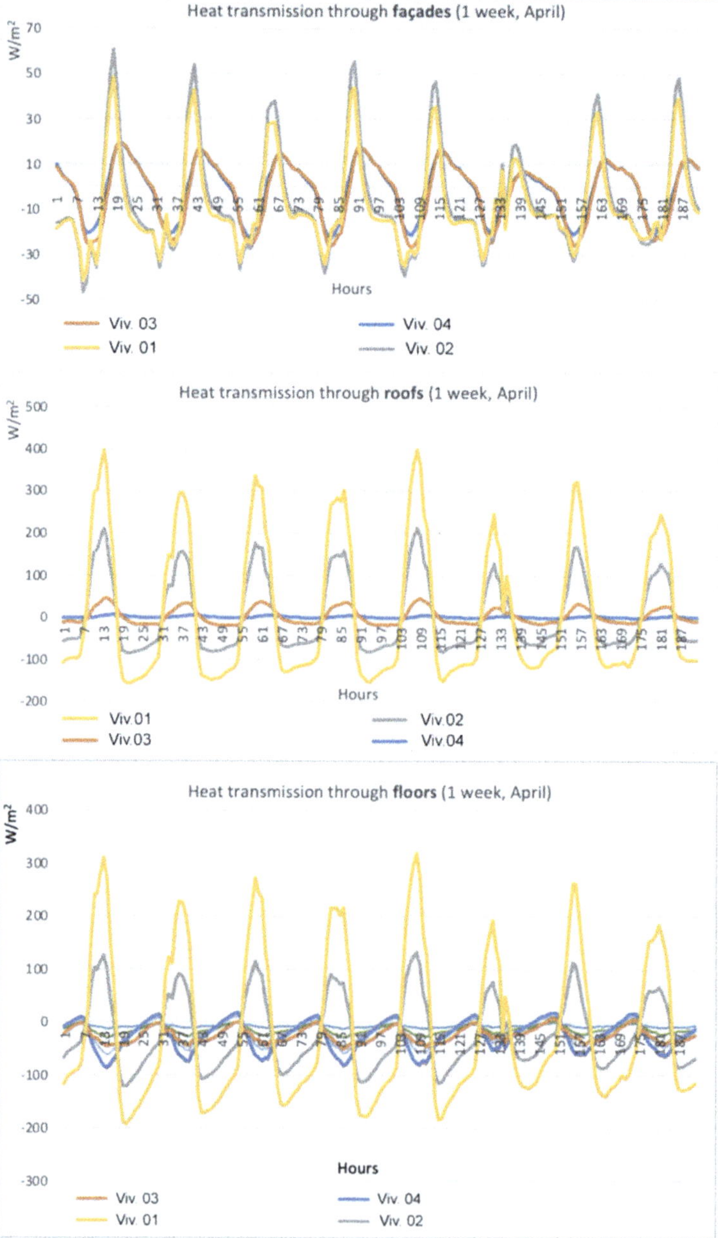

Source: authors.

- **Cooling demand:**

During the warmest week, Viv 04 (improved vernacular) had the lowest cooling demand, with a reduction of up to 48% compared to Viv 01 (block and aluzinc), which had the highest demand. Viv 03 (traditional vernacular) also demonstrated a significant reduction, while Viv 02 (brick and aluzinc) was in an intermediate position but still with high demand.

In the annual analysis, Viv 04 showed a reduction in cooling demand of around 60% compared to Viv 01. Viv 03 (traditional vernacular) achieved also a notable reduction, close to 54% compared to Viv 01. Conventional dwellings (Viv 01 and Viv 02) exhibited the highest annual demand, with concrete blocks performing the worst.

Figure 12. Graphs of Cooling demand for the for the four Housing models for the year warmest week, April.

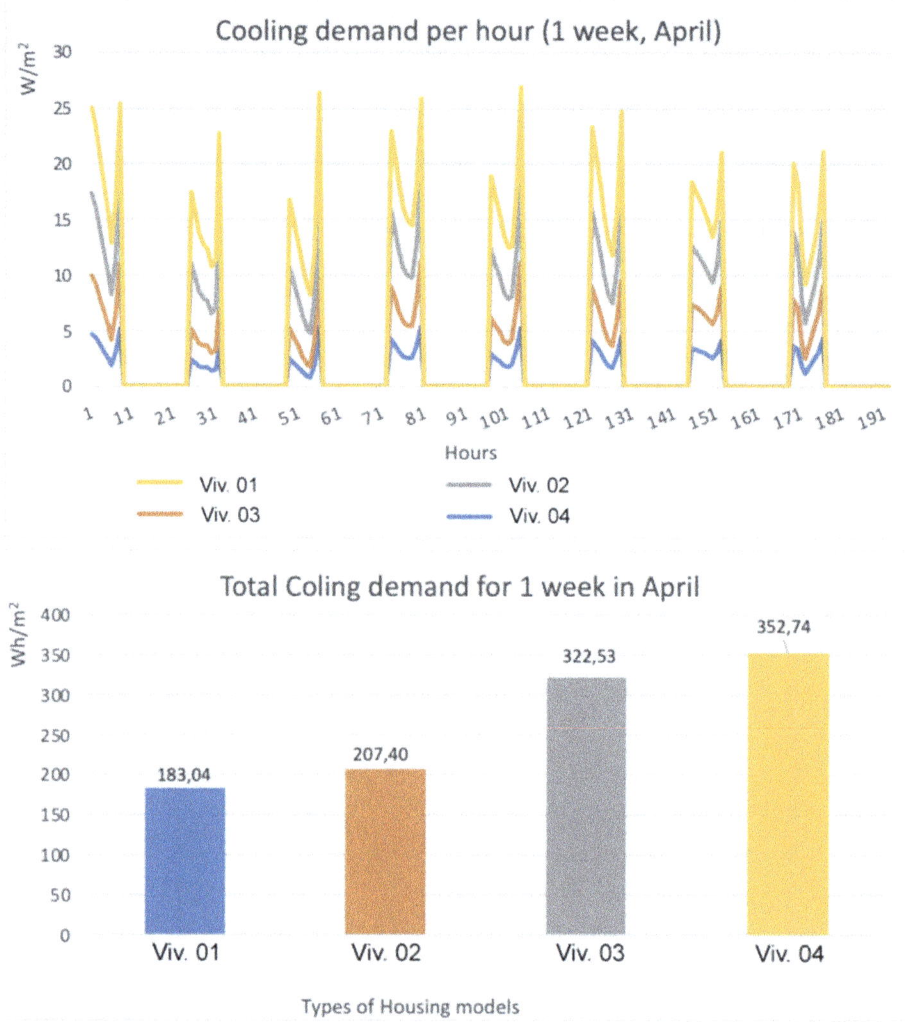

Source: authors.

- **Cooling electric consumption**

Patterns of cooling electric consumption for closely followed demand trends. During the hottest week, Viv 04 (the improved vernacular model) achieved a 48% reduction in electricity consumption compared to Viv 01. Annually, Viv 04 achieved a 60% reduction in cooling electricity consumption compared to Viv 01. Viv 03 (traditional vernacular) also demonstrated a substantial reduction of approximately 55%.

Figure 13. Graphs of Cooling demand per month and for the whole year for the for the four Housing models.

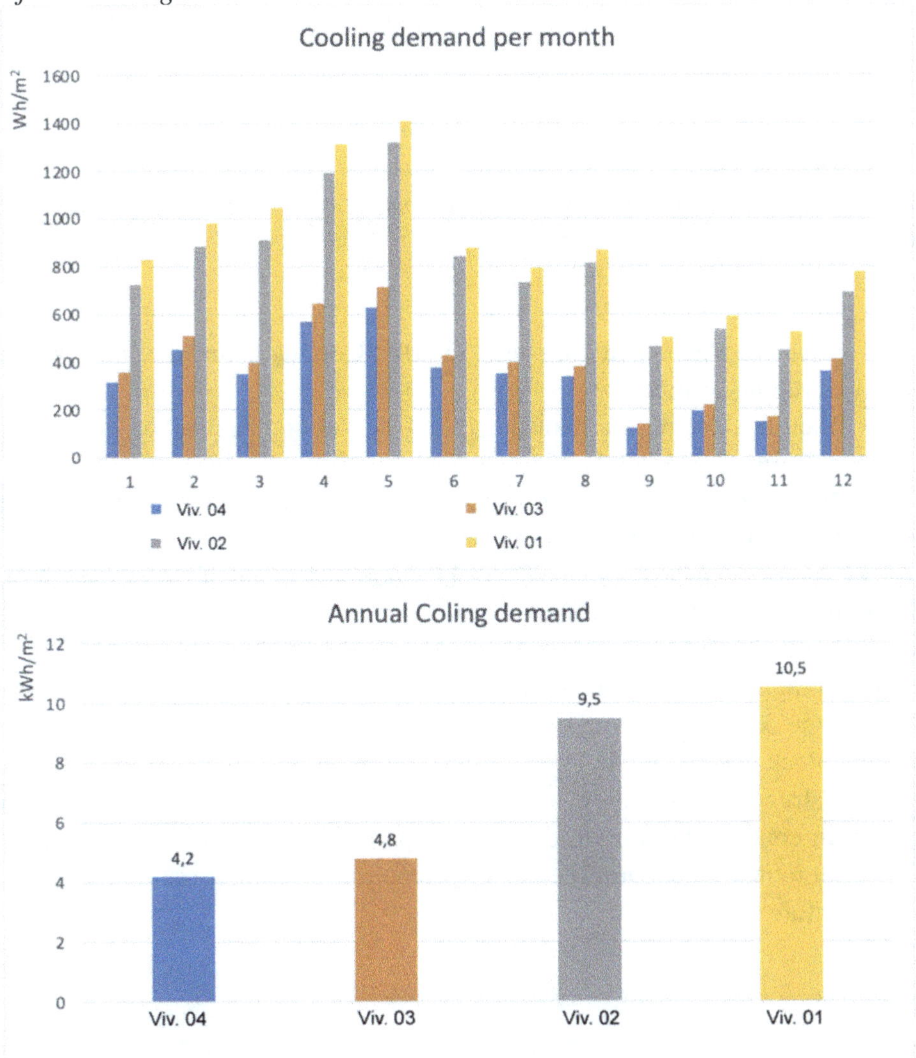

Source: authors.

The quantitative data summarised in Table 4 provide compelling evidence that adopting vernacular materials and bioclimatic design principles can lead to substantial energy savings and considerably improved thermal comfort. These results call into question the suitability of modern construction methods commonly used in social housing in the Honduran climate, demonstrating that traditional solutions, especially when optimised (as in Viv 04), are more energy efficient.

Energy Rating Proposal and Benchmarks

To contextualise the results of the simulations and communicate their meaning more clearly to the general public, an energy rating estimate has been made for the four Housing typologies studied. This rating is based on the Spanish Energy Efficiency Certificate (CEE) system and its reference values for climate zone α2 (Las Palmas de Gran Canaria), which is considered climatically comparable to Tegucigalpa, although slightly cooler. The main indicator used for the rating was annual cooling demand (kWh/m^2·year).

The results of this assigned rating were as follows:

- Enhanced Vernacular Housing (Viv 04): Cooling demand of 4.2 kWh/m^2-year. Rating assigned: A.
- Traditional Vernacular Housing (Viv 03): Cooling demand of 4.8 kWh/m^2-year. Rating assigned: A.
- Brick and aluzinc housing (Viv 02): Cooling demand: 9.5 kWh/m^2-year. Rating assigned: A. Rating assigned: C.
- Concrete block and aluzinc (Viv 01): Cooling demand of 10.5 kWh/m^2/year. Rating assigned: C.

The average reference value for a new single-family dwelling in the Spanish α2 climate zone is 10.70 kWh/m^2 per year[1]. This figure is close to that obtained for concrete block and aluzinc housing (Viv 01), indicating that this type of conventional construction would only just meet basic efficiency standards in a regulated context. In contrast, both traditional and improved vernacular dwellings achieve the highest rating (A) according to this reference scale, demonstrating outstanding energy performance.

Table 4. Key Performance Indicators of Design Builder Simulations for the Housing Typologies Studied.

Housing Typology (brief description)	Annual Cooling Demand (kWh/ m²·year)	Annual Electricity Consumption for Cooling (kWh/ m²·year)	% Reduction in Cooling Demand (vs. Viv 01)	% Reduction in Electricity Consumption (vs. Viv 01)	Estimated Energy Rating (Spain CTE DB-HE α2)
Viv01: Concrete block, aluzinc roof	10.5	5.2	0% (Baseline)	0% (Baseline)	C
Viv 02: Perforated brick (*rafón*), aluzinc roof	9.5	4.8	9.5%	7.7%	C
Viv 03: Traditional adobe, clay tile roof	4.8	2.4	54.3%	53.8%	A
Viv 04: Improved adobe, improved tile roof, insulation, tile floor	4.2	2.1	60.0%	60.0%	A

Source: authors own elaboration.

It is important to note the limitations inherent in using an external rating system. The particular climatic conditions in Tegucigalpa, particularly with regard to the severity of the summer weather, may require adjustments to enable a more accurate comparison. Nevertheless, this benchmarking exercise effectively illustrates the performance gap between conventional construction and vernacular-bioclimatic solutions. Furthermore, it highlights the urgent need to develop and implement energy efficiency standards and appropriate rating systems in Honduras. As highlighted in an Inter-American Development Bank (IDB) report on resilience and sustainability in building codes in Latin America and the Caribbean, Honduras's building codes are 100% non-compliant with energy parameters (Serrano et al., 2014), demonstrating a critical regulatory gap that must be addressed to promote more sustainable buildings. Demonstrating that high levels of energy efficiency can be achieved using local approaches (such as Viv 03 and Viv 04) should encourage the development of these Honduran standards.

DESIGN PROPOSAL FOR A VERNACULAR-BIOCLIMATIC SOCIAL HOUSING MODEL IN HONDURAS.

Design Philosophy: Synthesising Tradition and Contemporary Needs

The Social Housing proposal presented in this chapter is based on a design philosophy that seeks to reconcile the ancestral wisdom of Lenca vernacular architecture with the needs and aspirations of contemporary Honduran families. The aim is to provide more than just shelter, offering homes that are affordable, environmentally sustainable, and culturally appropriate, and which promote the comfort and dignity of their inhabitants. This approach aligns with the work of pioneering architects in the fields of sustainable and vernacular architecture. Hassan Fathy's work, which emphasises "Architecture for the Poor" and uses local materials such as adobe, as well as techniques that empower the community (Fathy, 1973), resonates deeply with the intent of this project. Similarly, the research of Paul Oliver on the diversity and cultural appropriateness of vernacular architecture globally (Oliver, 2006) and the work of Bernard Rudofsky, who appreciated 'architecture without architects' as a source of ingenious and adaptive solutions (Rudofsky, 1964), inform the valuing of Lenca traditions as a starting point for contextually relevant innovation. Therefore, the proposal does not seek to nostalgically replicate the past, but rather to creatively synthesise the best of traditional knowledge with the possibilities of current bioclimatic design.

The Progressive Social Housing Model

The proposed design is a progressive social housing model conceived to adapt to families' changing needs and economic circumstances over time. This flexibility is crucial for social housing, where families tend to grow and their spatial requirements evolve.

- **Spatial Layout and Functionality:**

The housing is proposed in an initial stage with 25 m^2 of built area, which is suitable for a couple or small family. There is the possibility of vertical growth to approximately 65 m^2 in a second stage, which would allow more rooms to be added for larger families. This range of sizes complies with the size criteria for social housing established by National Regulations such as METROPLAN (*Plan de Desarrollo Metropolitano del Municipio del Distrito Central*) and organizations such as CHICO (*Cámara Hondureña de la Industria de la Construcción*). The architectural plans are organised to optimise the functional distribution of spaces, including living areas, a dining room, a kitchen, bedrooms and bathrooms. Particular attention has been given to Lenca architectural heritage, for example by incorporating wide corridors that serve not only as circulation spaces, but also as semi-outdoor living, transition and sun protection spaces. These elements are highly valued in the local tradition. Cross and longitudinal sections of the project illustrate spatial qualities such as generous ceiling heights, which promote warm air stratification and ease convection ventilation. The sections also demonstrate how solar shading elements are integrated and how natural ventilation paths are configured.

Figure 14. New proposal for progressively growing Social Housing.

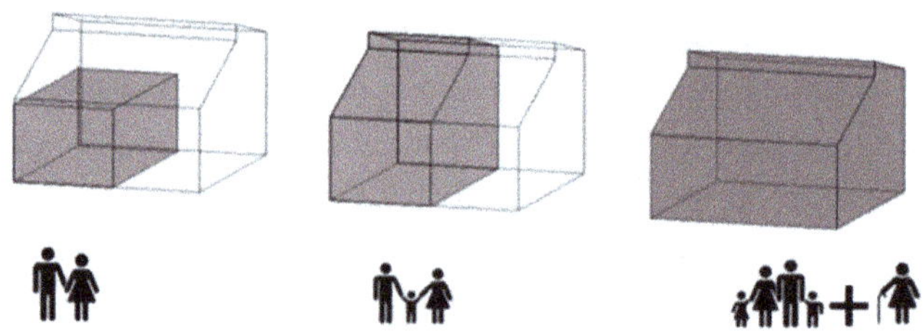

Proposal of a single level: 1 bedroom apartment for a couple. 25 m². L-shaped corridor.

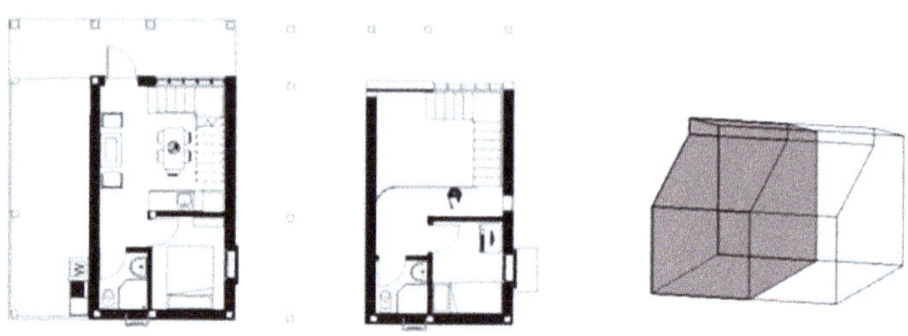

Two-level, 2-bedroom proposal, 3 people, 40 m2. L-shaped corridor.

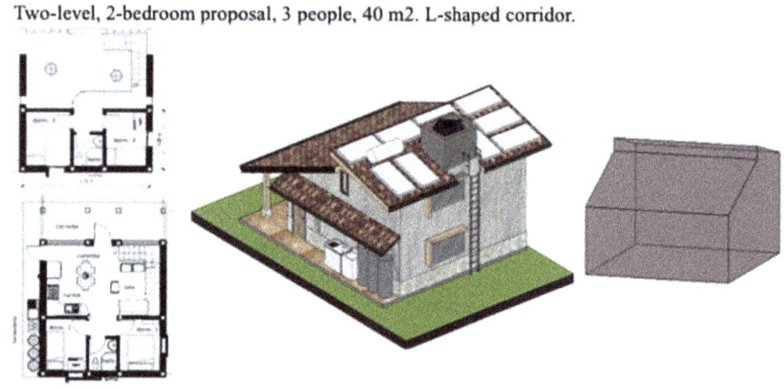

Two-level, 4-bedroom proposal, 4-5 people, 65 m². L-shaped corridor.

Source: authors.

- **Materials and Building construction systems specifications:**

Walls: The proposal is to use locally manufactured 40 cm-thick adobe bricks with a plastered finish on both sides, made from a mixture of earth and lime. This improves the bricks' durability and aesthetic appeal. The walls will rest on a stone masonry foundation, approximately 1 metre high, to protect them from soil moisture.

Roof: Traditional baked clay tiles mounted on a seasoned pine wood frame. To improve thermal performance, the roof incorporates a 15 cm ventilated air chamber and R19 thermal insulation. An interior suspended ceiling made of reeds, or a similar material adds aesthetic value and contributes to insulation and indoor micro-ventilation.

Floors: For the interior, fired clay tiles have been selected. These are a locally sourced material with good thermal inertia and will be laid on top of a layer of extruded polystyrene insulation (XPS) and a thin concrete slab.

Foundation: A cyclopean masonry foundation is proposed, made from local stone set in a mortar of sand, lime and cement in the correct proportions to ensure stability.

Windows: They are designed with operable wooden or similar lattices that allow precise control of ventilation and light entry. Mosquito nets are also provided for comfort and health.

Doors: Made of local wood with designs that allow them to open in half (top or bottom), improving ventilation while maintaining privacy and security, as seen in some traditional buildings.

- **Bioclimatic strategies integrated into design:**

Passive Solar Design: The orientation of the home and its openings is determined through a solar analysis of the site to optimise natural lighting and minimise excessive heat gain. Fixed or mobile wooden sunshades are incorporated to protect the most exposed facades and are sized according to shade mask calculations.

Natural ventilation: The design promotes cross ventilation through opposing openings and the chimney effect (convection), making use of the interior heights and roof design to expel hot air.

Thermal mass: The thick adobe walls and clay floors significantly contribute to the thermal inertia of the building envelope, stabilising interior temperatures.

Natural Lighting: The layout and size of the windows are optimized to maximize the entry of diffused natural light, reducing the need for artificial lighting during the day, without compromising thermal control.

Habitable outdoor spaces: Corridors and patios act as thermal transition zones and provide shade, creating pleasant and protected outdoor spaces for a variety of family and social activities.

Figure 15. Cross-sections of the proposal for progressively growing Social Housing.

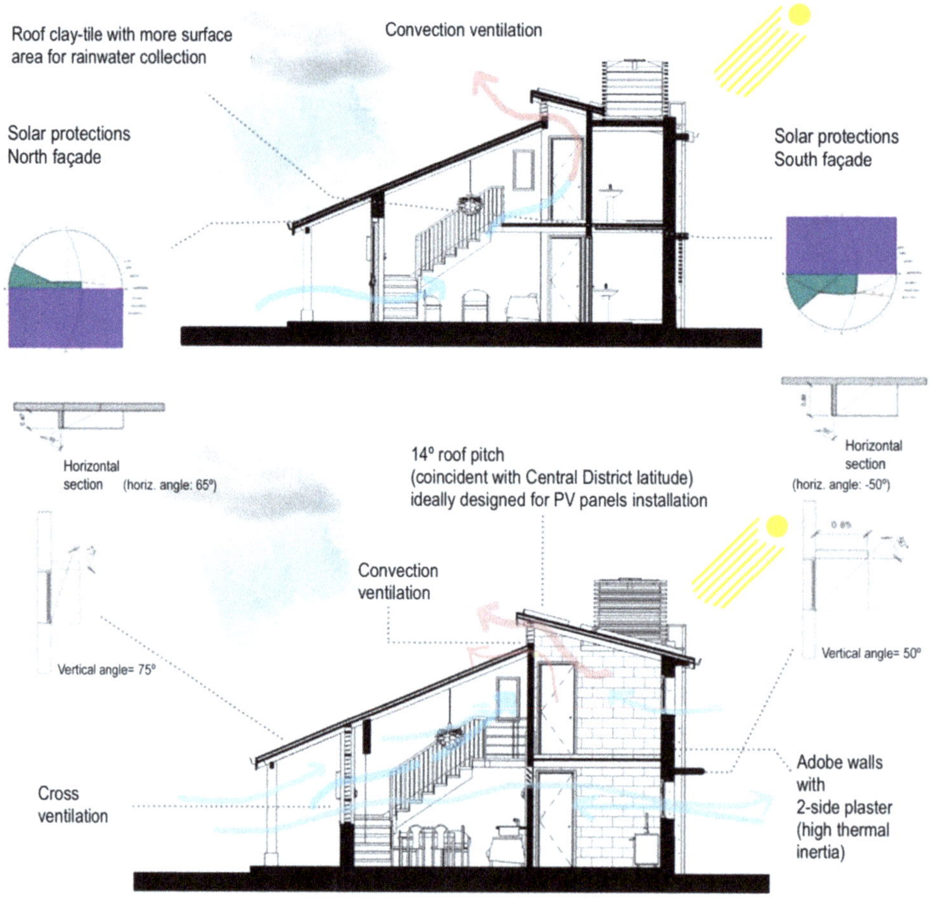

Source: authors.

• **Integration of ecotechnologies:**

Rainwater Harvesting Systems: Storage tanks are planned to collect and store rainwater, reducing dependence on the public network and promoting the efficient use of water resources.

Solar water heating: Installing solar thermal panels to supply domestic hot water is being considered as a way of reducing conventional energy consumption.

Photovoltaic solar energy: The design considers the feasibility of incorporating photovoltaic solar panels to generate electricity, with the aim of achieving greater energy autonomy and reducing operating costs.

Elevated water tank: Ensures a gravity-fed water supply, even in the event of utility service interruptions.

These features are visually presented in the axonometric view and diagrams of the building construction system (Figure16), which makes the proposal easier to understand. The 'progressive' nature of this housing model is fundamental. It directly addresses the dynamic needs of low-income families who often have to make informal and unstable additions to standard social housing, which quickly becomes insufficient. The proposal reflects a deep understanding of how homes evolve by planning for growth in a manner that is safe, resource-efficient and consistent with the original design principles. This principle is also observed in the way traditional housing adapts and grows organically over time.

Figure 16. Cross-sections of the proposal for progressively growing Social Housing.

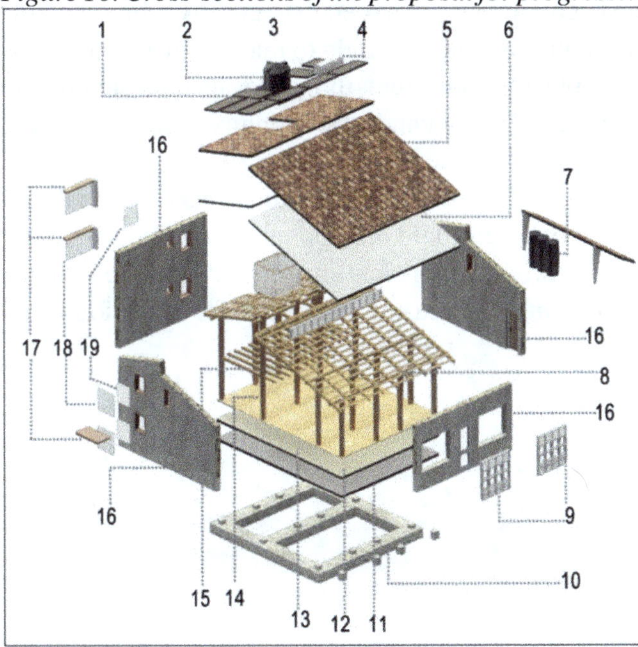

1. PV solar panels
2. 2,500-litre elevated tank (5-day)
3. Solar panel for water heating
4. Hot water storage tank
5. Fired-clay tiled roof
6. Thermal insulation (R19)
7. Rainwater storage tanks
8. Pine wood structure, tiled roof with 2"x6" beams and 1"x3" battens.
9. Lattice windows
10. Foundation: 60% stone, 30% sand, 10% cement and lime
11. Select material foundation.
12. XPS thermal insulation
13. Floor fired-clay tiles
14. 4"x4" pine wood columns
15. Pine wood mezzanine with 2"x6" beams and 1"x10" planks.
16. Plastered adobe walls on both sides with a 1-metre, 40-cm-thick stone wall foundation.
17. Wooden sunshades (dimensions based on sun protection calculations).
18. Lattice windows.
19. Mosquito-proof metal mesh.

Source: authors.

Site Planning and Community Integration

The success of a Social Housing proposal depends not only on the design of the individual units themselves, but also on how they are grouped together and integrated into the wider community and environmental context. Site planning for this vernacular bioclimatic housing model should consider the following:

- Plot layout and orientation: The optimal orientation of each unit to maximise passive solar energy and natural ventilation should be a guiding principle in the subdivision's design. We will seek to learn from Lenca settlement patterns, which often favour community life and a relationship with the natural environment without the rigid property divisions of modern urban developments.
- Creation of communal spaces: The design of the complex should include shared green spaces and recreational and gathering areas that foster social interaction and a sense of community, aspects that are valued in Lenca culture.
- Integration of vegetation: Strategic planting of trees and other native vegetation around homes and in communal areas is encouraged. This improves the complex's aesthetics and contributes significantly to microclimatic control by providing shade, reducing ambient temperatures through evapotranspiration and acting as a barrier against cold or dusty winds. Community or family gardens for food production could also be considered.
- Sustainable water management on the site: In addition to harvesting rainwater at household level, the site design could incorporate Sustainable Drainage Systems (SUDS) to manage stormwater, prevent erosion and recharge aquifers. Community systems for treating and reusing greywater for irrigation could also be explored.

Figure 17. Exterior view of the progressive Social Housing proposal (final stage) and its environment.

Source: authors

A collaborative design approach that involves future residents in planning community spaces and defining environmental features can strengthen their sense of belonging and ensure the final design meets their needs and aspirations. When social housing is conceived in this way, it becomes a catalyst for building resilient and sustainable communities rather than just a collection of houses.

Comparisons With Regional Good Practices and Case Studies

The proposal for vernacular-bioclimatic Social Housing in Honduras forms part of a wider movement across Latin America that seeks to revive earth construction techniques and climate-adapted design principles in response to housing and environmental issues. Examining successful case studies in similar socio-climatic contexts reinforces the viability and relevance of this approach.

Mexico: The 'Adobe Housing Prototype with Renewable Energy' project in Raíces, State of Mexico, was developed for a marginalised population living in a protected natural area. It demonstrates how technified adobe, bioclimatic design (cross-ventilation, insulation, and sunscreens) and renewable energy (solar and wind) can be used to improve living conditions and reduce environmental impact. This project is stands out for its self-construction system and the use of a basic 40 m^2 module that can be adapted (Méndez & Pineda, 2017).
Nicaragua: The low-income housing construction project in the *El Pantanal* neighbourhood of Granada involved stabilising adobe bricks with bamboo fibres. Laboratory tests demonstrated that the bamboo fibre-enhanced adobes exceeded regulatory standards for compressive strength. The project also involved community members, particularly women, in the construction process, thereby transferring technical skills (Vásquez et al., 2020).
Peru: Research on the 'Bioclimatised Rural Housing Module (*Vivienda Rural Bioclimatizado*, VRB)' in the high Andean region of Puno, which uses local adobe and a passive solar strategy (attached greenhouse), showed an improvement of 37% in indoor temperature compared to the outdoor temperature, and an improvement of 27% compared to a typical adobe house. This reached comfort levels according to the ASHRAE 55 standard (Rincón et al., 2020).
Colombia: The '*ECO-FUTURO* Prototype' is a prototype for sustainable bioclimatic Social Housing in Quibdó for an Afro-descendant community in a warm, humid climate. It proposes the use of local wood (*'cativo'*) and bamboo (*'barilaque'*), with an elevated design to prevent humidity, internal patios for ventilation and flexible adaptation to different family sizes. The project aims to reduce energy consumption by 30–50% and significantly improve quality of life (Correa Giraldo et al., 2014).
Honduras: Architect Angela Stassano's projects, such as *'Las Casitas'* in San Pedro Sula, apply bioclimatic principles and local materials to create energy-efficient housing in the country's tropical climate. This work serves as an important local benchmark for the viability of these approaches (ArchDaily, n.d.).

These examples, alongside many others in the region, demonstrate the growing trend of revaluing earth construction and bioclimatic design (Neila González, 2004). These examples share common principles, such as using local, low-impact materials; adapting to the climate; pursuing energy efficiency; and, in many cases, encouraging community participation. Situating the Honduran proposal within this Latin American context not only reinforces its credibility, but also demonstrates that it is part of a regional trajectory towards more equitable, sustainable, and culturally relevant housing solutions.

EXPANDED IMPACTS: SOCIOECONOMIC, ENVIRONMENTAL AND POLITICAL DIMENSIONS

Adopting and promoting a social housing model based on vernacular and bioclimatic principles, as proposed here, could have a positive impact that extends far beyond simply providing housing. These repercussions can be seen in the economic, environmental, social, and political spheres.

Economic Benefits and Local Development

One of the most significant advantages of this approach is its potential to promote local and endogenous economic development. By prioritising the use of locally sourced materials such as adobe, clay tiles, sustainably harvested wood and stone, the project reduces reliance on industrial materials, many of which are imported and subject to the volatility of international markets. This translates into lower initial construction costs, particularly when transportation costs and the embodied energy of conventional materials are considered and ensures that a greater proportion of housing investment remains and circulates within the local economy.

The production and processing of vernacular materials, such as adobe and tiles, stone quarrying and carving, and traditional carpentry, stimulates local economies by supporting artisans, small producers, and family workshops, which often struggle to compete with mass industrial production. These activities generate employment in rural and peri-urban communities where these resources and skills are present. Furthermore, while labour-intensive, construction using these techniques offers training and employment opportunities for local workers, helping to preserve traditional trades that are at risk of disappearing.

In the long term, the inherent energy efficiency of well-designed bioclimatic homes translates into significantly lower operating costs for families, primarily due to reduced or eliminated reliance on air conditioning or artificial heating. The resulting savings on energy bills free up financial resources that low-income families can allocate to other basic necessities, such as food, healthcare or education. Therefore, shifting towards vernacular bioclimatic social housing is not just a construction strategy; it is also a tool for sustainable local economic development that creates value in the community and improves the economic resilience of its inhabitants.

Environmental Sustainability and Climate Resilience

From an environmental perspective, the benefits are equally substantial. Building houses with locally sourced, natural materials such as sustainably managed earth and wood results in a considerably lower carbon footprint than conventional construction methods. The production of cement, steel and fired bricks is highly energy-intensive and generates significant CO_2 emissions. In contrast, sun-dried adobe and wood have much lower embodied energy.

Resource efficiency is another key advantage. These materials are often renewable, such as the wood and straw used in adobe bricks or bahareque, or easily recyclable — the earth in adobe bricks, for example, can be reused indefinitely. This reduces the pressure on ecosystems resulting from the massive extraction of non-renewable raw materials.

In the context of climate change, the consequences of which are already being keenly felt in Honduras in the form of more intense hurricanes, floods and droughts, vernacular bioclimatic architecture offers greater resilience. The thermal stability provided by earth walls helps homes to buffer extreme temperature fluctuations. Passive design is inherently more robust against power outages, which are common during and after natural disasters, as it does not rely on complex mechanical systems or a constant power supply. Furthermore, when applied correctly and properly reinforced, some traditional construction techniques (e.g. well-designed *wattle and daub*) can offer good seismic response. Therefore, promoting these construction solutions is a climate change adaptation strategy that reduces community vulnerability.

Social and Cultural Benefits

Adopting a vernacular-bioclimatic approach to social housing has profound and multifaceted social and cultural impacts. Firstly, there is a tangible improvement in the living conditions of occupants. The superior thermal comfort resulting from passive temperature regulation and improved indoor air quality — due to the breathability of natural materials and adequate ventilation — has a direct positive effect on health. This could potentially reduce the incidence of respiratory diseases and other conditions associated with damp, poorly ventilated homes, or extreme temperatures.

This type of architecture can play a crucial role in strengthening cultural identity and community empowerment, beyond physical comfort. Using designs, materials and techniques that stem from local traditions - the Lenca tradition in this case - validates and gives continuity to a Cultural Heritage that has often been marginalised or undervalued. This can foster a greater sense of pride, belonging and ownership of homes among their inhabitants.

Building with earth and traditional techniques often lends itself to community participation in the construction process. Similar projects in Latin America - e.g. Nicaragua, (Vásquez et al., 2020)- have shown that involving future residents in the construction of their homes can reduce costs, transfer valuable skills, strengthen social cohesion and ensure the homes better respond to the community's needs and preferences. Thus, social housing ceases to be an externally imposed product and becomes a dignifying and empowering co-creation process.

Political and Regulatory Transformation: Scalability

Despite its many proven advantages, the widespread adoption of vernacular and bioclimatic architecture in Honduras's Social Housing sector faces significant political and regulatory obstacles. Current building Codes and Regulations often fail to recognise traditional materials and techniques such as adobe and *wattle and daub*, implicitly or explicitly favouring conventional industrialised materials and methods. The BID report (Serrano et al., 2014) confirms this shortcoming by highlighting the absence of sustainability and energy efficiency parameters in Honduran Codes. While the 'Green Regulation for the Central District' (*Reglamento Verde para el Distrito Central*) raises awareness of sustainable construction, its non-mandatory nature and lack of robust oversight severely limits its impact. For vernacular-bioclimatic solutions to be implemented on a large scale and contribute significantly to alleviating the housing shortage, a profound transformation of this regulatory and policy framework is essential.

Table 5. Proposed policy measures to promote vernacular and bioclimatic social housing in Honduras.

Policy Area	Proposed Specific Measure	Key Stakeholders Involved	Expected Outcome
Regulatory and Normative Framework	Develop and formalize national technical standards for earthen construction (adobe, rammed earth, bahareque) and other vernacular materials, including seismic-resistant and bioclimatic performance criteria.	SEVIAH, Professional Associations (Engineers, Architects), Universities, IHAH, PVAH.	Legalization and standardization of vernacular construction techniques, facilitating their approval and financing.
	Mandatorily incorporate energy efficiency and bioclimatic design criteria into building codes for all new social housing. Update and make the Green Regulation of the DC binding.	AMDC, SEVIAH, PVAH.	Reduction of energy consumption in the housing sector, improvement of thermal comfort.
Financial and Economic Incentives	Create preferential credit lines and specific subsidies (green bonds) through BANHPROVI and FONHVIVIENDA for social housing projects that use vernacular materials and bioclimatic design.	BANHPROVI, FONHVIVIENDA, SEFIN (Secretariat of Finance), PVAH.	Increased financial viability and attractiveness of sustainable housing projects for developers and families.
	Establish tax exemptions or reductions (property tax, building permits) for homes that meet high standards of sustainability and energy efficiency.	Municipalities, National Congress, SEFIN.	Reducing the total cost of sustainable housing to encourage its adoption.
Capacity building and training	Implement training and certification programs for architects, engineers, builders, and artisans in sustainable vernacular construction techniques and bioclimatic design.	Universities (UNAH), Vocational Technical Institutes, Professional Colleges, IHAH, NGOs.	Increased availability of professionals and skilled labor, improving construction quality.
Research and development	Fund applied research on the performance and optimization of local vernacular materials and techniques, and the development of prototypes adapted to different regions of Honduras.	Universities, Research Centers, IHAH, PVAH.	Generation of local knowledge, innovation in construction solutions, adaptation of techniques to diverse contexts.
Public awareness and promotion	Launch information and awareness campaigns to highlight the benefits (comfort, health, economy, culture, environment) of vernacular-bioclimatic housing and combat prejudices about its quality or modernity.	PVAH, SEVIAH, Media, NGOs, Community Leaders.	Increased demand and social acceptance of this type of Housing.

continued on following page

Table 5. Continued

Policy Area	Proposed Specific Measure	Key Stakeholders Involved	Expected Outcome
Integración Institucional y Proyectos Piloto	Mandate the PVAH and other social housing implementation bodies to develop and implement pilot projects for vernacular bioclimatic housing on different scales. These projects should be documented, and the results disseminated.	PVAH, SEVIAH, Municipalities.	Demonstration of practical viability and benefits, generation of replicable models, and institutional learning.
	Promote the creation of local material banks and networks of artisanal producers to ensure the supply and quality of local inputs.	Cooperatives, Producer Associations, NGOs, Chambers of Commerce.	Strengthening local value chains, improving access to quality materials.

Source: authors own elaboration.

These policy proposals are not just technical; they imply a shift in mindset that recognises the intrinsic value of local knowledge and the need for development that is more harmonious with the environment and culture. The transition towards more sustainable social housing in Honduras requires a concerted effort from a variety of stakeholders, including central and local government, the private sector, academia, and the communities themselves. Without this systemic shift in Policies and Regulations, innovative, contextually appropriate solutions such as those derived from this study risk remaining isolated experiments rather than becoming the norm.

CONCLUSIONS: RESCUING TRADITION IN THE FUTURE HONDURAN HOUSING

This chapter has examined the critical situation of Social Housing in Honduras's Central District in depth. This situation is characterised by a significant quantitative and qualitative deficit, which is exacerbated by conventional construction practices that are often unsuitable for the local climate and contribute to socio-environmental vulnerability. In contrast, this chapter has explored the rich legacy of Lenca vernacular architecture, identifying its materials, techniques, and design principles as a valuable source of sustainable, culturally rooted solutions. Dynamic energy simulations using Design Builder software have provided a scientific validation of the bioclimatic superiority of Lenca-inspired housing typologies. Specifically, the 'improved vernacular dwelling' model demonstrated a reduction in annual cooling demand and associated electricity consumption of up to 60%, compared to homes constructed with concrete blocks and aluzinc sheets, which are common in the current social housing landscape. Translated into an estimated energy rating, these findings place the vernacular proposals in the highest efficiency category (A), whereas the conventional designs only achieve a modest rating (C). Based on this analysis, a design proposal for progressive social housing has been presented which integrates not only vernacular bioclimatic materials and strategies, but also responds to the need for flexibility and adaptability for low-income families over time.

The evidence presented throughout this chapter conclusively supports the central argument that a paradigm shift in the design and provision of Social Housing in Honduras is imperative. Conventional models, which are often imported and poorly adapted, must be replaced with solutions that are deeply rooted in local construction traditions, environmental materials, and a sensitive understanding of the regional climate. Far from being a relic of the past, vernacular architecture emerges as a fundamental path towards genuine sustainability, the preservation of cultural identity and a substantial improvement in the quality of life of the most vulnerable populations. Rescuing and intelligently adapting this ancestral knowledge offers efficient technical solutions and promotes a more equitable, resilient, and respectful development model for Honduras' cultural and environmental diversity.

Transforming social housing to make it more sustainable and culturally relevant requires the commitment and coordinated action of various stakeholders:

- For policymakers and government agencies (such as PVAH, SEVIAH and AMDC): It is crucial to review and update regulatory frameworks and Building Codes in order to recognise and validate vernacular materials and techniques. Financial and fiscal incentives should be implemented to actively promote vernacular bioclimatic construction in national social housing programmes, with these approaches being prioritised rather than merely considered as alternatives.
- For architects, engineers, and urban planners: They have a professional responsibility to reconnect with traditional knowledge, innovate with local materials and design buildings and communities that respond sensitively to climate, culture and the real needs of users. They should prioritise human well-being and the health of the planet over ephemeral aesthetic trends or short-term economic efficiency.
- For academic and research institutions (such as UNAH and IHAH): It is essential to intensify the research, documentation and teaching of vernacular architecture, bioclimatic design and sustainable construction technologies adapted to the Honduran context. Training new generations of professionals in these skills is crucial for the future.
- For local communities, artisans, and builders: Their active participation is essential. Specific spaces must be created to enable the transmission of traditional knowledge and skills, and to empower communities to take ownership

of the design and construction processes of their homes. This will ensure that these reflect their identity and meet their aspirations.

Although this study is focused on the Central District of Honduras and the Lenca tradition, the principles and methodologies presented here have much broader applicability. The lessons learned regarding the valuation of local knowledge, climate adaptation and the use of local materials could inspire similar solutions in other regions of Honduras with different cultural and climatic contexts, as well as in other developing countries that are facing comparable Housing and sustainability challenges. Successful implementation of the proposed housing model could set a powerful precedent by demonstrating that sustainable, culturally relevant solutions are often found by looking to local wisdom and adapting it creatively to the present.

To consolidate and expand this approach, the following directions for future research are suggested:

- Post-occupancy monitoring and evaluation: Conduct long-term follow-up studies of the thermal, energy and structural performance of constructed vernacular bioclimatic housing prototypes and assess occupant satisfaction and well-being.
- Life Cycle Assessment (LCA): Develop complete life cycle assessments for the proposed vernacular housing typologies to quantify their total environmental impact more accurately, from material extraction to final disposal compared to conventional alternatives.
- Socio-cultural research: Investigate the socio-cultural acceptance, adaptation and appropriation of these dwellings by communities in more depth and explore effective community participation models in the design and construction processes.
- Development of local value chains: Research and promote the strengthening of value chains for vernacular construction materials to improve their availability, quality, and market competitiveness.
- Adaptation to diverse contexts: Investigate the application of vernacular bioclimatic principles to different ethnic groups, ecoregions within Honduras and building types beyond housing.

Ultimately, incorporating the rich tradition of construction into the future of Honduran housing is not only desirable, but a pressing necessity to build a more just, resilient, and sustainable habitat for all.

REFERENCES

Agurcia, R., & Soto, R. (1984). La Casa de Don Calecho: Un ejemplo de la arquitectura vernácula en Honduras. *Yaxkin*, 7(1), 47–59. https://cdihh.ihah.hn/revistayaxkin/1984_01/LaCasadeDonCalechoUnEjemplodelaArquitecturaVernaculaenHonduras.pdf

ArchDaily. *Architecture from Honduras*. Last access May 27, 2025. https://www.archdaily.com/country/honduras

Castañeda Moncada, M. A. (2012). *Construyendo Pueblitos, Arquitectura para comunidades pequeñas*. Castañeda Moncada, Mauricio Alejandro. (author website: http://www.macastaneda.com/)

Center for Gender & Refugee Studies (CGRS). (2023). *Honduras: La Emergencia Climática como Motor de Desplazamiento y Violaciones de Derechos Humanos*. CGRS, UC Law SF. Last access May 7, 2025. https://cgrs.uclawsf.edu/sites/default/files/Honduras%20Report_12.18.2023_FINAL_SP.pdf

Chapman, A. (1985-1986). *Los hijos del copal y la candela* (Vols. I & II). UNAM/CEMCA.

Correa Giraldo, V. M., Queiros, M., Ordoñez Candelaria, V. R., López Muñoz, L. F., Flores Méndez, E., & Zapata López, J. (2014, November). *El bahareque: Un sistema constructivo sismoresistente y sustentable para soluciones de vivienda social en México. XIX National Congress of Structural Engineering*, Puerto Vallarta, Jalisco, Mexico.

Fathy, H. (1973). *Architecture for the Poor: An Experiment in Rural Egypt*. University of Chicago Press. DOI: 10.7208/chicago/9780226239149.001.0001

Gaceta, L. (2020, 18 de febrero). *Ley Marco de Vivienda y Asentamientos Humanos de Honduras* (Decreto legislativo n.º 173-2019, La Gaceta n.º 35 178). Diario Oficial de la República de Honduras. Last access May 7, 2025. https://enag.sgjd.gob.hn/mfhandler.php?field=archivo&file=20200218+-+DECRETO+LEGISLATIVO++No.+173-2019+-+publicado+en+el+Diario+Oficial+La+Gaceta+35178.pdf

Government of Honduras. *Planes Estratégicos y Operativos del PVAH*. Unified Transparency Portal of Honduras. Last access May 27, 2025. https://portalunico.iaip.gob.hn/424/

Habitat for Humanity Honduras. (2017). *Barreras de acceso al suelo para vivienda social en Honduras*. Last access May 2, 2025. https://www.habitathn.org/wp-content/uploads/2017/08/Barreras-de-Acceso-al-Suelo-para-Vivienda-Social-en-Honduras.pdf

Honduran Chamber of the Construction Industry (CHICO). (2024). *Datos sobre el déficit habitacional*. El Heraldo/Tiempo.hn. Last access May 27, 2025. https://www.elheraldo.hn/honduras/deficit-vivienda-honduras-casi-un-millon-personas-sin-casa-AH20274756 and https://tiempo.hn/de-un-1-millon-600-mil-viviendas-es-el-deficit-habitacional-honduras/

Honduran Institute of Anthropology and History (IHAH). *Publicaciones sobre etnoarquitectura y cultura Lenca*. Last access May 20, 2025. https://portalunico.iaip.gob.hn/410/

Inter-American Development Bank (IDB). (2023). *Resiliencia y sostenibilidad en los códigos de construcción de América Latina y el Caribe: análisis regional comparativo y oportunidades de fortalecimiento*. IDB. Last access May 2, 2025. https://publications.iadb.org/publications/spanish/document/Resiliencia-y-sostenibilidad-en-los-codigos-de-construccion-de-America-Latina-y-el-Caribe-analisis-regional-comparativo-y-oportunidades-de-fortalecimiento.pdf

Méndez, M., & Pineda, L. (2017). Estabilización de adobes con fibra de bambú para la construcción de viviendas de interés social en el barrio "El Pantanal", Granada. *Revista de Arquitectura*, 13(1), 18-29. https://camjol.info/index.php/arquitectura/article/download/8981/10145/31194

Municipal del Distrito Central, A., & de Ordenamiento Territorial, D. (2020). *Reglamento verde para la construcción.* [Acuerdo municipal n.° 024, Acta n.° 016 del 14 de diciembre de 2020]. *La Gaceta: Diario Oficial de la República de Honduras*, 35 (493), 1-43. Last access May 22, 2025. https://www.tsc.gob.hn/web/leyes/Acuerdo-024-AMDC.pdf

Muñoz Torres, L. C. (2015). *Análisis tipológico y bioconstructivo de las viviendas vernáculas en el municipio San Antonio de Oriente, Honduras* [thesis, Zamorano University, Escuela Agrícola Panamericana]. https://bdigital.zamorano.edu/handle/11036/4550

National Congress of Honduras. (2020). *Ley Marco de Vivienda y Asentamientos Humanos* (Decree No. 173-2019). Last access May 7, 2025. https://dinero.hn/congreso-nacional-aprobo-ley-marco-de-vivienda-y-fondo-para-construccion/

Neila González, F. J. (2004). *Arquitectura Bioclimática en un entorno sostenible*. Munilla-Lería Publisher.

Oliver, P. (2006). *Built to meet needs: cultural issues in vernacular architecture*. Architectural Press. Last access May 2, 2025, https://www.routledge.com/Built-to-Meet-Needs-Cultural-Issues-in-Vernacular-Architecture/Oliver/p/book/9780750666572

Presidency of the Republic of Honduras. (2022). *Creación del Programa de Vivienda y Asentamientos Humanos (PVAH)* (PCM-005-2022). Last access May 27, 2025. https://pvah.gob.hn/

Rincón, J., Manrique, A., & Galvis, F. (2020). Prototipo de vivienda social bioclimática y sostenible en Quibdó, Chocó, Colombia. *Revista de Arquitectura (Bogotá)*, 22(1), 86-98. https://repository.ugc.edu.co/server/api/core/bitstreams/94473d6f-3f53-4428-bc40-952ac9dc1f98/content

Rudofsky, B. (1964). *Architecture Without Architects: A Short Introduction to Non-Pedigreed Architecture*. Museum of Modern Art. Last access May 2, 2025. https://www.blinkist.com/en/books/architecture-without-architects-en

Salinas, I. M. (1991). *Arquitectura de los grupos étnicos de Honduras*. Guaymuras Publishing.

Serrano, H., Barba, E., & Del Real, C. (2014). Prototipo de vivienda de adobe con energías renovables. *Ra Ximhai, 10*(4), 187–200. https://dialnet.unirioja.es/descarga/articulo/5035106.pdf

UNIAPRAVI. (2020). *Datos del déficit habitacional*. Last access May 2, 2025. https://www.uniapravi.org/objetos/publicacion/MTE=/04092024111559.pdf

United Nations Environment Programme & Global Alliance for Buildings and Construction. (2025). *Global Status Report for Buildings and Construction 2024/2025: Not just another brick in the wall*. Last access May 2, 2025. https://www.unep.org/resources/report/global-status-report-buildings-and-construction-20242025

Urban and Cities Platform - ECLAC. *Derecho a la vivienda en Honduras*. Last access May 2, 2025. https://plataformaurbana.cepal.org/es/derecho-la-vivienda-en-honduras

Vásquez, V., Puyen, Z., & Valdivia, P. (2020). *Módulo de vivienda rural bioclimatizado (VRB): mejora del confort térmico en la zona altoandina de Puno, Perú*. INUDI Publishing. Acccesed May 27, 2025. https://editorial.inudi.edu.pe/index.php/editorialinudi/catalog/book/30

KEY TERMS AND DEFINITIONS

Adobe: A traditional earthen building material made from clay-rich soil, sand, straw, and water, moulded into bricks and sun-dried rather than kiln-fired. Adobe walls provide high thermal mass, buffering indoor temperatures against daytime heat and night-time cooling, thereby improving comfort and reducing reliance on mechanical conditioning.

Bioclimatic design: An architectural approach that seeks to achieve optimal thermal comfort for occupants and maximum energy efficiency by working with the local climate. This is done by utilising passive strategies such as solar orientation, natural ventilation, thermal mass and shading in order to minimise dependence on mechanical systems.

Gable roof: A roof form consisting of two pitched planes that meet along a central ridge, creating a triangular gable at each end wall. The simple geometry efficiently sheds rain and snow, allows generous attic ventilation, and facilitates straightforward framing, making it a cost-effective solution for diverse climates—especially where seasonal precipitation demands rapid runoff.

Hip roof: A roof form in which all sides slope downwards from a common ridge toward the exterior walls, eliminating vertical gable ends. The inward-leaning planes enhance aerodynamic stability under high winds, distribute structural loads uniformly, and provide generous eave overhangs for consistent shading and rain protection, suiting hurricane-prone or exposed sites.

Lenca Architecture: A set of traditional construction practices of the Lenca indigenous people of Honduras and El Salvador, typically characterized using materials such as adobe, bahareque, wood, and clay tiles, with designs adapted to the regional climate and cultural norms.

Progressive housing: A housing model designed to be built in stages, allowing occupants to gradually expand or improve their homes based on their evolving needs and financial circumstances, offering flexibility and adaptability.

Qualitative housing deficit: The extent to which existing housing units fail to meet basic standards of habitability, structural safety, access to services (e.g. water, sanitation and electricity) and security of tenure, regardless of whether a physical structure exists.

Quantitative Housing Deficit: The numerical disparity between the number of existing adequate housing units and the number of households requiring adequate, independent housing.

Social Housing: Affordable housing provided by the government or non-profit organisations for low-income individuals and families who cannot access adequate housing on the open market. It often addresses issues of housing shortage and quality.

Thermal Mass: The property of a building material to absorb, store and subsequently release significant amounts of heat energy. Materials with high thermal mass, such as adobe or stone, help to moderate indoor temperature fluctuations.

Vernacular architecture: A set of building traditions that have developed over time. They are characterised using local materials and techniques in response to the specific climatic conditions and cultural practices of a particular region or community. These traditions often emerge without the intervention of formal architects.

Wattle and daub: A composite wall construction method that weaves flexible wooden strips (wattle) into a lattice and then coats both sides with a viscous mix of earth, clay, straw, and sometimes lime (daub). The intertwined lattice offers tensile strength while the earthen render supplies thermal mass and airtightness, yielding a low-cost, locally sourced envelope with moderate insulation and moisture-regulating properties.

Chapter 7
Sustainable Thermal-Acoustic Panels from Agricultural and Paper Waste: Alternative Materials for Construction and Environmental Engineering

Siwat Lawanwadeekul
https://orcid.org/0000-0002-0827-0078
Lampang Rajabhat University, Thailand

ABSTRACT

This chapter presents the development of binderless thermal-acoustic insulation panels using rice straw, water hyacinth, and wastepaper as locally sourced waste materials. Utilizing Persea kurzii powder as a natural adhesion enhancer, the panels were produced without synthetic binders or energy-intensive curing, aligning with sustainable manufacturing. Thermal conductivity values ranged from 0.040 to 0.067 W/m·K, and Noise Reduction Coefficients (NRC) between 0.48 and 0.51, comparable to conventional materials like expanded polystyrene (EPS) and glass wool. Economic analysis showed production costs 25–40% lower than imported synthetic options. Additional environmental advantages include biodegradability, low embodied carbon, and compatibility with circular economy models. The chapter discusses performance data, material economics, and long-term sustainability implications, concluding with recommendations for life cycle assessment and policy integration to enable broader market adoption.

DOI: 10.4018/979-8-3373-2414-2.ch007

1. INTRODUCTION

The construction sector plays a pivotal role in the global transition toward sustainability (Marconi et al., 2020; Ali et al., 2022). Responsible for nearly 40% of global carbon emissions and approximately 36% of global energy consumption, it exerts significant pressure on the planet's ecological balance (Ricciardi et al., 2021; Tiza, 2023; Zucaro & Agostinho, 2025). With the escalating urgency of climate change, resource depletion, and environmental degradation, the demand for innovative and environmentally responsible construction practices has become increasingly critical. Within this context, conventional building materials—such as concrete, gypsum boards, and synthetic insulation—are under scrutiny due to their substantial environmental footprints (Bamigboye et al., 2019). These materials are not only resource-intensive to produce but also contribute to long-term environmental burdens through high emissions, energy consumption, and post-use waste generation.

Beyond emissions, the construction industry remains entrenched in a linear resource consumption model—extract, manufacture, build, demolish, and discard—which perpetuates unsustainable patterns of material use (Dixit et al., 2010; Roque et al., 2021). Concurrently, millions of tons of agricultural and industrial waste are generated globally each year, much of which remains underutilized or is discarded through harmful practices such as open-air burning and landfill disposal. This dual crisis of material inefficiency and waste accumulation necessitates solutions that transcend simple material substitution, instead embracing the principles of the circular economy, wherein waste is transformed into valuable resources for regenerative production systems (Ben Hassen et al., 2025).

Developing countries such as Thailand exemplify this convergence of rapid construction growth and substantial agricultural waste generation. As an agriculturally driven economy, Thailand produces approximately 25 million tons of rice straw annually (Land Development Department, 2025), much of which is openly burned, contributing significantly to hazardous PM2.5 air pollution. In parallel, invasive aquatic weeds such as water hyacinth (Eichhornia crassipes) proliferate across national waterways, worsening environmental degradation and impeding water management systems (Philip & Rakendu, 2020; Jirawattanasomkul et al., 2021). Compounding these challenges, the country's vibrant artisanal sectors, including handmade paper and packaging industries, generate considerable waste streams that often lack formal recycling systems or valorization pathways (Yeon et al., 2014).

However, these environmental burdens can be reimagined as latent resources for sustainable innovation. Agricultural and industrial by-products—particularly those rich in cellulose and low in lignin—offer significant potential for engineered composite materials (Elsacker et al., 2020; Ojo, 2023). Over the past decade, researchers have increasingly explored the use of waste biomass—such as rice husk, jute, coconut coir, and banana peels—for developing building panels, boards, and bio-based thermal and acoustic insulation materials (Saxena et al., 2008; Hasan et al., 2021). Nevertheless, most bio-based materials still rely on chemical adhesives such as urea-formaldehyde, melamine, or phenolic resins, compromising environmental integrity by emitting volatile organic compounds (VOCs) and introducing toxicity into indoor environments (Mulder et al., 2011; Maiti et al., 2020; Mirindi, 2024).

Why The Circular Economy Matters in Construction

Life Cycle Assessment (LCA) studies have increasingly demonstrated that sustainability construction materials can significantly reduce embodied carbon and energy use compared to conventional products (Abera, 2024). These tools enable transparent comparisons across a product's lifecycle and are essential for validating circularity and sustainability claims. The construction industry's dependence on virgin raw materials has led to the depletion of critical resources such as sand and limestone, while exacerbating biodiversity loss and land degradation (Zhong et al., 2022). Transitioning toward a circular economy offers a pathway to decouple economic growth from resource consumption by extending material life cycles, enhancing recyclability, and promoting design-for-disassembly strategies. For developing nations like Thailand, circular economy practices can reduce reliance on costly imported materials while fostering localized supply chains for bio-based composites, thereby supporting economic self-reliance and sustainable community development (Sáez-de-Guinoa et al., 2022).

Health Impacts of PM2.5 From Open Agricultural Burning

The widespread practice of open-field burning of agricultural waste, particularly rice straw, is a major contributor to hazardous air pollutants, including fine particulate matter (PM2.5). Elevated PM2.5 levels are directly linked to increased rates of respiratory diseases, cardiovascular complications, and premature mortality (WHO, 2022). Amnuaylojaroen et al. (2020) estimated that in Thailand alone, exposure to PM2.5 pollution causes over 50,000 premature deaths annually. Reducing open burning through the value-added utilization of agricultural residues can therefore yield substantial public health benefits, aligning with SDG Goal 3: Good Health and Well-being.

Introduction to Binderless Technology

In response to these environmental and societal challenges, this study introduces a novel fabrication methodology centered on binderless composite technology.

Notably, several developing countries have also explored similar binderless or low-emission panel systems to address sustainability and health concerns associated with VOCs (Antov et al., 2020; Calvez et al., 2024). Here, the term "binderless" refers to materials manufactured without the use of synthetic adhesives or chemical binders. Instead, internal cohesion is achieved through the inherent bonding capabilities of natural fibers—facilitated by the high cellulose content in wastepaper and the polysaccharide-rich nature of Persea kurzii powder. These natural components promote hydrogen bonding and physical interlocking among fibers, eliminating the need for harmful chemical additives while maintaining sufficient mechanical stability (Lawanwadeekul et al., 2024; Lawanwadeekul & Chindaprasirt, 2025).

This low-energy, low-emission approach ensures environmental safety and cost-effectiveness. It is particularly suited to decentralized, small-scale production in rural and peri-urban communities. From a materials science perspective, natural fibers exhibit favorable characteristics for composite panel production, including high aspect ratios, low density, high porosity, and excellent intrinsic sound-dampening properties (Mondem & Balunaini, 2025). These features make them especially suitable for applications such as indoor partitions, wall linings, and thermal insulation. Additionally, their fire resistance can be enhanced through simple surface treatments, expanding their applicability across a broader range of construction uses. Unlike chemically bonded panels, which often require high temperature curing and emit formaldehyde or other VOCs, binderless panels minimize toxic emissions and reduce embodied energy throughout the production cycle. This distinction enhances their suitability for green-certified building systems and improves indoor environmental quality.

Importantly, the approach presented in this study directly supports several United Nations Sustainable Development Goals (SDGs), including Goal 11 (Sustainable Cities and Communities), Goal 12 (Responsible Consumption and Production), and Goal 13 (Climate Action). By transforming waste streams into construction materials, this project contributes to reducing environmental pollution, improving public health outcomes, and stimulating local economies through employment opportunities in material recovery and panel production.

Recent market studies forecast that the global green construction sector will surpass USD 610 billion by 2025, driven by increasing regulatory pressures, the adoption of green building certifications, and rising consumer demand for sustainable construction practices (Hasanov, 2023; Le et al., 2023). These trends underscore the urgent need for alternative, low-carbon construction materials, particularly in fast-growing economies where urbanization is accelerating.

In alignment with these developments, leading economies have begun integrating circular economy principles directly into their national development agendas, promoting material efficiency, waste reduction, and the valorization of agricultural residues (Tang et al., 2025). For example, the European Commission (2020) and recent initiatives across Southeast Asia emphasize the shift from resource-extractive models to regenerative, low-impact material cycles (Yang et al., 2024).

Moreover, the rising interest in carbon credit markets and Environmental, Social, and Governance (ESG) investment strategies further highlights the economic potential of sustainable construction materials (Razali et al., 2024). Companies pursuing carbon neutrality and net-zero emission targets are increasingly adopting green materials to meet both regulatory compliance and stakeholder expectations, creating significant market opportunities for innovative, low-impact products such as binderless thermal-acoustic panels.

In conclusion, this study lays the foundation for a transformative construction paradigm—one in which agricultural and industrial waste streams are no longer viewed as environmental liabilities but are instead converted into high-performance, low-impact construction materials. Through a process that is environmentally responsible, economically viable, and socially inclusive, this approach aims to catalyze meaningful change in how materials are sourced, fabricated, and applied within the built environment.

2. AGRICULTURAL WASTE SITUATION IN THAILAND

Thailand's agricultural sector generates a diverse range of biomass by-products that present both significant environmental management challenges and substantial opportunities for material recovery and reuse within a circular economy framework. The concept of a circular economy—promoted globally through initiatives such as the European Union's Circular Economy Action Plan (European Commission, 2020) and the Ellen MacArthur Foundation (2019)—advocates reintegrating waste materials into productive economic cycles, thereby minimizing resource extraction and environmental impacts. Among these by-products, rice straw, water hyacinth, and wastepaper from artisanal industries stand out due to their unique material characteristics, which align well with sustainable construction practices and low-carbon material development strategies.

Rice straw is a principal by-product of rice cultivation, which remains a cornerstone of Thailand's agricultural economy, particularly across the central, northern, and northeastern regions. According to the Land Development Department (2025), over 25 million tons of rice straw are produced annually. Despite its abundance, a considerable portion of this biomass is disposed of through open-field burning, primarily due to the lack of cost-effective alternative disposal methods and limited rural waste management infrastructure. Although this practice offers logistical convenience for farmers, it significantly contributes to environmental degradation, notably through elevated concentrations of PM2.5 particulate matter, which adversely affects air quality during the post-harvest season (Sarapin et al., 2024). The World Health Organization (2022) further highlights that PM2.5 exposure is directly associated with increased risks of cardiovascular and respiratory diseases, underscoring the critical need for effective biomass waste management to safeguard public health.

This unsustainable disposal method not only exacerbates air pollution but also results in the loss of valuable organic material that could otherwise be redirected into value-added industrial applications. Studies such as Wahyuni et al. (2019) have demonstrated that rice straw possesses several desirable characteristics for composite material production, including high porosity, low thermal conductivity, and excellent compatibility with other natural fibers. However, as Goodman (2020) notes, the utilization rate of rice straw remains disproportionately low, largely due to logistical barriers in collection and storage, limited technical expertise among smallholder farmers, and the absence of supportive institutional mechanisms for biomass valorization.

Parallel to terrestrial agricultural residues, Thailand also faces considerable challenges regarding aquatic biomass waste, particularly the aggressive proliferation of water hyacinth (*Eichhornia crassipes*). This invasive species thrives in Thailand's nutrient-rich freshwater environments, disrupting irrigation canals, reservoirs, and natural waterways, thereby impeding water management systems and exacerbating flood risks. According to the Department of Public Works and Town & Country Planning (2025), over 7 million tons of water hyacinth were removed from national water bodies between 2022 and 2023 alone, underscoring the magnitude of the problem.

Traditionally regarded as an ecological nuisance, water hyacinth's structural and fibrous properties make it highly suitable for sustainable material applications. Salas-Ruiz et al. (2019) reported that water hyacinth petioles exhibit favorable microstructural characteristics, including high porosity and low thermal conductivity, making them ideal for bio-based insulation panels. Further, Martwong et al. (2024) demonstrated the successful integration of water hyacinth fibers into green polyurethane composites for housing insulation, effectively enhancing both mechanical strength and thermal performance. These findings highlight the untapped potential of this invasive species in reducing environmental waste and contributing to green construction solutions.

Another underutilized waste stream originates from Thailand's artisanal paper industry, particularly in northern provinces such as Chiang Mai and Lampang, where handmade mulberry-based paper (*sa paper*) remains a culturally and economically significant craft. However, this traditional production process generates substantial quantities of fibrous offcuts, trimmings, and rejected sheets. As these wastepapers are typically uncoated and chemically untreated, they are unsuitable for conventional industrial recycling systems (Tanasombat et al., 2005). Consequently, these materials often accumulate in informal waste disposal sites, contributing to localized environmental degradation and missed opportunities for resource recovery.

Despite this underutilization, artisanal wastepaper possesses several favorable properties, such as high cellulose content and excellent absorption capacity, making it particularly well-suited for blending with lignocellulosic agricultural residues. Recent research by Lawanwadeekul and Chindaprasirt (2025) demonstrated the feasibility of incorporating such paper waste with water hyacinth to produce thermal-acoustic insulation panels without the use of chemical adhesives. The resulting materials achieved sound absorption and thermal insulation performances comparable to conventional synthetic products while remaining biodegradable, non-toxic, and aligned with circular economy principles.

Collectively, rice straw, water hyacinth, and artisanal wastepaper represent renewable, biodegradable, and regionally abundant resources for sustainable material innovation. Each waste stream contributes distinctive material properties that enable their integration into low-carbon construction solutions. Rice straw contributes tensile strength and thermal insulation capacity, water hyacinth enhances porosity and structural bulk, and artisanal wastepaper provides bond strength and internal cohesion. The synergistic combination of these materials, particularly in binderless composite panels, closely aligns with the global shift toward Nature-Based Solutions (NBS) and sustainable construction practices (Parmesan et al., 2022; Illankoon et al., 2023).

Despite these advantages, the widespread implementation of such innovations faces significant systemic challenges. Thailand's agricultural production is highly decentralized, and waste material availability varies considerably by season and region. Infrastructure for waste collection, preprocessing, and storage—such as drying facilities, shredding equipment, and baling stations—remains inadequate. Moreover, as Gaspar et al. (2020) observe, policy frameworks across Southeast Asia often prioritize biomass-to-energy conversion (e.g., combustion, biogas production) over material valorization strategies, despite the latter offering greater long-term environmental and economic benefits through life cycle extension and carbon footprint reduction.

Nonetheless, encouraging developments are beginning to emerge. Several municipal and regional authorities have initiated pilot programs to fabricate rice straw-based wall panels and decorative tiles using water hyacinth pulp, indicating growing awareness of the latent value embedded in agricultural and artisanal waste streams. Ben Hassen et al. (2025) underscores the importance of promoting scalable, low-cost technologies to support decentralized material processing initiatives. At the same time, policy integration that connects environmental, economic, and industrial development goals is crucial to unlocking the full potential of circular economy strategies in Thailand's construction sector.

To illustrate the scale and potential utilization of agricultural waste in Thailand, Fig. 1 presents the composition of unused biomass resources based on recent data from Krungthai COMPASS (2022). This analysis focuses specifically on the volumes of agricultural residues that remain unutilized and offer significant opportunities for value-added material development.

Figure 1. Agricultural waste composition in Thailand (Unused biomass per year)

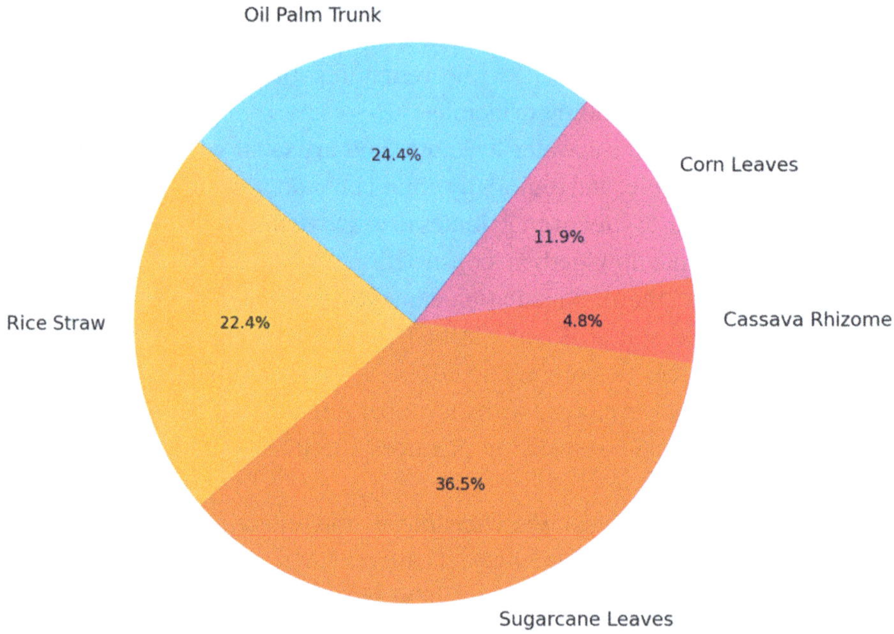

Source: Krungthai COMPASS (2022)

The data reveals that sugarcane leaves account for the largest share of unused biomass, representing approximately 45.2 million tons per year, followed by oil palm trunks (30.2 million tons) and rice straw (27.7 million tons). Although rice straw is frequently discussed in academic and industrial research, it ranks only third in terms of actual underutilized volume. This highlights the considerable untapped potential of sugarcane and oil palm waste in sustainable material development, particularly for bio-composites and green construction applications.

Furthermore, the substantial volumes of corn leaves (14.8 million tons) and cassava rhizomes (6.0 million tons) underscore the diversity of available biomass resources that remain largely unexplored. Integrating these waste streams into Thailand's circular economy policies—consistent with the European Union's Circular

Economy Action Plan (European Commission, 2020)—could unlock significant environmental and economic benefits.

The inclusion of this updated waste composition analysis reinforces the need for broader policy reforms and technological advancements to support the valorization of underutilized biomass. By expanding research efforts beyond rice straw and water hyacinth, Thailand can diversify its portfolio of sustainable raw materials for green construction and promote more balanced, resource-efficient industrial development.

Despite these opportunities, critical systemic challenges persist, including fragmented waste management systems, seasonal biomass availability, and the absence of standardized biomass-to-material conversion protocols. Overcoming these barriers will require targeted policy reforms, investments in decentralized processing infrastructure, and the promotion of cross-sectoral partnerships to unlock the full potential of Thailand's biomass resources.

Thailand's experience with agricultural and artisanal waste management vividly illustrates both the challenges and opportunities associated with sustainable resource utilization. While the country continues to face significant issues related to air pollution, invasive species proliferation, and underutilized organic waste, it also holds immense potential to transform these environmental liabilities into economic assets. Through coordinated investments in research and development (R&D), improved waste management infrastructure, and targeted policy reforms, Thailand can position itself as a leader in sustainable materials innovation—contributing to global efforts toward climate resilience, circular economy adoption, and the advancement of green construction practices across the Global South.

3. MATERIALS AND METHODS

3.1 Agricultural and Paper Waste in Conventional Binder-Based Applications

Agricultural by-products such as rice straw and water hyacinth, along with cellulosic waste from artisanal paper production, have increasingly been explored as renewable fillers and reinforcements in the construction sector. Their fibrous morphology, low density, and local availability make them attractive candidates for producing insulation panels, lightweight boards, and composite materials. However, most prior studies have incorporated chemical binders—typically synthetic adhesives or inorganic matrices—to stabilize these bio-based components (Alavéz-Ramírez et al., 2012; Ormondroyd, 2015; Hýsek et al., 2019).

Common chemical adhesives include urea-formaldehyde (UF), phenol-formaldehyde (PF), melamine-based resins, and polyvinyl acetate (PVA). These polymers enhance mechanical integrity and water resistance but simultaneously raise significant environmental and health concerns (Borges et al., 2022). UF and PF are particularly problematic, as they emit formaldehyde, a volatile organic compound (VOC) classified as a human carcinogen by the World Health Organization. Despite efforts to reduce free formaldehyde content, these binders continue to pose occupational health risks for producers and degrade indoor air quality for end users (Mulder et al., 2011; Maiti et al., 2021).

Recent research has further demonstrated that prolonged exposure to VOC emissions from UF- and PF-based panels significantly increases the risk of Sick Building Syndrome (SBS) and respiratory complications in indoor environments (Zhou et al., 2023). These concerns, combined with increasingly stringent environmental regulations—such as the European Union's Restriction of Hazardous Substances (RoHS) Directive—have reduced the market acceptance of chemically bonded composites in modern green building projects (European Commission, 2020).

In addition to synthetic resins, agricultural residues have been incorporated into inorganic matrices such as cement and gypsum. For example, rice straw fibers have been mixed with Portland cement to produce lightweight construction blocks (Saxena et al., 2008), while water hyacinth has been blended with gypsum to improve porosity and acoustic absorption in panel products (Philip & Rakendu, 2020). While these combinations enhance structural rigidity and fire resistance, they compromise biodegradability and limit end-of-life recyclability. Moreover, these materials often require high temperature curing or alkaline activation, resulting in increased energy consumption and higher carbon emissions (Sambucci et al., 2021).

Thermoplastic polymers such as polypropylene (PP), polyethylene (PE), and polylactic acid (PLA) have also been used to produce fiber-reinforced composites through extrusion or hot pressing. Wahyuni et al. (2019) reported that rice straw–polypropylene composites achieve acceptable mechanical strength and thermal insulation properties; however, these systems typically demand specialized equipment and energy-intensive processing, resulting in higher production costs and reduced feasibility for deployment in low-income or rural areas.

Taken together, these approaches underscore a fundamental trade-off: while binder-based systems often deliver superior mechanical performance, they introduce significant environmental, economic, and technical barriers that hinder large-scale adoption—particularly in contexts where affordable, low-impact materials are urgently required. This study, therefore, seeks to overcome these limitations by developing binderless, chemical-free composite panels using locally available Thai agricultural waste and ambient-temperature fabrication techniques.

3.2 Development of Chemical-Free Binderless Panels

This study proposes an alternative fabrication strategy that exploits the inherent bonding properties of lignocellulosic fibers and physical compaction to produce thermal-acoustic panels without the use of synthetic or chemical binders. The primary materials employed include rice straw, water hyacinth, and handmade wastepaper, all sourced locally from the Chiang Mai–Lampang region. In addition, *Persea kurzii* powder was introduced as a natural adhesion enhancer, owing to its high starch content and fiber-binding capability (Lawanwadeekul et al., 2024; Lawanwadeekul & Chindaprasirt, 2025).

Persea kurzii, a plant species native to Southeast Asia, is traditionally recognized for its medicinal applications. Although direct studies on its adhesive properties remain limited, its high starch content—rich in amylose and amylopectin—suggests considerable potential for natural adhesion enhancement through hydrogen bonding, a mechanism well-documented in starch-based bio-adhesives (Moini et al., 2022). These natural binders offer significant advantages in terms of biodegradability and reduced environmental toxicity compared to conventional synthetic adhesives.

The raw materials underwent a series of preparation processes to ensure uniformity and optimize their performance. Rice straw was collected post-harvest, thoroughly cleaned, sun-dried for seven days, and ground into fibers using a mechanical grinder. Water hyacinth was harvested from irrigation canals, washed to remove impurities, air-dried, and milled to a target fiber length of 10–40 mm. Wastepaper, obtained from local craft workshops, was soaked for 24 hours and mechanically pulped to improve fiber flexibility. To achieve particle size consistency, a sieve analysis was performed, and the final moisture content was maintained below 10% to prevent microbial degradation during storage.

Characterization of the raw materials was performed using Scanning Electron Microscopy (SEM) and Fourier Transform Infrared Spectroscopy (FTIR). SEM images revealed that rice straw fibers possess rigid, tubular structures conducive to thermal insulation, while water hyacinth fibers exhibit a loose, porous surface favorable for sound absorption. Wastepaper samples displayed a dense, compact matrix with residual sizing agents that enhance bonding, and *Persea kurzii* particles appeared irregular with a high surface area, facilitating effective pore filling and improved cohesion (Fig. 2).

Figure 2. Microstructure analysis of raw materials (a) rice straw, (b) water hyacinth, (c) wastepaper and (d) Persea kurzii.

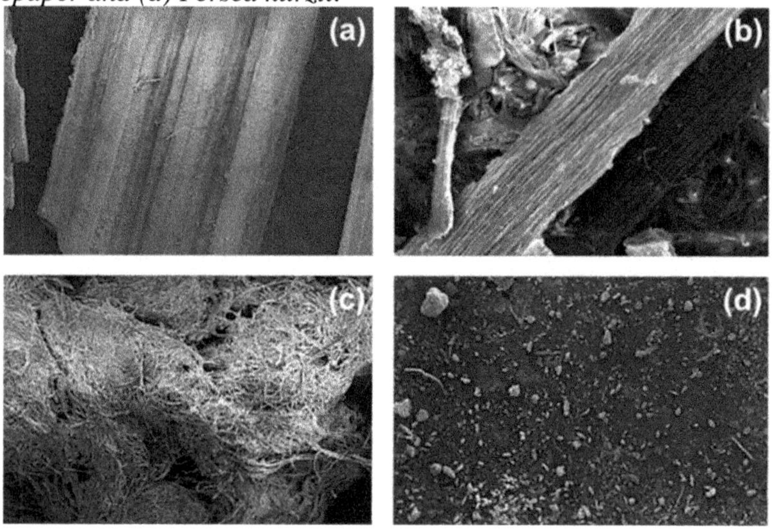

Source: Lawanwadeekul et al. (2024); Lawanwadeekul and Chindaprasirt (2025)

FTIR analysis confirmed the lignocellulosic composition of rice straw, water hyacinth, and wastepaper. Major absorption bands appeared at approximately 3330 cm^{-1} (O–H stretching), 2920 cm^{-1} (C–H stretching), 1730 cm^{-1} (C=O stretching of hemicellulose), and 1030 cm^{-1} (C–O–C stretching of cellulose). In wastepaper, aromatic ring vibrations detected between 1500–1600 cm^{-1} indicated residual lignin, which contributes to improved thermal stability (Sheng et al., 2014; Lawanwadeekul & Chindaprasirt, 2025). Notably, no signals corresponding to synthetic additives or chemical contaminants were observed.

Similarly, *Persea kurzii* powder exhibited distinct spectral features associated with its high starch and polysaccharide content. A strong absorption peak near 3400 cm^{-1}, corresponding to O–H stretching, confirmed the abundance of hydroxyl groups capable of hydrogen bonding. The band around 1640 cm^{-1} was attributed to H–O–H bending of absorbed water, while a sharp peak at approximately 1150 cm^{-1} confirmed the presence of C–O stretching associated with carbohydrate structures (Iftikhar et al., 2019; Kumar & Cumbal, 2016). These characteristics support the role of *Persea kurzii* as an effective natural binder by promoting hydrogen bonding between lignocellulosic fibers, thereby enhancing panel cohesion without requiring synthetic adhesives.

Panel fabrication was performed by varying the composition ratios of the raw materials. Rice straw and water hyacinth served as the primary structural framework, while wastepaper and *Persea kurzii* powder were incorporated as natural binding agents. The dry materials were manually mixed to ensure uniform distribution, followed by the addition of distilled water at 10% by weight to promote fiber interaction. The blended mixture was then compressed using a hydraulic press at 5 bars for 5 minutes. Panels were produced in dimensions suitable for characterization: Ø100 × 25 mm and Ø29 × 25 mm for acoustic testing, and 300 × 300 × 25 mm for thermal conductivity and mechanical testing, as shown in Fig. 3.

Figure 3. Some raw materials, production process and final product.

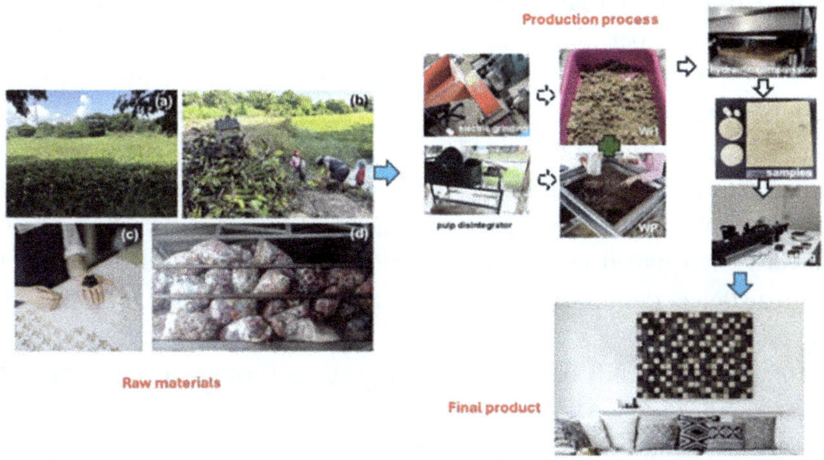

Following molding, the panels were air-dried under ambient air conditioning for 5 to 7 days until a constant weight was achieved. Importantly, no external heating, chemical curing, or resin impregnation was employed throughout the production process. This approach aligns with low-energy consumption principles and chemical-free production practices, supporting environmentally sustainable material innovation suitable for decentralized, small-scale manufacturing.

4. TESTING RESULTS, COMPARATIVE ANALYSIS, AND ECONOMIC EVALUATION

4.1 Physical and Mechanical Properties of Fabricated Panels

The binderless thermal-acoustic panels were assessed for their suitability in interior, non-load-bearing applications by evaluating three key indicators: density, porosity, and flexural strength—all critical to structural and functional performance. Panel densities ranged from 90 to 170 kg/m^3, influenced by material composition.

Higher wastepaper content led to denser panels due to its compact fiber network and residual sizing agents, enhancing internal cohesion. In contrast, panels with greater rice straw or water hyacinth content exhibited lower densities, which increased porosity and improved thermal insulation.

Porosity levels (70–85%) were optimal for sound absorption, with water hyacinth contributing significantly to this due to its inherently open-cell structure (Merli et al., 2021; Sangmesh et al., 2023). These results align with existing studies on agricultural residue-based composites (Maderuelo-Sanz et al., 2022).

Flexural strength values ranged from 0.09 to 0.48 MPa, sufficient for applications such as interior wall partitions, ceiling linings, and decorative panels. Incorporating Persea kurzii powder improved fiber-matrix interaction via hydrogen bonding, enhancing fracture resistance (Lawanwadeekul et al., 2024; Lawanwadeekul & Chindaprasirt, 2025). As shown in Fig. 4, wastepaper content (up to 50%) significantly increased flexural modulus—especially in 2-inch-thick panels—due to higher cellulose content and stronger fiber bonding.

Thickness also played a crucial role; thicker panels consistently exhibited higher bending resistance, demonstrating that geometric configuration is as important as material composition.

For context, chemically bonded panels (e.g., urea-formaldehyde) achieve higher flexural strengths (5.5–8 MPa) but at the cost of higher density, lower porosity, and adverse environmental impacts such as VOC emissions and non-biodegradability (Hussein et al., 2019). A comparative summary of the physical and mechanical properties of the fabricated panels and conventional chemically bonded panels is presented in Table 1.

Figure 4. Flexural strength of binderless panels at varying wastepaper content and panel thicknesses.

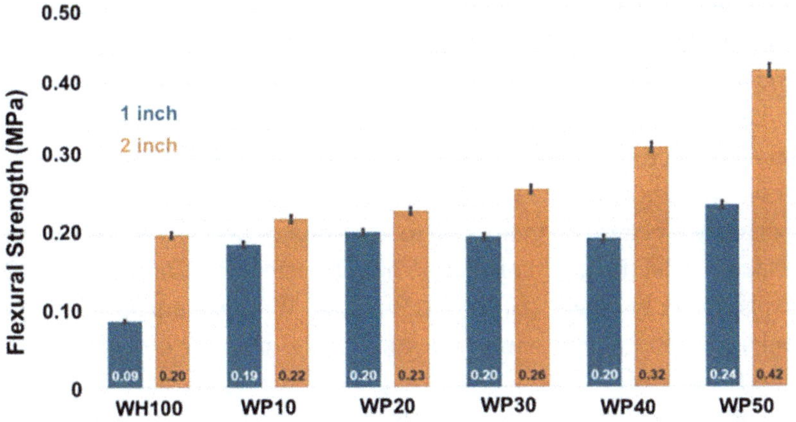

Source: Lawanwadeekul and Chindaprasirt (2025).

Table 1. Physical and mechanical properties of fabricated and conventional panels

Material Composition	Density (kg/m^3)	Porosity (%)	Flexural Strength (MPa)	Ref.
Rice Straw + Wastepaper	90–170	70–82	0.09–0.48	Lawanwadeekul et al., 2024
Rice Straw + Persea kurzii	95–160	72–80	0.09–0.48	Lawanwadeekul et al., 2024
Rice Straw + UF Resin	250–350	45–60	5.5–8.0	Hussein et al., 2019
Water Hyacinth + Wastepaper	85–160	75–85	0.09–0.42	Lawanwadeekul & Chindaprasirt, 2025
Water Hyacinth + UF Resin	180–300	~55	4.0–6.5	Salas-Ruiz et al., 2019

In summary, the fabricated panels offer a viable non-toxic and biodegradable alternative to conventional panels, with sufficient mechanical integrity for light construction applications. The use of Persea kurzii enhanced bonding without chemical binders, and the composition-thickness relationship provides design flexibility for various performance needs.

4.2 Thermal and Acoustic Performance of Fabricated Panels

Thermal and acoustic performance is a key determinant of an insulation material's viability in real-world building applications. The fabricated binderless panels were tested using standardized procedures to assess thermal conductivity and sound absorption capabilities—two critical functions for occupant comfort.

Thermal conductivity values ranged from 0.040 to 0.067 W/m·K, depending on material composition and density. Panels with higher rice straw and water hyacinth content performed better thermally due to their higher porosity and air entrapment, which reduce conductive and convective heat transfer (Ricciardi et al., 2021). Although expanded polystyrene (EPS) (0.035–0.040 W/m·K) remains the benchmark in thermal insulation (Ouakarrouch et al., 2022), it lacks biodegradability and acoustic performance.

Increasing wastepaper content slightly increased thermal conductivity due to densification of the fiber network, though insulation remained within acceptable ranges, particularly up to 30% wastepaper (Fig. 5a). This trade-off is counterbalanced by improvements in mechanical integrity and ease of panel handling—relevant for on-site installation.

Noise Reduction Coefficient (NRC) values ranged between 0.48 and 0.51, comparable to standard materials like glass wool and rock wool (0.50–0.60) (Maderuelo-Sanz et al., 2022; Lawanwadeekul & Chindaprasirt, 2025). As illustrated in Fig. 5b, panels with 10–20% wastepaper content provided the best acoustic performance-to-strength ratio. This positions them as well-suited for applications in classrooms, offices, and residential walls, where both sound absorption and material durability are essential.

Figure 5. Thermal conductivity and noise reduction coefficient of binderless panels at varying wastepaper content and panel thickness.

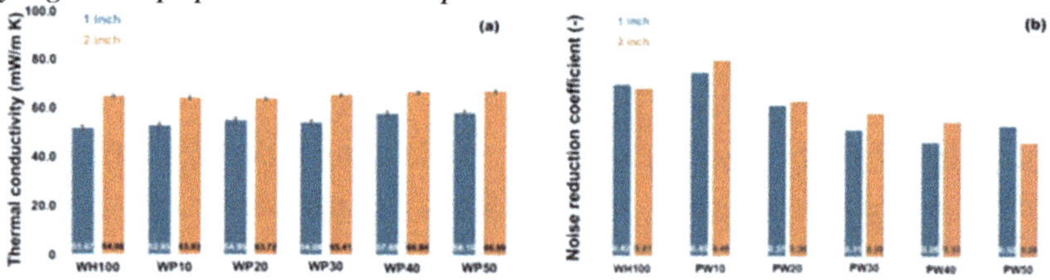

Source: Lawanwadeekul and Chindaprasirt (2025).

Frequency-dependent absorption analysis (Fig. 6) further highlights this potential. A panel with water hyacinth and wastepaper (90:10) reached a peak absorption coefficient near 0.75 at 800 Hz, matching typical conversational frequencies. While absorption performance declined at frequencies above 1000 Hz—where glass wool still dominates—the binderless panels provide balanced performance across mid-range frequencies, which are more relevant for indoor acoustic comfort.

Figure 6. Frequency-dependent sound absorption coefficients of the fabricated binderless panel (water hyacinth: wastepaper at 90:10) compared to conventional glass wool and rock wool panels at 25 mm thickness.

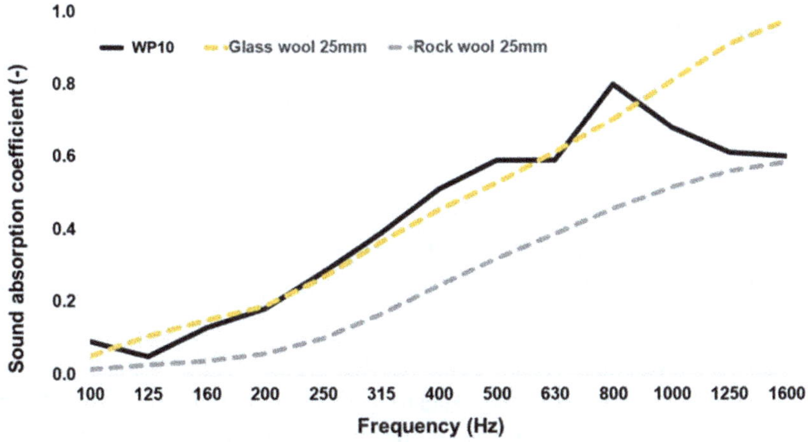

A comparative summary of the thermal and acoustic performance of the fabricated panels and conventional materials is presented in **Table 2.**

Table 2. Thermal and Acoustic Performance of Fabricated and Conventional Panels

Material Composition	Thermal Conductivity (W/m·K)	Noise Reduction Coefficient (-)	Ref.
Rice Straw + Wastepaper	0.042–0.065	0.48–0.50	Lawanwadeekul et al., 2024
Rice Straw + Persea kurzii	0.040–0.062	0.50–0.51	Lawanwadeekul & Chindaprasirt, 2025
Rice Straw + UF Resin	0.045–0.060	0.35–0.42	Hussein et al., 2019
Water Hyacinth + Wastepaper	0.043–0.067	0.49–0.51	Lawanwadeekul & Chindaprasirt, 2025
Water Hyacinth + UF Resin	0.046–0.065	0.36–0.44	Salas-Ruiz et al., 2019
Expanded Polystyrene (EPS)	0.035–0.040	~0.20	Ricciardi et al., 2021

continued on following page

Table 2. Continued

Material Composition	Thermal Conductivity (W/m·K)	Noise Reduction Coefficient (-)	Ref.
Glass Wool	0.033–0.040	0.50–0.60	Thai Watsadu. (n.d.); HomePro. (n.d.)
Rock Wool	0.035–0.045	0.50–0.60	Thai Watsadu. (n.d.); HomePro. (n.d.)

In practical terms, these findings suggest that binderless panels can substitute synthetic insulators in non-load-bearing partitions and ceilings in moderate to warm climates, especially in buildings prioritizing indoor acoustic quality and sustainability. Overall, the balance between thermal insulation, sound absorption, and biodegradability makes these panels highly promising for sustainable architecture.

4.3 Degradation Behavior Under High Humidity and Water Contact

In real-world settings, the long-term durability under humid conditions is a pivotal consideration for bio-based thermal-acoustic panels, particularly in tropical regions where water exposure is frequent. As lignocellulosic materials such as rice straw, water hyacinth, and wastepaper are inherently hydrophilic, they readily absorb moisture—initiating microbial activity and hydrolytic degradation (Yang et al., 2014; Kliem et al., 2020).

Experimental results showed up to a 30% reduction in flexural strength after four weeks at 85% RH and 30°C, and a 20% mass loss after two weeks of water immersion. These effects stem from the leaching of soluble compounds and microbial degradation (Deroiné et al., 2014; Celestine et al., 2020), underscoring moderate water sensitivity but also confirming controlled biodegradability—an essential trait for end-of-life sustainability.

While moisture degradation may seem like a drawback, it is strategically beneficial for temporary structures, post-disaster shelters, compostable partitions, and applications where materials are expected to return safely to the biosphere without landfill impact.

Persea kurzii, used as a natural binder, moderately improves water resistance via polysaccharide gelation, which partially blocks capillary ingress (Moini et al., 2022). Yet, this effect is not permanent, as gel networks weaken with prolonged hydrolysis.

In contrast, urea-formaldehyde and phenol-formaldehyde resin panels offer higher water resistance but at environmental costs—poor biodegradability, formaldehyde emissions, and disposal complications (Pizzi & Papadopoulos, 2020; Ramesh et al.,

2022). These features violate circular economy principles by prolonging environmental persistence and introducing hazardous waste streams.

Overall, binderless bio-panels present a balanced trade-off: acceptable short-term performance for indoor or semi-dry zones, and full compostability post-use, aligning with green construction and zero-waste objectives. Their limitations in exterior or structural uses can be addressed by selective application design or complementary surface treatments, while their biodegradation offers a long-term ecological advantage.

This degradation profile—while limiting in some scenarios—enhances circularity, compliance with biocycle-based standards, and supports scalable, low-impact construction models in developing regions.

4.4 Thermal Degradation Behavior and Flammability Assessment

The fire behavior of sustainable insulation materials is a key determinant of their safe integration into buildings, especially for interior applications where exposure to elevated temperatures or electrical heat sources may occur. To evaluate this, thermal degradation and flammability properties of the binderless panels were analyzed using Thermogravimetric Analysis (TGA) and Differential Scanning Calorimetry (DSC), as shown in Fig 7.

The results revealed a three-stage degradation profile common to lignocellulosic composites. The first stage (below 150°C) involved evaporation of moisture and low-volatility compounds, particularly in panels with high water hyacinth content due to its porosity. The second stage (200–350°C) represented the main pyrolysis phase, where hemicellulose and cellulose decomposed. Notably, panels with higher wastepaper content showed a delayed decomposition peak—up to 320°C—indicating better thermal shielding and slower breakdown under heat stress. The third stage (above 350°C) involved lignin degradation and formation of carbonaceous char, which served as a natural barrier to further combustion.

DSC data confirmed this enhanced stability, showing exothermic peaks corresponding to cellulose decomposition and delayed secondary peaks in panels with 40–50% wastepaper, pointing to stronger char layer formation. Although these panels are not engineered as flame-resistant, this passive charring effect significantly slows fire spread, a valuable trait for interior materials not in direct flame contact.

Figure 7. TGA and DSC thermograms of binderless panels with varying wastepaper content.

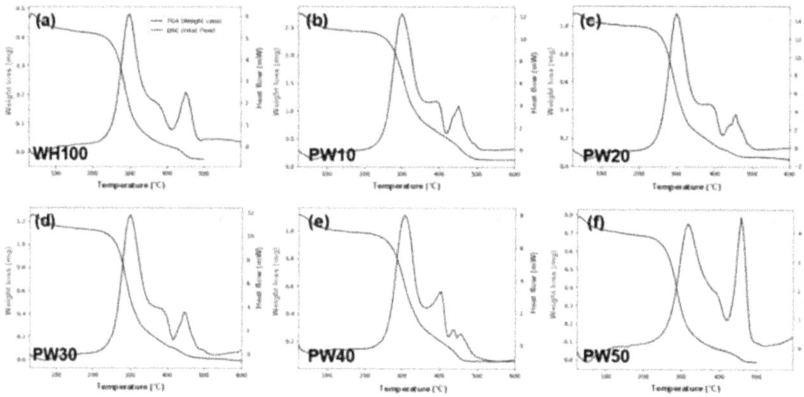

Source: Lawanwadeekul and Chindaprasirt (2025)

The use of *Persea kurzii* as a natural binder modestly enhanced thermal performance by reducing oxygen penetration into the matrix via gelation (Moini et al., 2022). However, its effect diminishes under prolonged heat and does not eliminate the need for fire-conscious design choices.

In comparison, synthetic adhesive panels, such as those using urea-formaldehyde (UF) or phenol-formaldehyde (PF), exhibit stronger fire resistance but pose significant sustainability drawbacks, notably the release of formaldehyde and other toxins during thermal breakdown (Pizzi & Papadopoulos, 2020; Ramesh et al., 2022).

Ultimately, panels with 30–50% wastepaper content offer a well-balanced thermal profile—delaying degradation, promoting protective charring, and maintaining eco-friendly performance. While they are not substitutes for certified fire-rated systems in high-risk areas, they are well-suited for low-risk interior applications where thermal protection, sustainability, and biodegradability are equally valued.

4.5 Economic Evaluation

The economic feasibility of producing chemical-free, binderless thermal-acoustic panels was systematically assessed by analyzing key cost components, including raw materials, energy consumption, labor, and logistics. A critical economic advantage of this sustainable production approach lies in the complete elimination of synthetic adhesives and the utilization of ambient-temperature processing. These factors significantly reduce both operational costs and environmental burdens, making the panels particularly suitable for low-resource settings and decentralized manufacturing models.

The estimated production costs for the binderless panels range between 540 and 690 THB per square meter, depending on material composition and production scale. In comparison, conventional imported synthetic insulation materials, such as expanded polystyrene (EPS) and glass wool, are priced considerably higher, typically ranging between 800 and 1,200 THB per square meter. This substantial cost differential is primarily attributable to three factors:

- **Utilization of Locally Sourced Agricultural and Industrial Waste**: Agricultural residues such as rice straw, water hyacinth, and wastepaper are abundant and underutilized in many regions of Thailand, significantly reducing raw material procurement costs.
- **Low-Energy Fabrication Processes**: The elimination of energy-intensive chemical curing and thermal treatment processes minimizes production energy demands, resulting in lower operational expenditures.
- **Simplified Production Techniques and Equipment**: The fabrication process requires only basic mechanical equipment, reducing initial capital investment and facilitating **decentralized, small-scale production units** that are ideal for rural and peri-urban communities.

In addition to direct cost savings, the adoption of **localized production models generates broader socio-economic benefits**. These include the creation of employment opportunities within local communities, reduced transportation costs associated with raw material and product logistics, and a corresponding reduction in greenhouse gas emissions from transportation activities. Such localized systems also enhance resilience in supply chains, especially in regions where centralized infrastructure is underdeveloped.

However, successful implementation at scale may face challenges. These include:

- **Logistics and Quality Control**: Variability in agricultural waste properties may require region-specific processing calibration, and quality consistency across distributed production sites must be ensured.
- **Regulatory Acceptance**: Binderless panels are relatively new to market, and their use in regulated construction projects may require material certification, performance standard validation, and approval under national building codes.
- **Market Adoption**: Awareness among contractors and building code officers remains limited. A lack of precedents for field deployment can delay adoption unless pilot demonstration projects or public procurement incentives are introduced.

These issues can be strategically addressed through community-based cooperatives, technical training programs, and collaboration with local governments to establish "green building innovation zones" where new materials can be piloted and refined.

Table 3. Cost comparison per square meter of thermal-acoustic panels

Material Type	Estimated Production Cost (USD/ m^2)	Ref.
Chemical-Free Binderless Panels	15.00 – 19.20	Lawanwadeekul & Chindaprasirt, 2025
Imported EPS Panels	22.20 – 33.30	Belloni et al., 2021
Glass Wool Panels	23.60 – 31.90	Ouakarrouch et al., 2022
Rock Wool Panels	25.00 – 36.10	Sangmesh et al., 2023

Note: *Conversion rate applied at 1 USD = 36 THB (2024 average).*

The economic evaluation clearly demonstrates that chemical-free binderless panels are highly cost-competitive when compared to conventional insulation materials. With production costs approximately 25%–40% lower than those of imported alternatives such as EPS and glass wool, these panels present a financially viable solution for scaling up sustainable construction practices, particularly in developing countries.

Moreover, decentralized manufacturing aligns with the Bio-Circular-Green (BCG) economic model, supporting inclusive growth, environmental stewardship, and circularity. These factors collectively underscore the potential of binderless panels to play a transformative role in transitioning the construction sector toward low-impact, locally embedded material systems.

4.6 Long-Term Environmental and Economic Benefits

The adoption of sustainable, chemical-free binderless insulation materials is not only a response to immediate environmental challenges but also a strategic investment in long-term economic and regulatory resilience. This transition aligns directly with international green building frameworks such as LEED (Leadership in Energy and Environmental Design), WELL Building Standard, and EDGE (Excellence in Design for Greater Efficiencies). These certification systems increasingly prioritize materials that demonstrate low embodied carbon, non-toxic composition, and minimal environmental impact throughout their life cycles, making bio-based materials an attractive choice for both policymakers and industry leaders.

Beyond short-term cost efficiencies, the integration of these materials into mainstream construction practices offers substantial long-term benefits. These include enhanced corporate competitiveness, alignment with evolving environmental legislation, and opportunities to participate in emerging carbon credit markets.

Key Long-Term Benefits

- **Reduced Regulatory Compliance Costs**: As global and national regulatory frameworks increasingly restrict the use of hazardous materials and impose stricter waste management protocols, bio-based materials inherently comply with such requirements. This reduces the need for costly compliance adjustments and mitigates financial risks associated with future policy changes.
- **Support for Corporate ESG Targets**: Companies are under increasing pressure from investors and stakeholders to adopt Environmental, Social, and Governance (ESG) strategies. The use of bio-based, chemical-free materials demonstrates a tangible commitment to sustainable development goals (SDGs), enhancing brand reputation and improving access to sustainability-focused investment funds.
- **Improved Market Competitiveness**: With the green construction sector projected to surpass USD 610 billion globally by 2025 (Hasanov, 2023; Le et al., 2023), early adopters of sustainable materials can secure competitive advantages in markets that increasingly demand green-certified buildings. This also supports compliance with Thailand's national sustainability agenda under the Bio-Circular-Green (BCG) Economy Model.
- **Participation in Carbon Credit and Incentive Markets**: The global shift towards carbon-neutral economies has created opportunities for organizations to monetize emission reductions through participation in carbon credit markets. By adopting low-carbon, biodegradable materials, developers and manufacturers can reduce their operational carbon footprints and qualify for carbon offset credits and tax incentives under emerging regulatory mechanisms.

Socio-Economic and Environmental Synergies

In addition to financial and regulatory advantages, the transformation of underutilized agricultural and industrial waste into high-value construction materials contributes directly to inclusive economic growth, particularly in marginalized rural communities. Decentralized production models facilitate localized employment opportunities, reduce reliance on imported materials, and promote technological self-sufficiency.

From an environmental perspective, this approach significantly reduces lifecycle emissions, minimizes waste sent to landfills, and prevents harmful practices such as open burning of agricultural residues, which contribute to PM2.5 pollution and associated public health risks.

In conclusion, the large-scale adoption of chemical-free, binderless insulation panels offers a multifaceted pathway toward sustainable development. By aligning with green building certifications, supporting ESG commitments, and enabling participation in carbon credit markets, these materials provide both environmental resilience and long-term economic value. As regulatory pressures and market demands for sustainable construction intensify, organizations that embrace these innovations early will be better positioned to achieve financial success, regulatory compliance, and enhanced corporate reputation in an increasingly sustainability-conscious global economy.

5. CONCLUSION

This study successfully demonstrated the technical, economic, and environmental viability of chemical-free, binderless thermal-acoustic panels fabricated from underutilized agricultural and industrial waste streams, including rice straw, water hyacinth, and artisanal wastepaper. Leveraging locally available biomass resources and simple, low-energy manufacturing processes, the research contributes directly to advancing sustainable material innovation within the framework of the circular economy.

The fabricated panels achieved competitive performance metrics, with thermal conductivity values ranging from 0.040 to 0.067 W/m·K and Noise Reduction Coefficient (NRC) values between 0.48 and 0.51. These results are comparable to those of conventional insulation materials such as expanded polystyrene (EPS), glass wool, and rock wool, while offering significant environmental advantages, including biodegradability, non-toxicity, and lower embodied carbon. The binderless production process—free from synthetic adhesives and energy-intensive curing—further reinforces the environmental benefits by minimizing volatile organic compound (VOC) emissions and reducing operational energy consumption.

From an economic perspective, the panels demonstrated clear cost competitiveness, with production costs estimated between USD 15 and 19.20 per square meter, representing a 25–40% reduction compared to imported synthetic insulation products. This economic advantage, combined with environmental sustainability, positions the developed panels as an ideal solution for affordable housing, public infrastructure projects, and green-certified building initiatives, particularly in emerging economies such as Thailand.

Despite these promising outcomes, the research also identified critical areas for further investigation. Specifically, while the panels exhibit moderate resistance to moisture and thermal degradation, long-term performance under extreme environmental conditions warrants additional study. Comprehensive Life Cycle Assessment (LCA) and Carbon Footprint Analysis, following standardized methodologies (ISO 14040:2006; ISO 14044:2006), are necessary to quantify the full environmental benefits and validate claims of carbon emission reductions over the product life cycle. Additionally, the integration of natural fire-retardant additives could further enhance material safety and broaden the range of applications.

In summary, this work establishes a strong foundation for the adoption of binderless, bio-based insulation materials in sustainable construction practices. By addressing both material circularity and functional performance, these panels represent a viable pathway for reducing environmental impacts in the built environment, supporting national sustainability goals, and contributing to global climate resilience initiatives.

6. FUTURE WORK AND RESEARCH DIRECTIONS

6.1 Life Cycle Assessment (LCA) and Carbon Footprint Analysis

Future research should adopt internationally recognized LCA methodologies, particularly ISO 14040:2006 and ISO 14044:2006, to rigorously assess the full environmental impacts associated with the production, use, and end-of-life stages of binderless thermal-acoustic panels. A comprehensive evaluation should encompass key indicators such as Global Warming Potential (GWP), embodied energy, water footprint, and greenhouse gas (GHG) emissions across the entire life cycle—from raw material acquisition and manufacturing to disposal or recycling.

Moreover, comparative LCA studies against conventional insulation materials, including expanded polystyrene (EPS), glass wool, and rock wool, are critical to providing quantitative evidence of the environmental advantages of bio-based alternatives. Such data will be invaluable for guiding material selection in green-certified building projects and informing stakeholders engaged in sustainable construction initiatives.

6.2 Policy Integration and Market Development

In addition to technical validation, the widespread adoption of these sustainable materials requires strategic alignment with national and international policy frameworks. Future research should explore how Thailand's Bio-Circular-Green (BCG)

Economic Model and the Climate Change Master Plan (2021–2030) can support the commercialization and policy endorsement of bio-based construction materials.

Comparative analyses of policy instruments from leading economies—such as the European Union's Circular Economy Action Plan and the United States' Inflation Reduction Act (IRA)—can offer valuable insights into effective mechanisms for accelerating the adoption of green materials. These may include the establishment of financial incentives, carbon credit trading schemes, green procurement policies, and public-private partnerships (PPPs) that stimulate market demand while supporting local industries.

6.3 Technical Optimization and Field Applications

To further enhance the viability of binderless bio-based panels, future experimental studies should focus on optimizing material formulations to improve mechanical strength, moisture resistance, and fire retardancy—without compromising biodegradability or environmental integrity. The integration of natural flame retardants, such as clay minerals or phosphorus-based bio-compounds, may offer a sustainable pathway to meet safety requirements for wider commercial applications.

Additionally, pilot-scale field trials in real-world construction projects are essential to validate the long-term durability and functional performance of these panels under diverse climatic and operational conditions. Such demonstrations will provide critical feedback for refining material properties and production processes while increasing stakeholder confidence in commercial-scale deployment.

In summary, while this study has established a solid foundation for the development of binderless, bio-based insulation materials, realizing their full potential will require a multidisciplinary research agenda. This includes comprehensive environmental impact assessments, policy integration strategies, technical optimizations, and real-world application trials. By addressing these critical areas, future research can accelerate the transition toward a more sustainable, circular, and resilient built environment at both local and global scales.

By integrating scientific innovation with regulatory foresight and grassroots implementation, the roadmap outlined above can transform agricultural waste into a mainstream construction asset. Through coordinated efforts between researchers, policymakers, and industry actors, bio-based binderless panels have the potential to become both a technological solution and a symbol of Thailand's commitment to sustainable development in the built environment.

REFERENCES

Abera, Y. A. (2024). Sustainable building materials: A comprehensive study on eco-friendly alternatives for construction. *Composites and Advanced Materials*, *33*, 26349833241255957. DOI: 10.1177/26349833241255957

Alavez-Ramirez, R., Chiñas-Castillo, F., Morales-Dominguez, V. J., & Ortiz-Guzman, M. (2012). Thermal conductivity of coconut fibre filled ferrocement sandwich panels. *Construction & Building Materials*, *37*, 425–431. DOI: 10.1016/j.conbuildmat.2012.07.053

Ali, M., Alabdulkarem, A., Nuhait, A., Al-Salem, K., Iannace, G., & Almuzaiqer, R. (2022). Characteristics of agro waste fibers as new thermal insulation and sound absorbing materials: Hybrid of date palm tree leaves and wheat straw fibers. *Journal of Natural Fibers*, *19*(13), 6576–6594. DOI: 10.1080/15440478.2021.1929647

Amnuaylojaroen, T., Inkom, J., Janta, R., & Surapipith, V. (2020). Long range transport of southeast asian PM2. 5 pollution to northern Thailand during high biomass burning episodes. *Sustainability (Basel)*, *12*(23), 10049. DOI: 10.3390/su122310049

Antov, P., Jivkov, V., Savov, V., Simeonova, R., & Yavorov, N. (2020). Structural application of eco-friendly composites from recycled wood fibres bonded with magnesium lignosulfonate. *Applied Sciences (Basel, Switzerland)*, *10*(21), 7526. DOI: 10.3390/app10217526

Bamigboye, G. O., Davies, I., Nwanko, C., Michaels, T., Adeyemi, G., & Ozuor, O. (2019, November). Innovation in construction materials-a review. []. IOP Publishing.]. *IOP Conference Series. Materials Science and Engineering*, *640*(1), 012070. DOI: 10.1088/1757-899X/640/1/012070

Ben Hassen, T., El Bilali, H., Daher, B., & Burkart, S. (2025). Sustainable and resilient food systems in times of crises. *Frontiers in Nutrition*, *12*, 1564950. DOI: 10.3389/fnut.2025.1564950

Borges, C. S., Jalali, S., Tsokanas, P., Marques, E. A., Carbas, R. J., & da Silva, L. F. (2022). Sustainable development approaches through wooden adhesive joints design. *Polymers*, *15*(1), 89. DOI: 10.3390/polym15010089

Calvez, I., Garcia, R., Koubaa, A., Landry, V., & Cloutier, A. (2024). Recent Advances in Bio-Based Adhesives and Formaldehyde-Free Technologies for Wood-Based Panel Manufacturing. *Current Forestry Reports*, *10*(5), 386–400. DOI: 10.1007/s40725-024-00227-3

Department of Public Works and Town & Country Planning. (2025, May 05). The preparation for water hyacinth and weed removal in 33 provinces to prepare for the rainy season of 2023. ThaiNews. https://thainews.prd.go.th/th/news/detail/TCATG230722105501843

Dixit, M. K., Fernández-Solís, J. L., Lavy, S., & Culp, C. H. (2010). Identification of parameters for embodied energy measurement: A literature review. *Energy and Building*, *42*(8), 1238–1247. DOI: 10.1016/j.enbuild.2010.02.016

Elsacker, E., Vandelook, S., Van Wylick, A., Ruytinx, J., De Laet, L., & Peeters, E. (2020). A comprehensive framework for the production of mycelium-based lignocellulosic composites. *The Science of the Total Environment*, *725*, 138431. DOI: 10.1016/j.scitotenv.2020.138431

European Commission. (2020). *Circular economy action plan: For a cleaner and more competitive Europe*. Publications Office of the European Union. https://ec.europa.eu/environment/circular-economy/

Gaspar, F., Bakatovich, A., Davydenko, N., & Joshi, A. (2020). Building insulation materials based on agricultural wastes. In *Bio-based materials and biotechnologies for eco-efficient construction* (pp. 149–170). Woodhead Publishing. DOI: 10.1016/B978-0-12-819481-2.00008-8

Goodman, B. A. (2020). Utilization of waste straw and husks from rice production: A review. *Journal of Bioresources and Bioproducts*, *5*(3), 143–162. DOI: 10.1016/j.jobab.2020.07.001

Hasan, K. F., Horváth, P. G., Bak, M., & Alpár, T. (2021). A state-of-the-art review on coir fiber-reinforced biocomposites. *RSC Advances*, *11*(18), 10548–10571. DOI: 10.1039/D1RA00231G

Hasanov, R. I. (2023). Promoting sustainability in Azerbaijan's energy sector: A green policy evaluation and future outlook. *Green Economics*, *1*(1), 62–69.

HomePro. (n.d.). *HomePro: Your home improvement store*. Retrieved December 25, 2024, from https://www.homepro.co.th

Hussein, Z., Ashour, T., Khalil, M., Bahnasawy, A., Ali, S., Hollands, J., & Korjenic, A. (2019). Rice straw and flax fiber particleboards as a product of agricultural waste: An evaluation of technical properties. *Applied Sciences (Basel, Switzerland)*, *9*(18), 3878. DOI: 10.3390/app9183878

Hýsek, Š., Neuberger, P., Sikora, A., Schönfelder, O., & Ditommaso, G. (2019). Waste utilization: Insulation panel from recycled polyurethane particles and wheat husks. *Materials (Basel)*, *12*(19), 3075. DOI: 10.3390/ma12193075

Iftikhar, M., Asghar, A., Ramzan, N., Sajjadi, B., & Chen, W. Y. (2019). Biomass densification: Effect of cow dung on the physicochemical properties of wheat straw and rice husk based biomass pellets. *Biomass and Bioenergy*, *122*, 1–16. DOI: 10.1016/j.biombioe.2019.01.005

Illankoon, W. A. M. A. N., Milanese, C., Karunarathna, A. K., Liyanage, K. D. H. E., Alahakoon, A. M. Y. W., Rathnasiri, P. G., Collivignarelli, M. C., & Sorlini, S. (2023). Evaluating sustainable options for valorization of rice by-products in Sri Lanka: An approach for a circular business model. *Agronomy (Basel)*, *13*(3), 803. DOI: 10.3390/agronomy13030803

International Organization for Standardization. (1991). *ISO 8302: Thermal insulation — Determination of steady-state thermal resistance and related properties — Guarded hot plate apparatus*. ISO.

International Organization for Standardization. (1998). *ISO 10534-2: Acoustics — Determination of sound absorption coefficient and impedance in impedance tubes — Part 2: Transfer-function method*. ISO.

International Organization for Standardization. (2006). *ISO 14040: Environmental management — Life cycle assessment — Principles and framework*. ISO.

International Organization for Standardization. (2006). *ISO 14044: Environmental management — Life cycle assessment — Requirements and guidelines*. ISO.

Jirawattanasomkul, T., Minakawa, H., Likitlersuang, S., Ueda, T., Dai, J. G., Wuttiwannasak, N., & Kongwang, N. (2021). Use of water hyacinth waste to produce fibre-reinforced polymer composites for concrete confinement: Mechanical performance and environmental assessment. *Journal of Cleaner Production*, *292*, 126041. DOI: 10.1016/j.jclepro.2021.126041

Kliem, S., Kreutzbruck, M., & Bonten, C. (2020). Review on the biological degradation of polymers in various environments. *Materials (Basel)*, *13*(20), 4586. DOI: 10.3390/ma13204586

Krungthai, C. O. M. P. A. S. S. (2022). *Biomass product market: Opportunities from agricultural waste* [Presentation slides]. Krungthai Bank Public Company Limited. Retrieved from https://krungthai.com/Download/news/MediaFile_471slide_upcycling_food_wastes.pdf

. Kumar, B., & Cumbal, L. (2016). UV-Vis, FTIR and antioxidant study of Persea americana (Avocado) leaf and fruit: a comparison.

Land Development Department. (2025, May 05). Plowing of stubble to improve soil and increase rice yield. https://www.ldd.go.th/menu_moc/POSTER/rice/rice.htm

Lawanwadeekul, S., & Chindaprasirt, P. (2025). Sustainable construction materials from leveraging water hyacinth and wastepaper for superior sound absorption and thermal insulation. *Journal of Building Engineering*, *103*, 112147. DOI: 10.1016/j.jobe.2025.112147

Lawanwadeekul, S., Jun-On, N., Kongthavorn, P., Sangkas, T., & Daothong, S. (2024). Chemical-free thermal-acoustic panels from agricultural waste for sustainable building materials. *Cleaner Materials*, *12*, 100245. DOI: 10.1016/j.clema.2024.100245

Le, A., Rodrigo, N., Domingo, N., & Senaratne, S. (2023). Policy mapping for net-zero-carbon buildings: Insights from leading countries. *Buildings (Basel, Switzerland)*, *13*(11), 2766. DOI: 10.3390/buildings13112766

Macarthur, E. L. L. E. N., & Heading, H. E. A. D. I. N. G. (2019). How the circular economy tackles climate change. *Ellen MacArthur Found*, *1*, 1–71.

Maderuelo-Sanz, R., García-Cobos, F. J., Sánchez-Delgado, F. J., Mota-López, M. I., Meneses-Rodríguez, J. M., Romero-Casado, A., Acedo-Fuentes, P., & López-Ramos, L. (2022). Mechanical, thermal and acoustical evaluation of biocomposites made of agricultural waste for ceiling tiles. *Applied Acoustics*, *191*, 108689. DOI: 10.1016/j.apacoust.2022.108689

Maiti, B., Abramov, A., Finn, M. G., & Díaz, D. D. (2021). Biopolymers as sustainable metal bio-adhesives. *Journal of Applied Polymer Science*, *138*(5), 49783. DOI: 10.1002/app.49783

Marconi, M., Barbanera, M., Calabrò, G., & Baffo, I. (2020). Reuse of leather scraps for insulation panels: Technical and environmental feasibility evaluation. *Procedia CIRP*, *90*, 55–60. DOI: 10.1016/j.procir.2020.01.053

Martwong, E., Minanandana, T., Puksuwan, K., Junthip, J., & Sukhawipat, N. (2024). Sound absorption and thermal insulation materials from waste palm oil for housing application: Green polyurethane/water hyacinth fiber sheet composite. *Construction & Building Materials*, *438*, 137007. DOI: 10.1016/j.conbuildmat.2024.137007

Merli, F., Belloni, E., & Buratti, C. (2021). Eco-sustainable wood waste panels for building applications: Influence of different species and assembling techniques on thermal, acoustic, and environmental performance. *Buildings (Basel, Switzerland)*, *11*(8), 361. DOI: 10.3390/buildings11080361

Mirindi, D. (2024). A review of particleboard development and performance using non-toxic and biodegradable adhesives. *International Journal of Engineering Trends and Technology*, *72*(5), 252–260. DOI: 10.14445/22315381/IJETT-V72I5P126

Moini, N., Khaghanipour, M., Faridani, F., & Jahandideh, A. (2022). Green adhesives—Past, present, and future outlook. In *Green Sustainable Process for Chemical and Environmental Engineering and Science* (pp. 341–372). Elsevier. DOI: 10.1016/B978-0-323-99643-3.00006-1

Mondem, N., & Balunaini, U. (2025). Optimal Blends of Artificial and Natural Aggregates for Enhanced Pavement Performance. In *Conference on GeoPractices towards Sustainable Infrastructure* (pp. 41-51). Springer, Singapore. DOI: 10.1007/978-981-96-3220-6_4

Mulder, W. J., Gosselink, R. J. A., Vingerhoeds, M. H., Harmsen, P. F. H., & Eastham, D. (2011). Lignin based controlled release coatings. *Industrial Crops and Products*, *34*(1), 915–920. DOI: 10.1016/j.indcrop.2011.02.011

Ojo, A. O. (2023). An overview of lignocellulose and its biotechnological importance in high-value product production. *Fermentation (Basel, Switzerland)*, *9*(11), 990. DOI: 10.3390/fermentation9110990

Ormondroyd, G. A. (2015). Adhesives for wood composites. In *Wood composites* (pp. 47–66). Woodhead Publishing. DOI: 10.1016/B978-1-78242-454-3.00003-2

Ouakarrouch, M., Bousshine, S., Bybi, A., Laaroussi, N., & Garoum, M. (2022). Acoustic and thermal performances assessment of sustainable insulation panels made from cardboard waste and natural fibers. *Applied Acoustics*, *199*, 109007. DOI: 10.1016/j.apacoust.2022.109007

Parmesan, C., Anshari, G., Buotte, P., Campbell, D., Castellanos, E., & Cowie, A. (2022). Nature-based solutions for climate change: Mitigation and adaptation. *Climatic Change*, •••, 303–309.

Philip, S., & Rakendu, R. (2020). Thermal insulation materials based on water hyacinth for application in sustainable buildings. *Materials Today: Proceedings*, *33*, 3803–3809. DOI: 10.1016/j.matpr.2020.06.219

Pizzi, A., Papadopoulos, A. N., & Policardi, F. (2020). Wood composites and their polymer binders. *Polymers*, *12*(5), 1115. DOI: 10.3390/polym12051115

Ramesh, M., Rajeshkumar, L., Sasikala, G., Balaji, D., Saravanakumar, A., Bhuvaneswari, V., & Bhoopathi, R. (2022). A critical review on wood-based polymer composites: Processing, properties, and prospects. *Polymers*, *14*(3), 589. DOI: 10.3390/polym14030589

Razali, N., Aziz, A. H. A., & Hassan, R. (2024). Governments' initiatives towards Islamic green finance. In *Islamic Green Finance* (pp. 135-143). Routledge.

Ricciardi, P., Belloni, E., Merli, F., & Buratti, C. (2021). Sustainable panels made with industrial and agricultural waste: Thermal and environmental critical analysis of the experimental results. *Applied Sciences (Basel, Switzerland)*, *11*(2), 494. DOI: 10.3390/app11020494

Roque, A. J., Paleologos, E. K., O'Kelly, B. C., Tang, A. M., Reddy, K. R., Vitone, C., Mohamed, A.-M. O., Koda, E., Goli, V. S. N. S., Vieira, C. S., Fei, X., Sollecito, F., Vaverková, M. D., Plötze, M., Petti, R., Podlasek, A., Puzrin, A. M., Cotecchia, F., Osiński, P., & Singh, D. N. (2021). Sustainable environmental geotechnics practices for a green economy. *Environmental Geotechnics*, *9*(2), 68–84. DOI: 10.1680/jenge.21.00091

Sáez-de-Guinoa, A., Zambrana-Vasquez, D., Fernández, V., & Bartolomé, C. (2022). Circular economy in the European construction sector: A review of strategies for implementation in building renovation. *Energies*, *15*(13), 4747. DOI: 10.3390/en15134747

Salas-Ruiz, A., del Mar Barbero-Barrera, M., & Ruiz-Téllez, T. (2019). Microstructural and thermo-physical characterization of a Water Hyacinth petiole for thermal insulation particle board manufacture. *Materials (Basel)*, *12*(4), 560. DOI: 10.3390/ma12040560

Sambucci, M., Sibai, A., & Valente, M. (2021). Recent advances in geopolymer technology. A potential eco-friendly solution in the construction materials industry: A review. *Journal of Composites Science*, *5*(4), 109. DOI: 10.3390/jcs5040109

Sangmesh, B., Patil, N., Jaiswal, K. K., Gowrishankar, T. P., Selvakumar, K. K., Jyothi, M. S., & Kumar, S. (2023). Development of sustainable alternative materials for the construction of green buildings using agricultural residues: A review. *Construction & Building Materials*, *368*, 130457. DOI: 10.1016/j.conbuildmat.2023.130457

Sarapin, P. Netpae, T., Phothi, R., Sangon, A. . ., & Chaothum, C. . . (2024). The Relationship Between Hotspots From Open Field Burning of Rice Straw and PM2.5 Change in Nakhon Sawan Province PSRU Journal of Science and Technology, 8(3), 54–68. retrieved from https://ph01.tci-thaijo.org/index.php/Scipsru/article/view/253255 (Original work published December 20, 2023)

Saxena, M., Morchhale, R. K., Asokan, P., & Prasad, B. K. (2008). Plant fiber—Industrial waste reinforced polymer composites as a potential wood substitute material. *Journal of Composite Materials*, *42*(4), 367–384. DOI: 10.1177/0021998307087014

Sheng, J., Ji, D., Yu, F., Cui, L., Zeng, Q., Ai, N., & Ji, J. (2014). Influence of chemical treatment on rice straw pyrolysis by TG-FTIR. *IERI Procedia*, *8*, 30–34. DOI: 10.1016/j.ieri.2014.09.006

Tanasombat, M., Okabayashi, Y., Sakurai, K., Thaiutsa, B., Thammincha, S., & Suekeaw, P. (2005). Silvicultural performance of paper mulberry in Thailand. *Tropics*, *14*(2), 149–162. DOI: 10.3759/tropics.14.149

Tang, L., Wu, T., & Li, Q. (2025). Construction and optimization strategy for collaborative governance of construction waste resource utilization. *KSCE Journal of Civil Engineering*, *29*(3), 100034. DOI: 10.1016/j.kscej.2024.100034

Thai Watsadu. (n.d.). *Thai Watsadu: Building materials and home improvement*. Retrieved December 25, 2024, from https://www.thaiwatsadu.com

Tiza, M. T., Ogunleye, E., Jiya, V. H., Onuzulike, C., Akande, E. O., & Terlumun, S. (2023). Integrating sustainability into civil engineering and the construction industry. *Journal of Cement Based Composites*, *4*(1), 1–11. DOI: 10.36937/cebacom.2023.5756

Wahyuni, N. L. E., Soeswanto, B., Aulia, H., & Fadhilah, R. A. (2019, September). The effects of rice straw weight fraction and particle size on thermal conductivity and mechanical properties of polypropylene composite. [). IOP Publishing.]. *Journal of Physics: Conference Series*, *1295*(1), 012067. DOI: 10.1088/1742-6596/1295/1/012067

World Health Organization (WHO). (2022). *WHO global air quality guidelines: Particulate matter (PM2.5 and PM10), ozone, nitrogen dioxide, sulfur dioxide and carbon monoxide*. Geneva: WHO Press. https://www.who.int/publications/i/item/9789240034228

. Yang, B., Blondeel, M., & Pattberg, P. (2024). The Evolution of China's Domestic Climate Policy Frames (2009–2024): Problems, Solutions, and Motivations. *Environmental Policy and Governance*.

Yang, J., Yang, Y., Wu, W. M., Zhao, J., & Jiang, L. (2014). Evidence of polyethylene biodegradation by bacterial strains from the guts of plastic-eating waxworms. *Environmental Science & Technology*, *48*(23), 13776–13784. DOI: 10.1021/es504038a

Yeon, J. O., Kim, K. W., Yang, K. S., Kim, J. M., & Kim, M. J. (2014). Physical properties of cellulose sound absorbers produced using recycled paper. *Construction & Building Materials*, *70*, 494–500. DOI: 10.1016/j.conbuildmat.2014.07.088

Zhong, Q., Li, H., Liang, S., Jetashree, , Wu, X., Qi, J., & Wang, S. (2022). Changes of production and consumption structures in coastal regions lead to mercury emission control in China. *Journal of Industrial Ecology*, *26*(5), 1760–1770. DOI: 10.1111/jiec.13314

Zucaro, A., & Agostinho, F. (2025). Urban sustainability: Challenges and opportunities for resilient and resource-efficient cities. *Frontiers in Sustainable Cities*, *7*, 1556974. DOI: 10.3389/frsc.2025.1556974

Chapter 8
Advancing Sustainable Construction With the Role of Next-Generation Eco-Friendly and High-Performance Materials

Pedro Miguel Gomes
https://orcid.org/0009-0005-4996-4534
Secretary General for Environment, Portugal

Tiago Jordão Cardoso
https://orcid.org/0009-0008-5223-2299
Secretary General for Environment, Portugal

ABSTRACT

The advancement of sustainable construction relies on next-generation eco-friendly and high-performance materials that minimize environmental impact while enhancing structural efficiency. Innovations in biodegradable composites, recycled materials, self-healing concrete, and energy-efficient insulation contribute to reducing carbon footprints and improving building longevity. Smart materials with adaptive properties, such as phase-change materials and nanotechnology-based solutions, further enhance energy conservation and resilience. This paper explores the integration of these cutting-edge materials into sustainable construction practices, highlighting their benefits, challenges, and potential for large-scale adoption. By leveraging technological advancements and green engineering principles, the construction industry can achieve greater sustainability, energy efficiency, and environmental responsibility.

DOI: 10.4018/979-8-3373-2414-2.ch008

1. INTRODUCTION

The construction industry stands at a critical crossroads in the 21st century. Rapid urbanization, population growth, and escalating climate change concerns have placed unprecedented pressure on the sector to shift from resource-intensive practices toward sustainable development models. Globally, buildings and construction are responsible for approximately 37% of energy-related CO_2 emissions, driven largely by the production and use of conventional materials such as Portland cement, steel, and glass. These materials, while essential for structural integrity and large-scale development, contribute significantly to environmental degradation, including carbon emissions, water pollution, and the depletion of non-renewable resources.

In parallel, the demand for infrastructure is surging, particularly in emerging economies where urban expansion is occurring at an unprecedented pace. As such, there is an urgent need for the construction sector to decouple economic growth from environmental impact. This requires a profound rethinking of how buildings are designed, constructed, and maintained — with materials playing a central role in this transformation.

Over the last two decades, advancements in material science, nanotechnology, and biotechnology have opened new pathways for reimagining the built environment. The emergence of eco-friendly and high-performance materials is enabling the construction of buildings that are not only structurally robust but also energy-efficient, climate-resilient, and environmentally restorative. These materials go beyond incremental improvements; they represent a paradigm shift in how we conceptualize the sustainability of the built environment.

For instance, biodegradable composites derived from renewable sources such as hemp, bamboo, and mycelium are redefining the boundaries of natural architecture. Similarly, recycled-content materials—including concrete with industrial by-products, reclaimed plastics, and recovered glass—support circular economy principles by extending the lifecycle of existing resources. Self-healing materials, notably microbial or microcapsule-based concretes, demonstrate the capacity to autonomously repair structural damage, reducing long-term maintenance costs and enhancing durability. Furthermore, phase-change materials (PCMs) offer passive thermal regulation by absorbing and releasing heat as ambient temperatures fluctuate, thereby reducing HVAC energy demand. At the cutting edge, nanotechnology-based materials improve performance at the molecular level, offering properties such as self-cleaning surfaces, enhanced mechanical strength, corrosion resistance, and air purification.

The convergence of these innovations is not merely technological but systemic. When integrated holistically into construction systems, these materials can significantly reduce embodied carbon, improve indoor environmental quality, and support buildings that are adaptive, smart, and net-zero ready. In fact, such materials are increasingly being linked with concepts such as regenerative architecture, low-impact design, and smart infrastructure.

Despite these promising developments, the widespread adoption of next-generation materials remains limited due to several challenges. Technical barriers such as lack of standardized testing protocols, economic factors like high initial costs, and regulatory uncertainties hinder their mainstream integration. Additionally, knowledge gaps among construction professionals and resistance to change within traditional building cultures further slow down the transition.

Given this context, this paper aims to provide a comprehensive and critical exploration of the role of next-generation eco-friendly and high-performance materials in sustainable construction. It seeks to answer key questions:

- What are the most promising classes of emerging materials for sustainable construction?
- How do these materials perform in terms of mechanical, thermal, and environmental properties?
- What are the barriers to their large-scale deployment?
- How can industry stakeholders, policymakers, and researchers accelerate their integration?

By drawing from recent scientific literature, case studies, and industry reports, this study contributes to the growing body of knowledge aimed at aligning construction practices with the broader goals of climate resilience, resource efficiency, and environmental stewardship. The findings offer valuable insights for engineers, architects, material scientists, and decision-makers committed to transforming the construction industry into a pillar of sustainability.

2. THE NEED FOR MATERIAL INNOVATION IN SUSTAINABLE CONSTRUCTION

The construction sector is undergoing a transformative shift in response to intensifying environmental concerns, regulatory pressure, and the urgent need to align with global climate targets. As one of the largest contributors to greenhouse gas emissions, resource depletion, and solid waste generation, the traditional construction model—based heavily on cement, steel, and virgin materials—is no longer sustainable.

Cement production alone accounts for approximately 8% of global CO_2 emissions, while construction and demolition waste (CDW) comprises over 35% of total solid waste in many developed economies (Sun et al., 2025; Naghibalsadati et al., 2025). These impacts are primarily rooted in the linear "take–make–dispose" model of material use, which fails to account for lifecycle performance, reusability, or environmental regeneration.

To address this, material innovation has become a core pillar in the transition to sustainable construction. Emerging materials are now designed not just for structural strength, but also for carbon reduction, durability, thermal performance, and compatibility with circular economy principles.

2.1 Strategic Drivers of Material Innovation

The impetus for innovation in construction materials is deeply rooted in the convergence of global challenges—climate change, urbanization, and resource scarcity—alongside emerging opportunities provided by digital transformation and scientific advancement. Governments, environmental agencies, and urban planners increasingly recognize that the materials selected for construction are not neutral choices; they have long-term implications for energy use, indoor environmental quality, and ecological footprint.

One of the most influential drivers is international climate policy, such as the Paris Agreement and the European Green Deal, which set ambitious targets for carbon neutrality by mid-century. These frameworks place increasing pressure on carbon-intensive sectors, including construction, to reduce emissions not only during building operation but also across the entire material lifecycle. Consequently, embodied carbon accounting has become a key metric in sustainability assessments.

In addition to regulatory forces, urban population growth and the need for climate-resilient infrastructure are accelerating demand for materials that are both high-performing and environmentally responsible. Traditional materials cannot fully meet this dual requirement. As a result, the industry is exploring new material paradigms—such as circular materials, bio-based systems, and smart composites—that are better suited to the environmental and social demands of contemporary construction.

Moreover, the digitalization of the design and planning process—particularly through the adoption of Building Information Modelling (BIM) and integrated lifecycle analysis—enables early-stage simulation and evaluation of environmental impacts. These tools are now being combined with artificial intelligence and big data analytics to optimize material choices based on performance, cost, and sustainability simultaneously.

Table 1. Comparative Overview: Traditional vs Innovative Construction Materials

Feature / Impact	Traditional Materials (e.g. Cement, Steel)	Innovative Materials (e.g. Geopolymers, Recycled Aggregates, Bio-composites)
Embodied Carbon (kg CO_2e/kg)	High (Cement ~0.9–1.0; Steel ~1.9–2.1)	Low to Negative (e.g., Geopolymer: ~0.2–0.3; Hempcrete: Negative)
Resource Origin	Non-renewable	Renewable / Secondary (waste-derived)
End-of-Life Disposal	Landfill, downcycling	Recyclable, biodegradable, or reusable
Thermal Performance	Often poor without insulation	High (e.g., phase-change, bio-insulators)
Durability / Maintenance	High maintenance over time	Self-healing or extended-life capabilities
Cost (Initial vs Lifecycle)	Low initial, high lifecycle cost	Moderate initial, lower lifecycle cost

Sources: Sun et al. (2025), Suriya Kumar et al. (2025), Akbulut et al. (2025)

2.2 Applications of Innovative Materials in Practice

Innovative materials are increasingly being implemented beyond the laboratory and are proving their viability in real-world construction scenarios. This transition from experimental research to field application is critical for validating their structural, economic, and environmental performance under practical conditions.

One of the most prominent examples is the development of ultra-high-performance concrete (UHPC) made entirely from industrial by-products. As shown by Sun et al. (2025), a combination of fly ash and steel slag can yield ASW-UHPC with mechanical strength comparable to conventional UHPC, while achieving a drastic reduction in embodied CO_2 emissions. This material is particularly suited to high-stress applications such as bridge decks, precast panels, and earthquake-resistant components.

Similarly, Suriya Kumar et al. (2025) introduced a fly ash-based geopolymer composite enhanced with oily sludge ash—a material previously regarded as hazardous waste. Their work demonstrates that this geopolymer not only reduces environmental impact but also maintains compressive strength suitable for load-bearing elements in modular construction. This reflects a broader trend where waste valorization is driving material circularity and innovation simultaneously.

In terms of digital tools, Naghibalsadati et al. (2025) illustrated how BIM-integrated lifecycle assessment helps decision-makers identify optimal material combinations based on carbon performance, recyclability, and end-of-life scenarios. This approach empowers project teams to compare design alternatives with real-time environmental feedback, shifting the paradigm from reactive compliance to proactive sustainability strategy.

In construction methodology, modular and prefabricated systems are emerging as natural complements to next-generation materials. Research by Khan et al. (2025) shows that modular systems incorporating innovative connection technologies not only enhance structural safety and earthquake resistance, but also enable easy disassembly, relocation, or reuse—which drastically reduces construction and demolition waste (CDW). These systems are increasingly used in housing developments, emergency shelters, and infrastructure in environmentally sensitive regions.

Collectively, these examples highlight a growing body of evidence that material innovation is viable at scale and can be tailored to diverse applications across climatic zones, building types, and economic contexts.

2.3 Critical Barriers to Adoption

Despite the growing interest in sustainable and innovative materials, their adoption at scale remains constrained by a series of structural, economic, and cultural barriers.

The most immediate challenge is the lack of standardization and certification frameworks. Many emerging materials—such as biocomposites, geopolymer concretes, and recycled aggregates—lack internationally recognized testing procedures or performance codes. This creates uncertainty for engineers and builders, especially in large-scale infrastructure projects where liability and compliance are major concerns. Without clear performance benchmarks, the risk perception associated with non-traditional materials remains high.

Another major barrier is cost, particularly initial investment. Although many sustainable materials offer reduced lifecycle costs due to lower maintenance, higher durability, or energy savings, their upfront price can be significantly higher than that of traditional alternatives. This cost differential is a deterrent in competitive markets, especially where short-term budgeting outweighs long-term thinking.

Supply chain limitations also affect adoption. Innovative materials often depend on localized sources of biomass, waste, or specialized processing, which may not be readily available or scalable in all regions. This uneven accessibility can hinder global uptake and increase transportation-related emissions—counteracting their environmental benefits.

Culturally, the construction sector is risk-averse and heavily reliant on established norms. Many professionals, including contractors, designers, and procurement officers, have limited exposure to new material technologies, and may lack the training to evaluate or implement them properly. This inertia is reinforced by conservative client preferences and fragmented regulatory environments.

To overcome these challenges, policy interventions are essential. These include:

- Government incentives for low-carbon materials

- Mandatory green procurement for public buildings
- Research funding for long-term performance testing
- Knowledge transfer platforms for training and certification

By addressing these barriers through coordinated action, stakeholders can create an enabling environment for sustainable material integration.

2.4 From Innovation to Transformation

Material innovation in construction must be viewed not as an isolated technological pursuit, but as a foundational element of systemic transformation within the built environment. It touches every dimension of sustainability—from carbon mitigation and climate adaptation to social equity and economic resilience.

The introduction of advanced materials enables a shift away from extractive and linear models of construction toward systems that are regenerative, circular, and climate-responsive. Bio-based materials store carbon instead of emitting it. Recycled aggregates reduce the need for raw material extraction. Nanomaterials extend the lifespan and functionality of building elements. Together, these innovations reconfigure the material metabolism of cities and reduce pressure on planetary boundaries.

Importantly, the transformation must also be institutional. The construction industry must evolve from a project-by-project mentality to one that is system-oriented and lifecycle-based. This implies:

- Embedding sustainability metrics in design, procurement, and delivery
- Valuing materials not only for their immediate function, but for their long-term ecological and social performance
- Integrating circular economy principles into policy, education, and business models

Ultimately, material innovation is a convergence point where science, policy, industry, and environmental ethics intersect. It empowers the construction sector to go beyond harm reduction and become a proactive force for environmental regeneration, economic sustainability, and human wellbeing in the Anthropocene.

3. TYPES OF NEXT-GENERATION ECO-FRIENDLY MATERIALS

The advancement of sustainable construction relies heavily on the identification and deployment of materials that are not only environmentally benign but also technically robust, adaptable, and scalable. These next-generation materials go beyond reducing environmental impact; they often enhance durability, promote energy efficiency, and enable circular design thinking. This section categorizes five key material types that are redefining the sustainability paradigm in construction: biodegradable and bio-based composites, recycled and waste-derived materials, self-healing concrete, phase-change materials, and nanotechnology-enabled solutions.

3.1 Biodegradable and Bio-Based Composites

The global shift towards sustainable construction practices has placed biodegradable and bio-based materials at the forefront of innovation. These materials respond directly to the environmental limitations of traditional construction systems by offering low-carbon, renewable, and end-of-life-friendly alternatives to conventional products. Their key appeal lies in their dual benefit: they reduce both embodied carbon emissions during production and environmental pollution at disposal, since many are designed to naturally degrade or be composted.

Biodegradable and bio-based materials have gained considerable momentum as alternatives to synthetic, carbon-intensive construction products. These materials are typically derived from renewable biological sources, such as plant fibres (e.g., hemp, flax, jute), fungal mycelium, bamboo, and cross-laminated timber (CLT). Their adoption aligns with circular economy principles by enabling biological cycling, reducing embodied carbon, and often allowing end-of-life biodegradability.

For example, hempcrete, a composite of hemp hurds and lime, offers excellent thermal insulation, low density, and carbon sequestration capacity, absorbing more CO_2 than it emits during its lifecycle (Martinez et al., 2023). Similarly, mycelium-based bricks are formed from fungal structures that grow within agricultural waste substrates, offering biodegradability, fire resistance, and zero-toxicity characteristics (Parhizi et al., 2025).

Bio-based composites have shown promise in non-structural applications such as insulation panels, cladding, and interior partitions, but their structural potential is growing through hybridization with other materials or natural resins. Nevertheless, challenges remain regarding standardization, moisture resistance, and scalability, which need further research and policy support.

Biological Carbon Storage

A significant advantage of these materials is their capacity for carbon sequestration. During growth, plants absorb atmospheric CO_2 via photosynthesis, which becomes chemically stored in the biomass. When harvested and processed into composites, this carbon remains locked in the material throughout its service life. For example, hemp can absorb up to 15 tonnes of CO_2 per hectare during growth, and when used in construction as hempcrete, it continues to sequester additional carbon through the carbonation of lime over time (Martinez et al., 2023).

Innovative Examples in Practice

Mycelium-based building blocks are a remarkable innovation within this category. These are produced by cultivating fungal mycelium in moulds filled with agricultural by-products such as sawdust or straw. The fungus binds the substrate into a dense, solid form with fire-resistant and self-extinguishing properties, all without the use of synthetic adhesives. Parhizi et al. (2025) demonstrated that mycelium composites achieve comparable insulation performance to polystyrene, with none of the associated toxicity or environmental persistence.

CLT and laminated bamboo panels, while not biodegradable in the strictest sense, are renewable and recyclable, and offer structural-grade performance with significantly lower embodied carbon compared to steel or concrete. They are increasingly used in multi-storey timber buildings, a trend supported by international green building certifications like LEED and BREEAM.

Applications and Technical Potential

Currently, bio-based materials are most widely used in non-structural applications, including:

- Internal partition walls
- Acoustic ceiling tiles
- Floor underlays
- Exterior cladding
- Thermal/acoustic insulation boards

However, technological advances in fibre treatment, natural resin binding, and lamination techniques are expanding their potential for load-bearing and structural roles. The use of bio-resins (e.g., lignin- or starch-based adhesives) instead of petrochemical binders is a key enabler of structural bio-composites that maintain sustainability integrity throughout the product.

Challenges and Limitations

Despite their ecological appeal and growing market acceptance, several barriers limit the widespread adoption of biodegradable and bio-based composites:

• Standardization and Certification: Lack of international standards for mechanical and fire performance makes it difficult to integrate these materials into conventional structural codes.
• Moisture Susceptibility: Many natural fibres are hygroscopic and vulnerable to biological degradation (e.g., mould, fungi) without protective treatment, especially in humid climates.
• Scalability and Cost: Production remains limited in scale, and costs can be 10–25% higher than mass-produced synthetic equivalents, especially when sourced from certified organic or regional inputs.
• Durability Concerns: While composite reinforcement and surface treatments are improving, long-term durability in high-stress environments remains under study.

The Role of Policy and Innovation

To overcome these constraints, policy support, investment in biomanufacturing infrastructure, and public procurement initiatives are essential. Several European and North American municipalities have begun to incorporate bio-based material quotas in public building projects, and life cycle-based taxation models are encouraging their adoption over synthetic materials with high environmental costs.

Moreover, interdisciplinary R&D—combining biotechnology, material science, and digital fabrication—is helping to close performance gaps. For example, additive manufacturing with bio-inks, or myco-fabrication in 3D moulds, is pushing the limits of what biodegradable materials can achieve in architectural and structural design.

3.2 Recycled and Waste-Derived Materials

The reuse of construction and demolition waste (CDW) and other industrial by-products plays a critical role in reducing the environmental burden of new construction. Materials such as recycled aggregates, fly ash, ground granulated blast furnace slag (GGBFS), and recycled plastics or glass are being widely incorporated into structural and non-structural components.

According to Akbulut et al. (2025), recycled aggregate concrete (RAC) offers comparable compressive strength to conventional concrete, especially when supplemented with pozzolanic materials. Additionally, using fly ash and oily sludge ash in geopolymer composites, as explored by Suriya Kumar et al. (2025), can reduce embodied carbon by up to 60% while maintaining mechanical stability.

To visualize the comparative sustainability benefits, the table below highlights key waste-derived materials and their applications:

Table 2. Common Waste-Derived Construction Materials and Their Environmental Benefits

Material	Source Waste	Application Area	Environmental Benefit
Recycled Aggregates	CDW	Structural concrete, base layers	Reduces need for virgin aggregates
Fly Ash & Slag	Coal plants, steel industry	Geopolymers, blended cement	Lowers embodied carbon, improves durability
Recycled Plastics	Packaging, municipal waste	Bricks, insulation, cladding	Diverts plastic from landfill, adds thermal value
Crushed Glass	Beverage industry waste	Tiles, paving blocks	Reduces sand extraction, aesthetic diversity

3.3 Self-Healing Concrete

As construction activity accelerates globally, the extraction of natural resources and disposal of construction and demolition waste (CDW) have reached unsustainable levels. In response, the industry is increasingly turning to recycled and waste-derived materials as a strategic avenue for reducing environmental impact, improving resource circularity, and complying with tightening sustainability standards. These materials are produced from post-industrial, post-consumer, or construction-derived waste streams, and are being integrated into structural, non-structural, and finishing applications.

The use of recycled materials contributes to waste diversion from landfills, reduction of embodied carbon, and conservation of virgin resources, while also supporting the principles of a circular economy. The environmental benefits are particularly significant when these materials replace high-impact virgin materials, such as natural aggregates, cement, and conventional polymers.

Concrete remains the most widely used construction material, but it is prone to cracking, which leads to water ingress, corrosion of reinforcement, and ultimately structural degradation. Self-healing concrete addresses this issue by incorporating materials that autonomously repair microcracks, enhancing the material's longevity and reducing maintenance costs.

There are primarily two approaches:

- Microbial-induced calcite precipitation (MICP) using bacteria such as Bacillus sphaericus that produce calcium carbonate to fill cracks.
- Encapsulated healing agents, such as polymers or minerals that are released upon cracking.

According to Singh et al. (2024), self-healing concrete can heal cracks up to 0.5 mm wide under favorable moisture conditions, extending the service life of structures by over 30%. While initial costs are higher, life-cycle cost analysis (LCCA) often reveals lower total expenditure due to decreased repair and downtime needs.

This technology is particularly promising for infrastructure applications where maintenance access is difficult or costly, such as bridges, tunnels, and marine structures.

Applications in Practice

Real-world implementations are growing across both infrastructure and building sectors:

- In road construction, recycled asphalt pavement (RAP) is widely used, lowering bitumen demand and emissions.
- In building envelopes, recycled PET-based panels have been applied for thermal and acoustic insulation in schools and public buildings.
    - Modular construction companies now incorporate waste-plastic bricks and panels into walls and partitions, especially in low-cost housing.

Furthermore, emerging technologies such as cold-bonded aggregates, alkali-activated binders, and additive manufacturing with recycled filaments are expanding the applicability of waste-based materials into high-value, customizable products.

Environmental and Economic Impacts

The substitution of virgin materials with recycled alternatives results in substantial sustainability gains:

- Recycled concrete aggregate (RCA) can reduce GHG emissions by 25–50% compared to natural aggregate concrete, depending on transport and processing distances (Akbulut et al., 2025).
- Geopolymers from fly ash have demonstrated 70–90% lower CO_2 footprint and better chemical resistance in corrosive environments (Suriya Kumar et al., 2025).
- Plastic-derived panels and blocks are lighter, reducing structural loads and transport emissions.

Economically, recycled materials often lower raw material costs, though processing and quality control expenses can offset some of these gains. Still, the overall life-cycle cost tends to be lower due to enhanced durability and reduced maintenance needs.

Barriers to Adoption

Despite promising developments, several barriers continue to limit widespread use:

- Lack of Technical Standards and Guidelines

Many recycled materials lack uniform performance data, especially regarding long-term durability, freeze–thaw resistance, and chemical stability. This undermines confidence among engineers and builders.

- Perception of Inferiority

There remains a stigma that recycled materials are "lower quality" or risky, particularly in structural applications.

- Regulatory and Procurement Constraints

Building codes and public procurement policies often favour conventional materials, making it difficult for recycled products to compete.

- Supply Chain Inconsistency

The availability and quality of waste-derived inputs are highly location-dependent and subject to fluctuations in collection and sorting infrastructure.

Policy and Innovation Opportunities

Governments and international agencies are beginning to incentivize circular material use. Policies such as recycled content mandates, green building certifications, and waste-to-resource tax credits are increasingly encouraging uptake. For instance:

- The EU Construction Product Regulation includes minimum performance benchmarks for recycled materials.
- LEED v4.1 provides points for building product disclosure and optimization, favouring materials with EPDs and recycled content.

On the innovation front, AI-powered material sorting, robotic deconstruction, and on-site crushing and recycling units are making recycled material use more feasible and economically attractive.

3.4 Phase-Change Materials (PCMs)

Phase-change materials (PCMs) have emerged as a promising class of thermal storage materials capable of passively regulating indoor temperatures in buildings. These substances operate based on the principle of latent heat storage: during a phase transition, typically from solid to liquid and vice versa, PCMs absorb or release significant amounts of thermal energy while maintaining a nearly constant temperature. This thermal buffering effect can reduce fluctuations in indoor environments, helping to maintain occupant comfort, decrease reliance on heating and cooling systems, and improve the overall energy efficiency of buildings.

The effectiveness of PCMs in the built environment stems from their ability to store and release heat precisely when indoor temperatures begin to exceed or fall below acceptable comfort levels. During the daytime, when solar radiation or internal gains increase room temperatures, the PCM absorbs excess heat by melting, thereby limiting the rise in temperature. At night, when temperatures drop, the material solidifies and releases the stored heat, maintaining thermal stability. This cyclical process reduces the need for active HVAC systems, particularly in climates with significant diurnal temperature variations.

Among the most common PCMs used in construction are paraffin waxes, salt hydrates, and fatty acids. Paraffin-based PCMs are favored for their chemical stability, non-corrosive nature, and long cycling life, although they exhibit low thermal conductivity and pose a flammability risk if not properly treated. Salt hydrates, on the other hand, offer higher latent heat storage capacity and better thermal conductivity, but can suffer from issues such as subcooling, phase separation, and corrosion. Fatty acids, often derived from renewable sources, present a bio-based alternative with moderate thermal performance and enhanced environmental compatibility.

In building applications, PCMs are typically integrated into construction materials such as gypsum boards, concrete panels, plasters, insulation foams, and prefabricated wall assemblies. These PCM-enhanced components are designed to function as part of the building envelope, absorbing and releasing heat throughout the day to maintain stable interior conditions. According to Li et al. (2022), incorporating PCMs into external wall systems can reduce peak cooling loads by up to 25% in warm climates. Their study demonstrated that rooms with PCM-enhanced walls experienced indoor temperature reductions of 3–5°C during peak summer hours, thus lowering air conditioning demand and extending thermal comfort periods without energy use.

PCMs are particularly suitable for passive solar buildings, zero-energy buildings (ZEBs), and highly glazed constructions where heat gains are significant. In these contexts, the use of PCM technology contributes to flattening the energy demand curve, which not only reduces utility costs but also minimizes stress on urban electrical grids. Furthermore, PCMs can be combined with solar thermal systems to store excess energy for night-time release, enhancing the performance of renewable energy systems in buildings.

Despite their benefits, several technical and economic challenges must be overcome for PCMs to achieve widespread adoption. One of the most significant limitations is thermal stability. Over multiple heating and cooling cycles, some PCMs—particularly inorganic types—experience degradation, such as phase separation or reductions in latent heat capacity, which compromise long-term performance. Leakage is another concern, particularly in formulations that lack adequate encapsulation. To address this, microencapsulation techniques are increasingly used, though these add complexity and cost to the manufacturing process. Furthermore, PCMs are still relatively expensive compared to conventional insulation or thermal mass materials, often costing two to five times more. This cost premium can extend payback periods unless energy prices are high or climate conditions particularly favorable. Another obstacle is the lack of universal performance standards and integration into building codes, which hampers the confidence of architects and engineers in specifying these materials for large-scale projects.

In recent years, research has focused on enhancing the performance of PCMs through the incorporation of additives and nanomaterials. For instance, graphite and carbon-based fillers are used to increase thermal conductivity, while bio-based PCMs derived from vegetable oils offer promising ecological advantages with improved biodegradability. Hybrid systems are also under investigation, such as combining PCMs with smart windows, photovoltaic panels, or responsive façade systems to optimize building performance. Moreover, advances in simulation tools and digital modeling, particularly through Building Information Modeling (BIM), are enabling designers to evaluate PCM performance during the early design phase, helping to determine optimal placement, thickness, and configuration based on the building's thermal profile.

In conclusion, PCMs represent a valuable advancement in the toolkit for sustainable building design. Their ability to passively regulate indoor temperatures, reduce energy consumption, and increase occupant comfort aligns strongly with the goals of decarbonization and energy efficiency in the built environment. While cost, performance degradation, and regulatory gaps remain obstacles, ongoing research and supportive policy frameworks are paving the way for more widespread use. As climate challenges intensify, the integration of PCMs in buildings offers a viable strategy to create resilient, low-energy, and climate-adaptive structures that contribute to global sustainability objectives.

3.5 Nanotechnology-Enabled Solutions

Nanotechnology in construction offers revolutionary enhancements by manipulating materials at the atomic or molecular scale (1–100 nm). These enhancements include improved mechanical strength, thermal insulation, fire resistance, water repellency, and even pollution mitigation through photocatalysis.

Key applications include:

- Nano-silica for higher concrete strength and reduced permeability.
- Titanium dioxide (TiO_2) coatings with self-cleaning and air-purifying properties (via photocatalysis).
- Carbon nanotubes in composite materials for advanced strength-to-weight ratios.

Liang et al. (2025) demonstrated that nano-enhanced concrete exhibited up to 40% higher compressive strength and reduced chloride penetration, improving its suitability for coastal and marine applications. In addition, graphene-based materials are being explored for their extraordinary conductivity and durability.

Despite the technical promise, adoption is hindered by concerns regarding health and environmental risks, as well as cost and regulatory gaps. Further research is required to standardize safety protocols and lifecycle impact assessment for nanomaterials.

4. PERFORMANCE METRICS AND ENVIRONMENTAL IMPACT

As sustainable construction evolves from concept to practice, the need for robust performance assessment frameworks becomes increasingly critical. Traditional materials have long been evaluated primarily based on mechanical strength, durability, and cost. However, in the context of climate change and finite resource constraints, new-generation construction materials must now be assessed using a broader set of environmental and functional metrics that reflect their life-cycle performance, resource efficiency, and circularity potential.

The most widely adopted tool in this context is the Life Cycle Assessment (LCA) methodology, standardized under ISO 14040 and 14044. LCA allows for a comprehensive evaluation of a material's environmental impact from raw material extraction ("cradle") to the end of its life, including manufacturing, transport, installation, use, and disposal ("grave" or "cradle-to-cradle" in circular systems). Within LCA, key indicators include embodied energy, global warming potential (GWP), acidification potential, eutrophication, and resource depletion.

One of the most pressing performance criteria is embodied carbon, which refers to the total greenhouse gas emissions associated with the material's production and life cycle. Materials such as cement, steel, and aluminum are known for their extremely high embodied carbon. In contrast, bio-based materials (e.g., hempcrete, CLT, mycelium composites) often show negative or near-zero carbon balances, due to carbon sequestration during biomass growth and lower manufacturing energy requirements. For instance, the study by Alrehaili et al. (2025) demonstrated that fiber-reinforced polymer composites made from bio-resins can reduce embodied carbon by over 60% compared to aluminum and steel in façade systems.

Another key metric is durability, which directly affects the life span of a building component. A longer-lasting material may require fewer replacements or repairs, thereby reducing life-cycle emissions even if its initial embodied carbon is higher. However, durability must be balanced against recyclability; many high-durability materials (such as epoxy-based composites) are difficult to disassemble and reuse. In contrast, modular construction materials designed with design-for-disassembly principles show strong performance in both durability and circularity metrics, especially when coupled with digital material passports that facilitate tracking and reuse.

Recyclability and material circularity are now central to environmental performance evaluation. Sandanayake et al. (2025) emphasized the importance of integrating circularity indices into building material selection frameworks. These indices combine quantitative factors such as recycled content, reuse potential, and end-of-life recovery rates to provide a holistic score for material circularity. For example, concrete reinforced with textile and cardboard fibers, when analyzed using life cycle costing and carbon assessments, demonstrated lower environmental impact and better circularity alignment than traditional Portland cement mixes (Sandanayake et al., 2025).

Water footprint and toxicity potential are also gaining relevance, particularly in materials containing synthetic additives or nanomaterials. Though nanotechnology enhances performance (e.g., in self-cleaning surfaces or ultra-high strength composites), studies are ongoing to quantify the eco-toxicological risks associated with nanoparticle release during degradation or disposal. As regulatory frameworks evolve, material suppliers will be increasingly required to provide Environmental Product Declarations (EPDs), offering third-party verified data on environmental impacts across life-cycle phases.

Additionally, thermal performance metrics such as thermal conductivity, specific heat capacity, and phase transition behavior are crucial in materials like PCMs and bio-insulation. These indicators affect not only energy consumption during use but also the building's ability to comply with nearly zero-energy building (nZEB) standards.

Importantly, performance metrics must not be analyzed in isolation. The trade-offs between criteria—for instance, between low embodied energy and shorter lifespan, or between high durability and limited recyclability—require multi-criteria decision-making tools, such as life cycle cost analysis (LCCA) and multi-attribute utility theory (MAUT), to guide material selection in sustainable construction.

The environmental performance of next-generation materials must be evaluated using a systems-thinking approach. It is no longer sufficient to consider only strength or cost; materials must demonstrate alignment with carbon reduction targets, circularity principles, and healthy indoor environments. With the emergence of comprehensive databases, advanced simulation tools, and growing regulatory demands for environmental transparency, these performance metrics are rapidly becoming central to construction material specification, procurement, and policy design.

5. APPLICATIONS AND CASE STUDIES

The transition toward sustainable construction is not limited to theoretical advancements or laboratory-scale innovation. Across the globe, numerous built environment projects have already incorporated next-generation eco-friendly and high-performance materials into their design, construction, and operational phases. These case studies provide concrete evidence of the technical feasibility, economic viability, and environmental benefits of these new materials, while also highlighting the context-specific challenges that can inform future deployment.

Real-world applications span residential, commercial, institutional, and infrastructure projects, where materials such as hempcrete, mycelium composites, self-healing concrete, phase-change materials, and nanotechnology-enhanced systems have been implemented. Their performance under operational conditions provides valuable insights into material durability, indoor comfort, embodied carbon reduction, and maintenance savings.

5.1 Hempcrete and Bio-Based Composites in Housing

One of the most widespread applications of bio-based composites is in low-carbon residential housing. The UK-based "Flat House" project, completed in 2020 by Practice Architecture, is a prominent example. Built using hempcrete blocks grown and processed on-site in Cambridgeshire, the house demonstrates the feasibility of creating carbon-negative, breathable buildings. According to post-occupancy evaluations, the house achieves U-values below 0.13 $W/m^2{\cdot}K$, significantly improving thermal performance while maintaining excellent humidity regulation and air quality without the need for mechanical ventilation systems.

Another notable project is the BioHome3D in Maine, USA—a fully 3D-printed house made from wood fiber and bio-resin composites. Developed by the University of Maine's Advanced Structures and Composites Center, this structure showcases the use of locally sourced, biodegradable materials in prefabrication, allowing for low-waste construction and high thermal efficiency.

5.2 Recycled and Waste-Based Materials in Public Infrastructure

Recycled materials have been widely adopted in large-scale infrastructure projects. In Oslo, Norway, the Økern Portal—a 50,000 m^2 commercial building completed in 2022—used recycled concrete aggregates and wood from deconstructed buildings in its structural and finishing elements. The project avoided over 200 tonnes of CO_2 emissions compared to standard material use, while achieving BREEAM Outstanding certification.

In India, the Gandhinagar Railway Station redevelopment used fly ash-based concrete blocks, plastic waste bricks, and recycled steel in structural and non-structural applications. Not only did this lower the cost of materials by 15%, but it also contributed to the project's eligibility under India's Green Rating for Integrated Habitat Assessment (GRIHA) system.

5.3 Self-Healing Concrete in Bridges and Tunnels

Infrastructure durability is a critical concern for civil engineers, and self-healing concrete has shown remarkable promise in mitigating maintenance costs. The M5 Motorway Bridge in the Netherlands, first trialed in 2017, integrated bacteria-based self-healing agents within concrete beams. Subsequent monitoring revealed that cracks up to 0.8 mm were sealed autonomously within six months, significantly extending service life and reducing the need for disruptive repairs.

Similarly, the Túnel de la Línea in Colombia—a 9-km road tunnel across the Andes—incorporated microencapsulated polymer-based healing agents into concrete linings to counteract thermal expansion cracks. Maintenance inspections confirmed a 30% reduction in crack propagation and longer intervals between repair cycles, demonstrating both functional and financial benefits.

5.4 Phase-Change Materials in Energy-Efficient Buildings

Several demonstration projects have highlighted the ability of phase-change materials (PCMs) to enhance thermal comfort and energy savings. In China, a comparative study of buildings in different climate zones using composite PCM wallboards revealed significant reductions in both heating and cooling demands, especially in climates with large daily temperature variations. The study by Li et al. (2022) showed that optimized PCM integration led to lower indoor temperature fluctuations and up to 23% energy savings, making it highly suitable for zero-energy building applications.

In China, the Shanghai Zero Energy Building pilot, supported by Tsinghua University, combined salt hydrate PCMs with radiant heating and solar PV. The building maintained stable indoor temperatures without active HVAC use for over 80% of the year, underscoring PCMs' contribution to net-zero energy performance in hot-summer/cold-winter climates.

5.5 Nanotechnology-Enhanced Materials in Urban Envelopes

Nanomaterials have primarily been applied in façade systems, coatings, and high-performance concrete. In Italy, the Torre Eurosky in Rome, one of the tallest green buildings in the country, utilized nano-TiO_2 façade coatings for self-cleaning and air purification. These photocatalytic surfaces degrade urban pollutants like NO_X and VOCs, contributing to healthier urban air while reducing maintenance costs.

In Qatar, stadiums constructed for the FIFA World Cup 2022 used nano-silica-enhanced concrete to withstand high thermal stress and chloride exposure, crucial for durability in desert climates. Field data indicated a 25–35% increase in compressive strength and a 60% reduction in permeability, improving lifespan and reducing water ingress.

Table 3. Summary of Case Studies on Sustainable Construction Materials

Project Name / Location	Material Used	Application Area	Key Impact / Benefit
Flat House, UK	Hempcrete	Residential walls and floors	Carbon-negative structure, breathable indoor air
BioHome3D, USA	Wood fiber bio-resin composite	3D-printed housing	Local sourcing, 100% compostable
Økern Portal, Norway	Recycled concrete and timber	Commercial building	200 t CO_2 avoided, BREEAM Outstanding

continued on following page

Table 3. Continued

Project Name / Location	Material Used	Application Area	Key Impact / Benefit
M5 Bridge, Netherlands	Bacterial self-healing concrete	Bridge elements	Cracks self-sealed, reduced maintenance
Lleida Classroom, Spain	Paraffin-based PCM in wallboards	Passive cooling	23% cooling energy saved, stable temperatures
Eurosky Tower, Italy	Nano-TiO_2 coatings	Façade system	Self-cleaning, pollution reduction

The above case studies demonstrate that next-generation eco-friendly and high-performance materials are not only technically viable but are already delivering measurable performance gains across multiple sectors. From carbon-negative houses to pollution-reducing façades and self-healing infrastructure, these materials are redefining the boundaries of what sustainable construction can achieve. While regional climates, regulatory environments, and material availability influence the outcomes, the lessons from these projects underscore the urgency and feasibility of scaling sustainable material adoption globally. These examples also reveal the importance of integrated design thinking, where material choice is aligned with performance goals from the earliest planning stages, ensuring that sustainability is embedded—not added—into construction systems.

6. BARRIERS TO ADOPTION

Although the environmental and functional benefits of next-generation eco-friendly construction materials are well-documented, their widespread adoption remains limited due to a combination of structural, technical, economic, and cultural barriers. These challenges manifest across the entire construction value chain—from design and specification to procurement, implementation, and post-occupancy evaluation—hindering the scalability and mainstream integration of innovative sustainable solutions.

One of the most significant obstacles is lack of standardization and certification. Unlike traditional materials, which benefit from decades of codification in national and international standards, many sustainable alternatives such as mycelium composites, geopolymer concretes, and phase-change materials lack harmonized performance benchmarks. According to Mahame et al. (2024), this absence of regulatory frameworks leads to uncertainty in structural integrity, fire resistance, and long-term durability, which in turn dissuades architects, engineers, and contractors from specifying these materials, particularly in projects with strict compliance and insurance requirements.

Closely related is the issue of technical performance uncertainty, especially under real-world conditions. While laboratory data on thermal conductivity, carbon sequestration, and mechanical strength may be promising, field data is often lacking. Materials like bio-based insulation or nano-enhanced coatings require further in-situ validation to assess how they perform under diverse climate conditions, moisture exposure, and load stress. This lack of longitudinal data complicates lifecycle cost forecasting and return-on-investment analysis—both critical for private developers and public procurement bodies.

Cost competitiveness is another persistent barrier. Although the lifecycle costs of sustainable materials are often lower due to energy savings and extended durability, the initial capital costs remain significantly higher in many regions. As noted by Alrehaili et al. (2025), materials such as self-healing concrete or encapsulated PCMs can cost two to five times more than their conventional counterparts, largely due to small-scale production, specialized additives, and limited supply chain maturity. In markets where project costs are tightly controlled, these premiums are difficult to justify without policy incentives or guaranteed payback models.

Moreover, the fragmentation of the supply chain further hinders adoption. Many innovative materials are not readily available through standard distribution channels. Builders may face long lead times, logistical challenges, or uncertain quality when sourcing bio-based or recycled content materials, especially in regions without developed circular economies or biomanufacturing infrastructure. This was emphasized by Sandanayake et al. (2025), who highlighted the lack of regional coordination in material reuse networks, which results in underutilization of viable construction waste.

Compounding these structural challenges is a more subtle but no less powerful factor: perceptual and cultural resistance. Decision-makers in the construction industry tend to prioritize tried-and-tested materials with known performance histories. New materials are often viewed as risky or non-conforming, especially by conservative stakeholders such as insurers, financial backers, or permitting authorities. Furthermore, there remains a widespread belief that sustainable materials are suitable only for niche or experimental projects, not for mainstream commercial or infrastructural applications. Changing these deeply embedded assumptions requires not only technical proof, but also targeted communication strategies, pilot demonstrations, and education at all levels of the construction workforce.

Policy fragmentation also plays a role. While some countries and municipalities have introduced green procurement guidelines, mandatory EPDs, or recycled content requirements, others continue to prioritize lowest-cost bidding without factoring in environmental externalities. This creates an uneven playing field where sustainable materials must compete with high-impact but inexpensive conventional alternatives.

Lastly, limited professional education and training on the use and integration of these materials exacerbates all the above barriers. Most architecture, engineering, and construction (AEC) curricula still focus on conventional systems, and lack updated content on emerging material science, circularity principles, or performance-based design. Without a skilled workforce capable of handling these technologies confidently, adoption will remain patchy and inconsistent.

In short, the transition to next-generation sustainable materials is constrained not by lack of innovation, but by a complex set of interrelated institutional, economic, regulatory, and behavioral factors. Overcoming these challenges requires a coordinated effort among industry stakeholders, researchers, governments, and educators to build a supportive ecosystem for sustainable material adoption.

7. FUTURE DIRECTIONS

As the urgency of the climate crisis intensifies and net-zero targets become binding through global policy frameworks, the construction industry is undergoing a profound transformation. Material innovation is no longer a peripheral concern—it is central to achieving carbon neutrality, resource efficiency, and circularity. Future pathways for sustainable construction materials will increasingly depend on four interlinked domains: smart adaptive materials, nanotechnology and sensing, biological systems, and digital lifecycle intelligence.

One of the most promising trajectories lies in the development of adaptive thermal materials capable of dynamically regulating building envelopes. These include phase-change materials (PCMs), thermochromic coatings, and passive thermal switches, which can adjust insulation capacity or reflectivity in response to outdoor conditions. For instance, Tajadod et al. (2025) used artificial intelligence to optimize the performance of climate-resilient buildings incorporating PCM-based façades, reporting significant reductions in cooling demand and CO_2 emissions. This synergy between materials and digital tools enhances building adaptability and energy performance.

In parallel, nanotechnology is enabling the next generation of multifunctional materials. These include cementitious matrices infused with nanoparticles to improve strength, impermeability, and durability, while also offering self-sensing capabilities. As Ahmadizadeh et al. (2024) explain, nanomaterials can also catalyze chemical reactions that repair micro-cracks, enhancing the lifespan of infrastructure and reducing maintenance costs. Furthermore, photocatalytic nanoparticles embedded in surfaces can contribute to air purification and self-cleaning facades, especially in dense urban environments.

Alongside nanotechnology, the rise of engineered living materials (ELMs) is shifting the boundaries of what building materials can be. These materials integrate fungal mycelium, microbial biofilms, or algal systems to create components that are self-growing, biodegradable, and regenerative. According to Liu et al. (2025), combining 3D printing with biofabrication offers possibilities for materials that not only heal themselves but can evolve in response to environmental changes. While still in experimental stages, ELMs hold vast potential for net-positive buildings—those that restore ecosystems rather than degrade them.

The digitalization of material decision-making is another key axis of future innovation. Building Information Modelling (BIM) integrated with Life Cycle Assessment (LCA) and Artificial Intelligence (AI) is transforming how materials are selected and managed. Yadav et al. (2024) illustrate how AI-enhanced BIM systems can optimize design choices by comparing materials based on cost, carbon footprint, recyclability, and lifespan—all in real time. These tools empower architects and engineers to simulate multiple scenarios and make data-driven sustainability decisions.

Equally critical is the embedding of circular economy principles at the materials level. Strategies such as design-for-disassembly, urban mining, material passports, and closed-loop supply chains are becoming operational, rather than aspirational. Iyiola (2025) presents an integrated framework combining LCA, carbon accounting, and circular design principles for building engineering systems. His work outlines the importance of tracking material flows from extraction to reuse, supported by blockchain and digital tagging.

Finally, systemic transformation will depend on regulatory alignment and education. Initiatives such as the EU Level(s) framework, Carbon Border Adjustment Mechanisms (CBAMs), and mandatory embodied carbon reporting are forcing industries to internalize environmental costs. Academic programs in engineering and architecture must now include modules on material circularity, biomimicry, and sustainable innovation to prepare the next generation of professionals for this evolving landscape.

8. CONCLUSION

The ongoing transformation of the construction industry toward sustainability is being driven by a convergence of material innovation, digital tools, and circular economy thinking. This study has demonstrated that next-generation eco-friendly and high-performance materials are not merely alternatives to conventional building systems, but essential pillars for achieving global environmental goals, particularly carbon neutrality, resource circularity, and infrastructure resilience.

The integration of biodegradable composites, waste-derived materials, self-healing concrete, and phase-change systems offers a substantial opportunity to reduce the sector's carbon footprint, both in terms of embodied and operational emissions. These materials can provide similar or superior performance while significantly reducing reliance on virgin raw materials, thus contributing to the decoupling of economic growth from resource depletion.

Moreover, emerging technologies such as nanotechnology and biofabrication have unlocked new classes of intelligent, multifunctional materials that not only perform structurally but also interact with their environment—through self-monitoring, self-healing, or adaptive thermal behavior. Coupled with AI-driven design tools and BIM–LCA integration, these innovations enable lifecycle-oriented decisions that improve environmental performance from the planning stage through to end-of-life.

Despite the technological progress, the path to large-scale adoption remains complex. As highlighted by Dipasquale et al. (2025), one of the critical challenges is aligning these advanced materials with broader circular economy strategies, including design-for-disassembly and material traceability across the building lifecycle. Implementing such strategies will require regulatory support, procurement reform, and industrial collaboration across supply chains.

From a policy perspective, Khatib and Chileshe (2025) argue that decarbonization efforts must be accompanied by stronger embodied carbon benchmarks, tax incentives, and circularity certifications that reward low-impact materials and systems. Education and workforce development will also play a central role—professionals must be equipped with skills in materials science, sustainability assessment, and data-driven design to implement these changes effectively.

Looking ahead, the evidence strongly supports the argument that next-generation sustainable construction materials are critical to building a climate-resilient future. However, their success hinges on systems-level alignment: between material science, design methodology, regulatory frameworks, and economic incentives.

Only through this integration can the construction sector evolve from a leading source of emissions to a net contributor to environmental restoration, actively supporting ecosystem health, reducing energy dependence, and enabling regenerative urban development.

As Yu et al. (2025) underscore in their work on graphene-enhanced recycled materials, performance gains must be matched with sustainable value chains, or the benefits of innovation will remain fragmented. The future of sustainable construction lies in coordinated action, where innovation is embedded not just in products—but in policies, partnerships, and practices.

REFERENCES

Ahmadizadeh, M., Heidari, M., Thangavel, S., Khashehchi, M., Rahmanivahid, P., Singh, V. P., & Kumar, A. (2024). Development of new materials for sustainable buildings. In *Advanced Materials for Energy-Efficient Buildings*. Taylor & Francis. DOI: 10.1201/9781003496656-2

Akbulut, Z. F., Guler, S., Yavuz, D., & Avcı, M. S. (2025). *Toward sustainable construction: A critical review of recycled aggregate concrete properties and future opportunities. Case Studies in Construction Materials*. Elsevier.

Alrehaili, E., Nurdiawati, A., & Al-Ghamdi, S. G. (2025). Sustainable Composites for Metal Replacement: Environmental Assessment and Material Selection of Fiber-Reinforced Polymer Across Industries. *Resources, Conservation & Recycling Advances*, *25*, 100051. DOI: 10.1016/j.rcradv.2025.200294

Dipasquale, L., Battisti, F., & Perdonò, M. (2025). Adopting Circular Economy Strategies in Building Regeneration Through a Holistic Approach. In *Transition Towards Carbon-Neutral Construction* (pp. 851–867). Springer. DOI: 10.1007/978-3-031-82323-7_58

Iyiola, O. O. (2025). *Integrated Lifecycle Assessment and Carbon Accounting of Building Engineering Systems Under Circular Economy Frameworks*. ResearchGate.

Khan, H., Zahid, Z. B., Malik, U. J., & Ahmad, J. (2025). Performance Evaluation of Innovative Connection Systems for Structural and Seismic Challenges in High-Rise Modular Construction: A State-of-the-Art Review. Journal of Building Engineering, Elsevier.DOI: 10.1016/j.jobe.2025.112808

Li, C., Wen, X., Cai, W., Wu, J., Shao, J. J., Yang, Y., Yu, H., Liu, D., & Wang, M. (2022). Energy performance of buildings with composite phase-change material wallboards in different climatic zones of China. *Energy and Building*, *272*, 112398. DOI: 10.1016/j.enbuild.2022.112398

Liang, H., Miao, C., Zhou, C., & Wang, Q. (2025). Norfloxacin–derived carbon dots modifying mechanical properties, impermeability, and chloride binding of cement–based materials. *Construction & Building Materials*, *409*, 134342. DOI: 10.1016/j.conbuildmat.2025.144191

Liu, Z. (2025). IoT for Innovation in the Building Process: IT for Design and Life Cycle Management of the Building Envelope [Doctoral Thesis]. University of Ferrara.

Mahame, C., Kikwasi, G. J., & Baruti, M. M. (2024). Barriers to the effective selection of sustainable materials for residential building projects: A qualitative study. *Sustainability (Basel), 16*(21), 9526. DOI: 10.3390/su16219526

Martínez, B., Bernat-Maso, E., & Gil, L. (2023). Applications and properties of hemp stalk-based insulating biomaterials for buildings. *Materials (Basel), 16*(8), 3245. DOI: 10.3390/ma16083245 PMID: 37110081

Naghibalsadati, F., Karimi, N., Mim, S. J., & Ng, K. T. W. (2025). Thematic evolution of Life Cycle Assessment in construction and demolition waste management: Before and after ISO 14040 and the Paris Agreement. Journal of Building Engineering, Elsevier.

Parhizi, Z., Dearnaley, J., Kauter, K., Mikkelsen, D., & Pal, P. (2025). Spore-Tacular Future Materials: Innovations and Applications of Mycelium-Based Composites. *Preprints*, 2025.

Sandanayake, M., Kraus, R., Haigh, R., Yaghoubi, E., & Vrcelj, Z. (2025). Tex-Crete—Carbon and Cost Assessment of Concrete with Textile and Cardboard Fibres—Case Studies Towards Circular Economy. *Applied Sciences (Basel, Switzerland), 15*(13), 6962. DOI: 10.3390/app15136962

Singh, S., Nateriya, R., & Raheem, J. (2024). Self-Healing Concrete: A Comprehensive Review. In *International Conference on Recent Advances in Structural Engineering and Construction Management* (pp. 285–297). Springer.

Sun, C., Yan, C., Li, J., Ma, J., & Li, Y. (2025). Ultra-high-performance concrete from 100% solid waste: Mechanical strength, microstructure and sustainability analysis. *Case Studies in Construction Materials, 24*, 100000. DOI: 10.1016/j.cscm.2025.e05482

Suriya Kumar, T., Adebanjo, A. U., & Shafiq, N. (2025). *Mechanical and sustainability assessment of the influence of oily sludge ash on fly ash-based geopolymer composites. Innovative Infrastructure Solutions*. Springer.

Tajadod, O. E., Ravanshadnia, M., & Ghanbari, M. (2025). Integrating circular economy and life cycle assessment strategies in climate-resilient buildings: An artificial intelligence approach to enhance thermal comfort and minimize CO_2. *Energy, 294*, 130468. DOI: 10.1016/j.energy.2025.135064

Yadav, J., Singh, V. P., & Kumar, A. (2024). Life Cycle Assessment of Sustainable Building Materials. In *Advanced Materials for Energy-Efficient Buildings*. Taylor & Francis. DOI: 10.1201/9781003496656-4

Yu, J., Sui, H., Wang, S., Ma, Y., Xu, X., & Liu, Y. (2025). Influence of graphene oxide dispersion strategy on recycled concrete powder for cementitious systems. SSRN Electronic Journal. DOI: 10.1016/j.conbuildmat.2025.144771

Compilation of References

Abdellatef, Y., Khan, M. A., Khan, A., Alam, M. I., & Kavgic, M. (2020). Mechanical, thermal, and moisture buffering properties of novel insulating hemp-lime composite building materials. *Materials (Basel), 13*(21), 5000. DOI: 10.3390/ma13215000 PMID: 33171950

Abed, J., Rayburg, S., Rodwell, J., & Neave, M. (2022). A review of the performance and benefits of mass timber as an alternative to concrete and steel for improving the sustainability of structures. *Sustainability (Basel), 14*(9), 5570. DOI: 10.3390/su14095570

Abera, Y. A. (2024). Sustainable building materials: A comprehensive study on eco-friendly alternatives for construction. *Composites and Advanced Materials, 33*, 26349833241255957. DOI: 10.1177/26349833241255957

Abhilash, P. P., Nayak, D. K., Sangoju, B., Kumar, R., & Kumar, V. (2021). Effect of nano-silica in concrete; a review. *Construction & Building Materials, 278*, 122347. DOI: 10.1016/j.conbuildmat.2021.122347

Abuelnuor, A. A., Omara, A. A., Saqr, K. M., & Elhag, I. H. (2018). Improving indoor thermal comfort by using phase change materials: A review. *International Journal of Energy Research, 42*(6), 2084–2103. DOI: 10.1002/er.4000

Adesina, A. (2021). Overview of the influence of waste materials on the thermal conductivity of cementitious composites. *Cleaner Engineering and Technology, 2*, 100046. Advance online publication. DOI: 10.1016/j.clet.2021.100046

Adesina, A. (2021). Performance of cementitious composites reinforced with chopped basalt fibres—An overview. *Construction & Building Materials, 266*, 120970. Advance online publication. DOI: 10.1016/j.conbuildmat.2020.120970

Adesina, A., & Das, S. (2021). Evaluation of the durability properties of engineered cementitious composites incorporating recycled concrete as aggregate. *Journal of Materials in Civil Engineering*, *33*(2), 04020439. Advance online publication. DOI: 10.1061/(ASCE)MT.1943-5533.0003563

Adesina, A., & Das, S. (2021). Sustainable utilization of recycled asphalt as aggregates in engineered cementitious composites. *Construction & Building Materials*, *283*, 122727. Advance online publication. DOI: 10.1016/j.conbuildmat.2021.122727

Adhikary, S. K., Ashish, D. K., & Rudžionis, Ž. (2021). Aerogel based thermal insulating cementitious composites: A review. *Energy and Building*, *245*, 111058. DOI: 10.1016/j.enbuild.2021.111058

Aggarwal, C., & Molleti, S. (2024). State-of-the-art review: Effects of using cool building cladding materials on roofs. *Buildings (Basel, Switzerland)*, *14*(8), 2257. DOI: 10.3390/buildings14082257

Aguacil, S., Duque, S., Lufkin, S., & Rey, E. (2024). Designing with building-integrated photovoltaics (BIPV): A pathway to decarbonize residential buildings. *Journal of Building Engineering*, *96*, 110486. DOI: 10.1016/j.jobe.2024.110486

Agurcia, R., & Soto, R. (1984). La Casa de Don Calecho: Un ejemplo de la arquitectura vernácula en Honduras. *Yaxkin*, 7(1), 47–59. https://cdihh.ihah.hn/revistayaxkin/1984_01/LaCasadeDonCalechoUnEjemplodelaArquitecturaVernaculaenHonduras.pdf

Ahmadizadeh, M., Heidari, M., Thangavel, S., Khashehchi, M., Rahmanivahid, P., Singh, V. P., & Kumar, A. (2024). Development of new materials for sustainable buildings. In *Advanced Materials for Energy-Efficient Buildings*. Taylor & Francis. DOI: 10.1201/9781003496656-2

Ahmed, G. H. (2023). A review of "3D concrete printing": Materials and process characterization, economic considerations and environmental sustainability. *Journal of Building Engineering*, *66*, 105863. DOI: 10.1016/j.jobe.2023.105863

Ahmed, H. U., Mahmood, L. J., Muhammad, M. A., Faraj, R. H., Qaidi, S. M., Sor, N. H., Mohammed, A. S., & Mohammed, A. A. (2022). Geopolymer concrete as a cleaner construction material: An overview on materials and structural performances. *Cleaner Materials*, *5*, 100111. DOI: 10.1016/j.clema.2022.100111

Ahmed, R., & Singh, A. (2025). Mechanical durability and recyclability of montmorillonite-reinforced polyester nanocomposites. *Heliyon*, *11*(4), e01403. DOI: 10.1016/j.heliyon.2025.e01403

Ajith, G., Kumar, S., & Eligehausen, R. (2021). Effect of mineral admixtures and manufactured sand on compressive strength of engineered cementitious composite. *Journal of Building Pathology and Rehabilitation*, *6*(1), 37. Advance online publication. DOI: 10.1007/s41024-021-00137-y

Akbulut, Z. F., Guler, S., Yavuz, D., & Avcı, M. S. (2025). *Toward sustainable construction: A critical review of recycled aggregate concrete properties and future opportunities. Case Studies in Construction Materials*. Elsevier.

Akın, N., & Mutlu Danacı, H. (2021). An investigation into the architectural use of nanotechnology in the context of the titanium dioxide. *Environmental Science and Pollution Research International*, *28*(45), 64130–64136. DOI: 10.1007/s11356-021-15268-5 PMID: 34247357

Alaloul, W. S., Musarat, M. A., Haruna, S., Law, K., Tayeh, B. A., Rafiq, W., & Ayub, S. (2020). Mechanical and deformation properties of rubberized engineered cementitious composite (ECC). *Case Studies in Construction Materials*, *13*, e00385. Advance online publication. DOI: 10.1016/j.cscm.2020.e00385

Alassaf, Y. (2024). Comprehensive review of the advancements, benefits, challenges, and design integration of energy-efficient materials for sustainable buildings. *Buildings (Basel, Switzerland)*, *14*(9), 2994. DOI: 10.3390/buildings14092994

Alavez-Ramirez, R., Chiñas-Castillo, F., Morales-Dominguez, V. J., & Ortiz-Guzman, M. (2012). Thermal conductivity of coconut fibre filled ferrocement sandwich panels. *Construction & Building Materials*, *37*, 425–431. DOI: 10.1016/j.conbuildmat.2012.07.053

Alazmi, S., Abdelmegid, M. A., Sarhan, S., Poshdar, M., Gonzalez, V., & Bidhendi, A. (2025). An integrated framework to improve waste management practices and environmental awareness in the Saudi construction industry. *Cleaner Waste Systems*, *10*, 100195. Advance online publication. DOI: 10.1016/j.clwas.2024.100195

Al-Homoud, M. S. (2005). Performance characteristics and practical applications of common building thermal insulation materials. *Building and Environment*, *40*(3), 353–366. DOI: 10.1016/j.buildenv.2004.05.013

Alhotan, A., Yates, J., Zidan, S., Haider, J., & Silikas, N. (2021). Assessing fracture toughness and impact strength of PMMA reinforced with nano-particles and fibre as advanced denture base materials. *Materials (Basel)*, *14*(15), 4127. DOI: 10.3390/ma14154127 PMID: 34361320

Ali, A., Issa, A., & Elshaer, A. (2024). A comprehensive review and recent trends in thermal insulation materials for energy conservation in buildings. *Sustainability (Basel)*, *16*(20), 8782. DOI: 10.3390/su16208782

Ali, M., Alabdulkarem, A., Nuhait, A., Al-Salem, K., Iannace, G., & Almuzaiqer, R. (2022). Characteristics of agro waste fibers as new thermal insulation and sound absorbing materials: Hybrid of date palm tree leaves and wheat straw fibers. *Journal of Natural Fibers*, *19*(13), 6576–6594. DOI: 10.1080/15440478.2021.1929647

Alrehaili, E., Nurdiawati, A., & Al-Ghamdi, S. G. (2025). Sustainable Composites for Metal Replacement: Environmental Assessment and Material Selection of Fiber-Reinforced Polymer Across Industries. *Resources, Conservation & Recycling Advances*, *25*, 100051. DOI: 10.1016/j.rcradv.2025.200294

Altomonte, S., & Schiavon, S. (2013). Occupant satisfaction in LEED and non-LEED certified buildings. *Building and Environment*, *68*, 66–76. DOI: 10.1016/j.buildenv.2013.06.008

Amir Ali, A., Kumar, R., Khan, M. A., & Chauhan, D. P. S. (2025). Biochar-amended high-strength engineered cementitious composites. *Cement and Concrete Composites*, *164*, 106219. Advance online publication. DOI: 10.1016/j.cemconcomp.2025.106219

Ammasi, A. K., & Ragul, G. (2018). Strength and durability of high volume fly ash in engineered cementitious composites. *Materials Today: Proceedings*, *5*(11), 24050–24058. DOI: 10.1016/j.matpr.2018.10.198

Amnuaylojaroen, T., Inkom, J., Janta, R., & Surapipith, V. (2020). Long range transport of southeast asian PM2. 5 pollution to northern Thailand during high biomass burning episodes. *Sustainability (Basel)*, *12*(23), 10049. DOI: 10.3390/su122310049

Amran, M., Onaizi, A. M., Fediuk, R., Danish, A., Vatin, N. I., Murali, G., Abdelgader, H. S., Mosaberpanah, M. A., Cecchin, D., & Azevedo, A. (2022). An ultra-lightweight cellular concrete for geotechnical applications–A review. *Case Studies in Construction Materials*, *16*, e01096. DOI: 10.1016/j.cscm.2022.e01096

Andersen, B., & Rasmussen, T. V. (2025). Biobased building materials: Moisture characteristics and fungal susceptibility. *Building and Environment*, *275*, 112720. DOI: 10.1016/j.buildenv.2025.112720

Andrade Neto, J., De la Torre, A. G., & Kirchheim, A. P. (2021). Effects of sulfates on the hydration of Portland cement—A review. *Construction & Building Materials*, *279*, 122428. Advance online publication. DOI: 10.1016/j.conbuildmat.2021.122428

Antov, P., Jivkov, V., Savov, V., Simeonova, R., & Yavorov, N. (2020). Structural application of eco-friendly composites from recycled wood fibres bonded with magnesium lignosulfonate. *Applied Sciences (Basel, Switzerland)*, *10*(21), 7526. DOI: 10.3390/app10217526

Anwar, A., Nasir, M., Hussain, I., & Shah, S. A. R. (2020). Enhanced properties of cementitious composite tailored with graphene oxide nanomaterial—A review. *Developments in the Built Environment*, *1*, 100002. Advance online publication. DOI: 10.1016/j.dibe.2019.100002

An, X., & Pivo, G. (2020). Green buildings in commercial mortgage-backed securities: The effects of LEED and Energy Star certification on default risk and loan terms. *Real Estate Economics*, *48*(1), 7–42. DOI: 10.1111/1540-6229.12228

ArchDaily. *Architecture from Honduras*. Last access May 27, 2025. https://www.archdaily.com/country/honduras

Arulanandam, P. M., Arunachalam, S., & Gobinath, R. (2020). Behavior of engineered cementitious composite structural elements—A review. *Indian Concrete Journal*, *94*(6), 5–28.

Arumugam, P., Ramalingam, V., & Vellaichamy, P. (2022). Effective PCM, insulation, natural and/or night ventilation techniques to enhance the thermal performance of buildings located in various climates–A review. *Energy and Building*, *258*, 111840. DOI: 10.1016/j.enbuild.2022.111840

Asayesh, S., Javid, A. A. S., Ziari, H., & Mehri, B. (2021). Evaluating fresh state, hardened state, thermal expansion and bond properties of geopolymers for the repairing of concrete pavements under restrained conditions. *Construction & Building Materials*, *292*, 123398. DOI: 10.1016/j.conbuildmat.2021.123398

Asdrubali, F., Baldinelli, G., Bianchi, F., & Sambuco, S. (2015). A comparison between environmental sustainability rating systems LEED and ITACA for residential buildings. *Building and Environment*, *86*, 98–108. DOI: 10.1016/j.buildenv.2015.01.001

Asdrubali, F., Ferracuti, B., Lombardi, L., Guattari, C., Evangelisti, L., & Grazieschi, G. (2017). A review of structural, thermo-physical, acoustical, and environmental properties of wooden materials for building applications. *Building and Environment*, *114*, 307–332. DOI: 10.1016/j.buildenv.2016.12.033

Asghari, N., & Memari, A. M. (2024). State of the art review of attributes and mechanical properties of hempcrete. *Biomass*, *4*(1), 65–91. DOI: 10.3390/biomass4010004

Asghari, Y., Alaghebandian, A., & Zhang, Y. (2024). 3D printing and implementation of engineered cementitious composites—A review. *Case Studies in Construction Materials, 21*, e03462. Advance online publication. DOI: 10.1016/j.cscm.2024.e03462

Ashraf, M., Khan, A. N., Ali, M., & Khan, M. A. (2022). Developing a sustainable concrete incorporating bentonite clay and silica fume: Mechanical and durability performance. *Journal of Cleaner Production, 337*, 130315. Advance online publication. DOI: 10.1016/j.jclepro.2021.130315

ASTM International. (2015). *ASTM C1012/C1012M-15: Standard test method for length change of hydraulic-cement mortars exposed to a sulfate solution*. ASTM International.

ASTM International. (2019). *ASTM C39/C39M-19: Standard Test Method for Compressive Strength of Cylindrical Concrete Specimens*. West Conshohocken, PA: ASTM International. https://www.astm.org/c0039_c0039m-19.html

ASTM International. (2020). *ASTM C642-13: Standard Test Method for Density, Absorption, and Voids in Hardened Concrete*. West Conshohocken, PA: ASTM International. https://www.astm.org/c0642-13.html

Awadh, O. (2017). Sustainability and green building rating systems: LEED, BREEAM, GSAS and Estidama critical analysis. *Journal of Building Engineering, 11*, 25–29. DOI: 10.1016/j.jobe.2017.03.010

Azhar, S., Carlton, W. A., Olsen, D., & Ahmad, I. (2011). Building information modeling for sustainable design and LEED rating analysis. *Automation in Construction, 20*(2), 217–224. DOI: 10.1016/j.autcon.2010.09.019

Bahraq, A. A., Maslehuddin, M., & Al-Dulaijan, S. U. (2020). Macro- and micro-properties of engineered cementitious composites (ECCs) incorporating industrial waste materials: A review. *Arabian Journal for Science and Engineering, 45*(10), 7869–7895. DOI: 10.1007/s13369-020-04729-7

Bamigboye, G. O., Davies, I., Nwanko, C., Michaels, T., Adeyemi, G., & Ozuor, O. (2019, November). Innovation in construction materials-a review. [). IOP Publishing.]. *IOP Conference Series. Materials Science and Engineering, 640*(1), 012070. DOI: 10.1088/1757-899X/640/1/012070

Barbaros, I., Yang, Y., Safaei, B., Yang, Z., Qin, Z., & Asmael, M. (2022). State-of-the-art review of fabrication, application, and mechanical properties of functionally graded porous nanocomposite materials. *Nanotechnology Reviews, 11*(1), 321–371. DOI: 10.1515/ntrev-2022-0017

Barletta, M., Aversa, C., Ayyoob, M., Gisario, A., Hamad, K., Mehrpouya, M., & Vahabi, H. (2022). Poly (butylene succinate)(PBS): Materials, processing, and industrial applications. *Progress in Polymer Science*, *132*, 101579. DOI: 10.1016/j.progpolymsci.2022.101579

Bas, Y., & Dike, A. S. (2024). Comprehensive characterizations of organoclay types based on their integration to poly (butylene terephthalate) nanocomposites. *Journal of Nanoparticle Research*, *26*(3), 60. DOI: 10.1007/s11051-024-05972-x

Bawono, A. A., Lechner, B., & Yang, E. H. (2019). Skid resistance and surface water drainage performance of engineered cementitious composites for pavement applications. *Cement and Concrete Composites*, *104*, 103387. Advance online publication. DOI: 10.1016/j.cemconcomp.2019.103387

Ben Hassen, T., El Bilali, H., Daher, B., & Burkart, S. (2025). Sustainable and resilient food systems in times of crises. *Frontiers in Nutrition*, *12*, 1564950. DOI: 10.3389/fnut.2025.1564950

Ben-Alon, L., & Rempel, A. R. (2023). Thermal comfort and passive survivability in earthen buildings. *Building and Environment*, *238*, 110339. DOI: 10.1016/j.buildenv.2023.110339

Berardi, U. (2018). Aerogel-enhanced systems for building energy retrofits: Insights from a case study. *Energy and Building*, *159*, 370–381. DOI: 10.1016/j.enbuild.2017.10.092

Berndt, M. L. (2009). Properties of sustainable concrete containing fly ash, slag and recycled concrete aggregate. *Construction & Building Materials*, *23*(7), 2606–2613. DOI: 10.1016/j.conbuildmat.2009.02.011

Bertino, G., Kisser, J., Zeilinger, J., Langergraber, G., Fischer, T., & Österreicher, D. (2021). Fundamentals of building deconstruction as a circular economy strategy for the reuse of construction materials. *Applied Sciences (Basel, Switzerland)*, *11*(3), 939. DOI: 10.3390/app11030939

Bheel, N., Mahro, S. K., & Adesina, A. (2023). Effects of graphene oxide on the properties of engineered cementitious composites: Multi-objective optimization technique using RSM. *Buildings (Basel, Switzerland)*, *13*(8), 2018. Advance online publication. DOI: 10.3390/buildings13082018

Bheel, N., Mahro, S. K., & Mohammed, B. S. (2025). Optimization of durability characteristics of engineered cementitious composites combined with titanium dioxide as a nanomaterial applying RSM modelling. *Scientific Reports*, *15*(1), 94382. Advance online publication. DOI: 10.1038/s41598-025-94382-7

Bikiaris, D. (2011). Can nanoparticles really enhance thermal stability of polymers? Part II: An overview on thermal decomposition of polycondensation polymers. *Thermochimica Acta, 523*(1-2), 25-45 Can nanoparticles really enhance thermal stability of polymers? Part II: An overview on thermal decomposition of polycondensation polymers. *Thermochimica Acta*, *523*(1-2), 25–45. DOI: 10.1016/j.tca.2011.06.012

Biswas, K., Desjarlais, A., Smith, D., Letts, J., Yao, J., & Jiang, T. (2018). Development and thermal performance verification of composite insulation boards containing foam-encapsulated vacuum insulation panels. *Applied Energy*, *228*, 1159–1172. DOI: 10.1016/j.apenergy.2018.06.136

Biyani, Y., Patil, L. G., & Chandrasekhar, C. N. K. (2019). Engineered cementitious composites (ECC). In *Engineered cementitious composites (ECC)* (pp. 206–214). Springer. https://doi.org/DOI: 10.1007/978-3-662-58438-5

Björklund, M., von Malmborg, F., & Nordensvärd, J. (2023). Lessons learnt from 20+ years of research on multilevel governance of energy-efficient and zero-carbon buildings in the European Union. *Energy Efficiency*, *16*(8), 98. DOI: 10.1007/s12053-023-10178-6

Boafo, F. E., Kim, J. H., Ahn, J. G., Kim, S. M., & Kim, J. T. (2021). Slim curtain wall spandrel integrated with vacuum insulation panel: A state-of-the-art review and future opportunities. *Journal of Building Engineering*, *42*, 102445. DOI: 10.1016/j.jobe.2021.102445

Bogue, R. (2011). Nanocomposites: A review of technology and applications. *Assembly Automation*, *31*(2), 106–112. DOI: 10.1108/01445151111117683

Booya, E., Gorospe, K., Dashti, A., & Ghaednia, H. (2020). The influence of utilizing slag in lieu of fly ash on the performance of engineered cementitious composites. *Construction & Building Materials*, *256*, 119412. Advance online publication. DOI: 10.1016/j.conbuildmat.2020.119412

Borges, C. S., Jalali, S., Tsokanas, P., Marques, E. A., Carbas, R. J., & da Silva, L. F. (2022). Sustainable development approaches through wooden adhesive joints design. *Polymers*, *15*(1), 89. DOI: 10.3390/polym15010089

Bostanci, S. C., Limbachiya, M., & Kew, H. (2018). Use of recycled aggregates for low carbon and cost effective concrete construction. *Journal of Cleaner Production*, *189*, 176–196. DOI: 10.1016/j.jclepro.2018.04.090

Brolinson, B., Palmer, K., & Walls, M. (2023). Does Energy Star certification reduce energy use in commercial buildings? *Journal of the Association of Environmental and Resource Economists*, *10*(1), 55–93. DOI: 10.1086/720952

Brzezicki, M. (2021). A systematic review of the most recent concepts in smart windows technologies with a focus on electrochromics. *Sustainability (Basel), 13*(17), 9604. DOI: 10.3390/su13179604

Busselli, M., Cassol, D., Prada, A., & Giongo, I. (2021). Timber-based integrated techniques to improve energy efficiency and seismic behaviour of existing masonry buildings. *Sustainability (Basel), 13*(18), 10379. DOI: 10.3390/su131810379

Cai, Z., Liu, X., Chen, H., & Yu, J. (2021). Development of ultra-high ductility engineered cementitious composites as a novel and resilient fireproof coating. *Construction & Building Materials, 288*, 123090. Advance online publication. DOI: 10.1016/j.conbuildmat.2021.123090

Calvez, I., Garcia, R., Koubaa, A., Landry, V., & Cloutier, A. (2024). Recent Advances in Bio-Based Adhesives and Formaldehyde-Free Technologies for Wood-Based Panel Manufacturing. *Current Forestry Reports, 10*(5), 386–400. DOI: 10.1007/s40725-024-00227-3

Cannavale, A., Ayr, U., Fiorito, F., & Martellotta, F. (2020). Smart electrochromic windows to enhance building energy efficiency and visual comfort. *Energies, 13*(6), 1449. DOI: 10.3390/en13061449

Cao, D., Xu, C., Lu, W., Qin, C., & Cheng, S. (2018). Sunlight-driven photo-thermochromic smart windows. *Solar RRL, 2*(4), 1700219. DOI: 10.1002/solr.201700219

Castañeda Moncada, M. A. (2012). *Construyendo Pueblitos, Arquitectura para comunidades pequeñas*. Castañeda Moncada, Mauricio Alejandro. (author website: http://www.macastaneda.com/)

Castro-Hoyos, A. M., Rojas Manzano, M. A., & Maury-Ramírez, A. (2022). Challenges and opportunities of using titanium dioxide photocatalysis on cement-based materials. *Coatings, 12*(7), 968. DOI: 10.3390/coatings12070968

Cavagnoli, S., Fabiani, C., Albuquerque Landi, F. F., & Pisello, A. L. (2024). Advancing sustainable construction through comprehensive analysis of thermal, acoustic, and environmental properties in prefabricated panels with recycled PET materials. *Energy and Building, 312*, 114218. DOI: 10.1016/j.enbuild.2024.114218

CEDENNA. (2024). *Nanotecnología aplicada a la construcción y la energía*. https://cedenna.cl/node/889

Center for Gender & Refugee Studies (CGRS). (2023). *Honduras: La Emergencia Climática como Motor de Desplazamiento y Violaciones de Derechos Humanos*. CGRS, UC Law SF. Last access May 7, 2025. https://cgrs.uclawsf.edu/sites/default/files/Honduras%20Report_12.18.2023_FINAL_SP.pdf

Chang, Y., Wilkinson, S., Potangaroa, R., & Seville, E. (2011). Identifying factors affecting resource availability for post-disaster reconstruction: A case study in China. *Construction Management and Economics*, *29*(1), 37–48. DOI: 10.1080/01446193.2010.521761

Chan, K., & Zinchenko, A. (2023). Design and synthesis of functional materials by chemical recycling of waste polyethylene terephthalate (PET): Opportunities and challenges. *Journal of Cleaner Production*, *139828*, 139828. Advance online publication. DOI: 10.1016/j.jclepro.2023.139828

Chapman, A. (1985-1986). *Los hijos del copal y la candela* (Vols. I & II). UNAM/CEMCA.

Chen, Y., He, S., Gan, Y., Çopuroğlu, O., Veer, F., & Schlangen, E. (2022). A review of printing strategies, sustainable cementitious materials and characterization methods in the context of extrusion-based 3D concrete printing. *Journal of Building Engineering*, *45*, 103599. DOI: 10.1016/j.jobe.2021.103599

Chen, Y., Zhang, Y., & Li, L. (2023). Biochar-derived nanocomposites for sustainable environmental remediation: A review. *Environmental Nanotechnology, Monitoring & Management*, *20*, 100673. DOI: 10.1016/j.enmm.2023.100673

Chindaprasirt, P., Kanchanda, P., Sathonsaowaphak, A., & Cao, H. T. (2007). Sulfate resistance of blended cements containing fly ash and rice husk ash. *Construction & Building Materials*, *21*(6), 1356–1361. DOI: 10.1016/j.conbuildmat.2005.10.005

Choudhary, M. K., & Eastes, W. (2024). Effective thermal conductivity of fiberglass insulation. *International Journal of Applied Glass Science*, *15*(3), 307–316. DOI: 10.1111/ijag.16652

CIRE2N. (2023). *Revista educativa de nanotecnología ambiental*. https://www.cire2n.upr.edu/wp-content/uploads/2023/12/Revista-IV_31_12_23.pdf

Colombo, M. A., Díaz, F. R., Kodali, D., Rangari, V., Güven, O., & Moura, E. A. (2023). Influence of Reinforcing Efficiency of Clay on the Mechanical Properties of Poly (butylene terephthalate) Nanocomposite. *Ceramics*, *6*(1), 58–73. DOI: 10.3390/ceramics6010005

Constructivo. (2023). *Nanotecnología aplicada a la construcción*. https://constructivo.com/noticia/nanotecnologia-aplicada-a-la-construccion-1689945783

Correa Giraldo, V. M., Queiros, M., Ordoñez Candelaria, V. R., López Muñoz, L. F., Flores Méndez, E., & Zapata López, J. (2014, November). *El bahareque: Un sistema constructivo sismoresistente y sustentable para soluciones de vivienda social en México. XIX National Congress of Structural Engineering*, Puerto Vallarta, Jalisco, Mexico.

Cuce, E., Cuce, P. M., Wood, C. J., & Riffat, S. B. (2014). Toward aerogel based thermal superinsulation in buildings: A comprehensive review. *Renewable & Sustainable Energy Reviews*, *34*, 273–299. DOI: 10.1016/j.rser.2014.03.017

da Cunha, S. R. L., & de Aguiar, J. L. B. (2020). Phase change materials and energy efficiency of buildings: A review of knowledge. *Journal of Energy Storage*, *27*, 101083. DOI: 10.1016/j.est.2019.101083

Dabaieh, M., Wanas, O., Hegazy, M. A., & Johansson, E. (2015). Reducing cooling demands in a hot dry climate: A simulation study for non-insulated passive cool roof thermal performance in residential buildings. *Energy and Building*, *89*, 142–152. DOI: 10.1016/j.enbuild.2014.12.034

Danish, A., & Mosaberpanah, M. A. (2021). Influence of cenospheres and fly ash on the mechanical and durability properties of high-performance cement mortar under different curing regimes. *Construction & Building Materials*, *279*, 122458. Advance online publication. DOI: 10.1016/j.conbuildmat.2021.122458

Danish, A., Ozbakkaloglu, T., Mosaberpanah, M. A., Salim, M. U., Bayram, M., Yeon, J. H., & Jafar, K. (2022). Sustainability benefits and commercialization challenges and strategies of geopolymer concrete: A review. *Journal of Building Engineering*, *58*, 105005. DOI: 10.1016/j.jobe.2022.105005

Das, P., Nayak, P. K., Muthusamy, S., & Kesavan, R. K. (2023). Foams for thermal insulation. In *Polymeric Foams: Applications of Polymeric Foams* (Vol. 2, pp. 145–165). American Chemical Society. DOI: 10.1021/bk-2023-1440.ch007

de Andrade Salgado, F., & de Andrade Silva, F. (2022). Recycled aggregates from construction and demolition waste towards an application on structural concrete: A review. *Journal of Building Engineering*, *52*, 104452. DOI: 10.1016/j.jobe.2022.104452

de Siqueira, A. A., & Cordeiro, G. C. (2022). Sustainable cements containing sugarcane bagasse ash and limestone: Effects on compressive strength and acid attack of mortar. *Sustainability (Basel)*, *14*(9), 5683. Advance online publication. DOI: 10.3390/su14095683

Deng, S., Wang, R. A., & Dai, Y. J. (2014). How to evaluate performance of net zero energy building–A literature research. *Energy*, *71*, 1–16. DOI: 10.1016/j.energy.2014.05.007

Deng, Z., & Liu, D. (2025). Impact resistance of engineered cementitious composites reinforced with hybrid fibers. *Journal of Advanced Concrete Technology*, *23*(7), 296–309. DOI: 10.3151/jact.23.296

Department of Public Works and Town & Country Planning. (2025, May 05). The preparation for water hyacinth and weed removal in 33 provinces to prepare for the rainy season of 2023. ThaiNews. https://thainews.prd.go.th/th/news/detail/TCATG230722105501843

Dhandapani, Y., Sakthivel, T., Santhanam, M., & Gettu, R. (2020). Perspectives on blended cementitious systems with calcined clay-limestone combination for sustainable low carbon cement transition. *Indian Concrete Journal*, *94*(2), 31–45.

Ding, D., & Xian, M. (2024). Research trends and design strategies for bamboo architecture: A bibliometric and systematic review. *Advances in Bamboo Science*, *8*, 100094. DOI: 10.1016/j.bamboo.2024.100094

Dipasquale, L., Battisti, F., & Perdonò, M. (2025). Adopting Circular Economy Strategies in Building Regeneration Through a Holistic Approach. In *Transition Towards Carbon-Neutral Construction* (pp. 851–867). Springer. DOI: 10.1007/978-3-031-82323-7_58

Divya, S., Thomas, B. S., & Anand, K. B. (2024). Pozzolanic attributes of hydraulic cement paste hybridized with agricultural by-product and nano-carbon. *Construction & Building Materials*, *411*, 134119. Advance online publication. DOI: 10.1016/j.conbuildmat.2023.134119

Dixit, M. K., Fernández-Solís, J. L., Lavy, S., & Culp, C. H. (2010). Identification of parameters for embodied energy measurement: A literature review. *Energy and Building*, *42*(8), 1238–1247. DOI: 10.1016/j.enbuild.2010.02.016

Duan, Q., Zhao, Y., & Wang, J. (2020). Thermal performance and condensation risk of single-pane glazing with low emissivity coatings. *MRS Advances*, *5*(50), 2555–2564. DOI: 10.1557/adv.2020.160

Du, H., Du, S., & Liu, X. (2014). Durability performances of concrete with nano-silica. *Construction & Building Materials*, *73*, 705–712. DOI: 10.1016/j.conbuildmat.2014.10.014

Dwi, M., Huang, W., & Hsu, H. (2024). Improving properties of engineered cementitious composite (ECC) using Bacillus subtilis immobilized in silica gel. *Case Studies in Construction Materials, 20*, e03165. Advance online publication. DOI: 10.1016/j.cscm.2024.e03165

Eastern Engineering Group. (2024). *Nanotecnología en la construcción: Nuevas perspectivas*. https://www.easternengineeringgroup.com/es/construyendo-el-futuro-explorando-la-nanotecnologia-en-la-construccion/

Elahi, M. M. A., Shearer, C. R., Reza, A. N. R., Saha, A. K., Khan, M. N. N., Hossain, M. M., & Sarker, P. K. (2021). Improving the sulfate attack resistance of concrete by using supplementary cementitious materials (SCMs): A review. *Construction & Building Materials, 281*, 122628. DOI: 10.1016/j.conbuildmat.2021.122628

Elaouzy, Y., & El Fadar, A. (2022). Energy, economic and environmental benefits of integrating passive design strategies into buildings: A review. *Renewable & Sustainable Energy Reviews, 167*, 112828. DOI: 10.1016/j.rser.2022.112828

Elmesalami, N., & Celik, K. (2022). A critical review of engineered geopolymer composite: A low-carbon ultra-high-performance concrete. *Construction & Building Materials, 346*, 128491. DOI: 10.1016/j.conbuildmat.2022.128491

Elsacker, E., Vandelook, S., Van Wylick, A., Ruytinx, J., De Laet, L., & Peeters, E. (2020). A comprehensive framework for the production of mycelium-based lignocellulosic composites. *The Science of the Total Environment, 725*, 138431. DOI: 10.1016/j.scitotenv.2020.138431

Elsayed, M. A., & El-Gendy, A. S. (2022). Development of cellulose nanofiber-reinforced PET membranes for petroleum wastewater treatment. *Journal of Hazardous Materials, 430*, 128415. DOI: 10.1016/j.jhazmat.2022.128415

Enshassi, A., Mohamed, S., & Abushaban, S. (2009). Factors affecting the performance of construction projects in the Gaza Strip. *Journal of Civil Engineering and Management, 15*(3), 269–280. DOI: 10.3846/1392-3730.2009.15.269-280

EPA. (2020). *Nanotechnology and the Environment*. U.S. Environmental Protection Agency. https://www.epa.gov/nanotechnology

European Commission. (2020). *Circular economy action plan: For a cleaner and more competitive Europe*. Publications Office of the European Union. https://ec.europa.eu/environment/circular-economy/

Fang, Y., Memon, S., Peng, J., Tyrer, M., & Ming, T. (2020). Solar thermal performance of two innovative configurations of air-vacuum layered triple-glazed windows. *Renewable Energy, 150*, 167–175. DOI: 10.1016/j.renene.2019.12.115

Farghali, M., Osman, A. I., Mohamed, I. M., Chen, Z., Chen, L., Ihara, I., Yap, P. S., & Rooney, D. W. (2023). Strategies to save energy in the context of the energy crisis: A review. *Environmental Chemistry Letters*, *21*(4), 2003–2039. DOI: 10.1007/s10311-023-01591-5 PMID: 37362011

Fathy, H. (1973). *Architecture for the Poor: An Experiment in Rural Egypt*. University of Chicago Press. DOI: 10.7208/chicago/9780226239149.001.0001

Ferrari, S., Zoghi, M., Blázquez, T., & Dall'O, G. (2022). New Level(s) framework: Assessing the affinity between the main international Green Building Rating Systems and the European scheme. *Renewable & Sustainable Energy Reviews*, *155*, 111924. DOI: 10.1016/j.rser.2021.111924

Firoozi, A. A., Firoozi, A. A., Oyejobi, D. O., Avudaiappan, S., & Flores, E. S. (2024). Emerging trends in sustainable building materials: Technological innovations, enhanced performance, and future directions. *Results in Engineering, 24*, 103521.

Flynn, M. A., Richman, R., Gorgolewski, M., Saunders, K., & Race, C. (2017). An investigation into the hygrothermal performance of a mineral wool based externally insulated enclosure in a cold climate. *Energy Procedia*, *132*, 345–350. DOI: 10.1016/j.egypro.2017.09.746

Freire, A. L., José, H. J., & Moreira, R. D. F. P. M. (2022). Potential applications for geopolymers in carbon capture and storage. *International Journal of Greenhouse Gas Control*, *118*, 103687. DOI: 10.1016/j.ijggc.2022.103687

Fujishima, A., & Zhang, X. (2006). Titanium dioxide photocatalysis: Present situation and future approaches. *Comptes Rendus. Chimie*, *9*(5-6), 750–760. DOI: 10.1016/j.crci.2005.02.055

Fukushima, K., Fina, A., Geobaldo, F., Venturello, A., & Camino, G. (2012). Properties of poly (lactic acid) nanocomposites based on montmorillonite, sepiolite and zirconium phosphonate. *Express Polymer Letters*, *6*(11), 914–926. Advance online publication. DOI: 10.3144/expresspolymlett.2012.97

Fu, Q., Han, Z., Chen, J., Lu, Y., Wu, H., & Wang, Y. (2022). Applications of reinforcement learning for building energy efficiency control: A review. *Journal of Building Engineering*, *50*, 104165. DOI: 10.1016/j.jobe.2022.104165

Fu, Y., Qiu, C., Ni, L., Ye, H., Zou, H., Luo, Y., & Liang, M. (2024). Cell structure control and performance of rigid polyurethane foam with lightweight, good mechanical, thermal insulation and sound insulation. *Construction & Building Materials*, *447*, 138068. DOI: 10.1016/j.conbuildmat.2024.138068

Gaceta, L. (2020, 18 de febrero). *Ley Marco de Vivienda y Asentamientos Humanos de Honduras* (Decreto legislativo n.º 173-2019, La Gaceta n.º 35 178). Diario Oficial de la República de Honduras. Last access May 7, 2025. https://enag.sgjd.gob.hn/mfhandler.php?field=archivo&file=20200218+-+DECRETO+LEGISLATIVO++No.+173-2019+-+publicado+en+el+Diario+Oficial+La+Gaceta+35178.pdf

Gad, M. M., Abualsaud, R., Alqarawi, F. K., Emam, A. N. M., Khan, S. Q., Akhtar, S., Mahrous, A. A., & Al-Harbi, F. A. (2021). Translucency of nanoparticle-reinforced PMMA denture base material: An in-vitro comparative study. *Dental Materials Journal, 40*(4), 972–978. DOI: 10.4012/dmj.2020-296 PMID: 33883327

Gan, J., Chen, M., Semple, K., Liu, X., Dai, C., & Tu, Q. (2022). Life cycle assessment of bamboo products: Review and harmonization. *The Science of the Total Environment, 849*, 157937. DOI: 10.1016/j.scitotenv.2022.157937 PMID: 35952867

Gao, M., Qian, D., Zhao, L., & Jin, R. (2025). Experimental study on the photothermal properties of thermochromic glass. *Buildings (Basel, Switzerland), 15*(2), 233. DOI: 10.3390/buildings15020233

Garces, J. I. T., Beltran, A. B., Tan, R. R., Ongpeng, J. M. C., & Promentilla, M. A. B. (2022). Carbon footprint of self-healing geopolymer concrete with variable mix model. *Cleaner Chemical Engineering, 2*, 100027. DOI: 10.1016/j.clce.2022.100027

García Aranda, M. Á., & Shirani, S. (2023). *Optimización de alternativas al cemento Portland mediante técnicas avanzadas de imagen con sincrotrón*. Universidad de Málaga.

Gaspar, F., Bakatovich, A., Davydenko, N., & Joshi, A. (2020). Building insulation materials based on agricultural wastes. In *Bio-based materials and biotechnologies for eco-efficient construction* (pp. 149–170). Woodhead Publishing. DOI: 10.1016/B978-0-12-819481-2.00008-8

Ghobadi, M., & Sepasgozar, S. M. (2023). Circular economy strategies in modern timber construction as a potential response to climate change. *Journal of Building Engineering, 77*, 107229. DOI: 10.1016/j.jobe.2023.107229

Ghosh, A., Norton, B., & Duffy, A. (2016). First outdoor characterisation of a PV-powered suspended particle device switchable glazing. *Solar Energy Materials and Solar Cells, 157*, 1–9. DOI: 10.1016/j.solmat.2016.05.013

Gilbert, P., Wilson, P., Walsh, C., & Hodgson, P. (2017). The role of material efficiency to reduce CO2 emissions during ship manufacture: A life cycle approach. *Marine Policy, 75*, 227–237. DOI: 10.1016/j.marpol.2016.04.003

Göçügenci, A., & Keskin, S. B. (2023). The effect of graphene nanoplatelet addition on the mechanical, durability and self-healing properties of engineered cementitious composites. *MATEC Web of Conferences, 378*, Article 02024. https://doi.org/DOI: 10.1051/matecconf/202337802024

Golbazi, M., & Aktas, C. B. (2018). Energy efficiency of residential buildings in the US: Improvement potential beyond IECC. *Building and Environment*, *142*, 278–287. DOI: 10.1016/j.buildenv.2018.06.029

Goodman, B. A. (2020). Utilization of waste straw and husks from rice production: A review. *Journal of Bioresources and Bioproducts*, *5*(3), 143–162. DOI: 10.1016/j.jobab.2020.07.001

Government of Honduras. *Planes Estratégicos y Operativos del PVAH*. Unified Transparency Portal of Honduras. Last access May 27, 2025. https://portalunico.iaip.gob.hn/424/

Grazieschi, G., Asdrubali, F., & Thomas, G. (2021). Embodied energy and carbon of building insulating materials: A critical review. *Cleaner Environmental Systems*, *2*, 100032. DOI: 10.1016/j.cesys.2021.100032

Grzegorzewska, M., & Kirschke, P. (2021). The impact of certification systems for architectural solutions in green office buildings in the perspective of occupant well-being. *Buildings (Basel, Switzerland)*, *11*(12), 659. DOI: 10.3390/buildings11120659

Guan, X., Li, M., & Yang, E. H. (2020). Effect of exposure conditions on self-healing behavior of engineered cementitious composite incorporating limestone powder. *Cement and Concrete Composites*, *114*, 103808. Advance online publication. DOI: 10.1016/j.cemconcomp.2020.103808

Gupta, D. K., Kumar, S., & Wani, M. Y. (2024). MOF magic: Zirconium-based frameworks in theranostic and bio-imaging applications. *Journal of Materials Chemistry. B*, *12*(11), 2691–2710. Advance online publication. DOI: 10.1039/D3TB02562D PMID: 38419476

Gupta, P., & Khatri, V. (2024). Eco-friendly polymer nanocomposite coatings for next-generation fire retardants in building materials. *Polymers*, *16*(14), 2045. DOI: 10.3390/polym16142045 PMID: 39065362

Habitat for Humanity Honduras. (2017). *Barreras de acceso al suelo para vivienda social en Honduras*. Last access May 2, 2025. https://www.habitathn.org/wp-content/uploads/2017/08/Barreras-de-Acceso-al-Suelo-para-Vivienda-Social-en-Honduras.pdf

Hallik, J., Gustavson, H., & Kalamees, T. (2019). Air leakage of joints filled with polyurethane foam. *Buildings (Basel, Switzerland)*, *9*(7), 172. DOI: 10.3390/buildings9070172

Han, F., Wang, Y., Feist, W., Cao, X., Yu, Z., Song, B., Benli, H., & Dermentzis, G. (2022). Exploring solutions to achieve carbon neutrality in China: A comparative study of a large-scale passive house district and a green building district in Qingdao. *Energy and Building*, *268*, 112224. DOI: 10.1016/j.enbuild.2022.112224

Han, H. J., Jeon, Y. I., Lim, S. H., Kim, W. W., & Chen, K. (2010). New developments in illumination, heating and cooling technologies for energy-efficient buildings. *Energy*, *35*(6), 2647–2653. DOI: 10.1016/j.energy.2009.05.020

Han, J., Zhang, J., Kim, D., & Lee, S. (2024). Development of self-sensing engineered cementitious composite sensors for monitoring flexural performance of reinforced concrete beam. *Developments in the Built Environment*, *18*, 100407. Advance online publication. DOI: 10.1016/j.dibe.2024.100407

Harilal, M., George, R. P., Philip, J., & Albert, S. K. (2021). Binary blended fly ash concrete with improved chemical resistance in natural and industrial environments. *Environmental Science and Pollution Research International*, *28*(22), 28107–28132. DOI: 10.1007/s11356-021-12453-4 PMID: 33533001

Hasan, K. F., Horváth, P. G., Bak, M., & Alpár, T. (2021). A state-of-the-art review on coir fiber-reinforced biocomposites. *RSC Advances*, *11*(18), 10548–10571. DOI: 10.1039/D1RA00231G

Hasanov, R. I. (2023). Promoting sustainability in Azerbaijan's energy sector: A green policy evaluation and future outlook. *Green Economics*, *1*(1), 62–69.

Hernando Castro, S. M., Rodríguez, R. C., López Arquillo, J. D., & Álvarez de Eulate, M. P. (2025). Comparative evaluation of construction insulation materials: Environmental performance across production, end-of-life, and beyond system boundaries. *Journal of Building Engineering*, *99*, 111570. Advance online publication. DOI: 10.1016/j.jobe.2024.111570

He, X. F., Nie, G. Z., Zhang, S., Xie, Y. Q., Ge, T. J., & Liu, P. H. (2024). Preparation and properties of modified zirconium dioxide reinforced poly (butylene adipate-co-terephthalate). *Journal of Thermoplastic Composite Materials*, *08927057241230877*(10), 3238–3251. Advance online publication. DOI: 10.1177/08927057241230877

He, Y., Lin, E. S., Tan, C. L., Yu, Z., Tan, P. Y., & Wong, N. H. (2021). Model development of Roof Thermal Transfer Value (RTTV) for green roof in tropical area: A case study in Singapore. *Building and Environment*, *203*, 108101. DOI: 10.1016/j.buildenv.2021.108101

HomePro. (n.d.). *HomePro: Your home improvement store*. Retrieved December 25, 2024, from https://www.homepro.co.th

Honduran Chamber of the Construction Industry (CHICO). (2024). *Datos sobre el déficit habitacional*. El Heraldo/Tiempo.hn. Last access May 27, 2025. https://www.elheraldo.hn/honduras/deficit-vivienda-honduras-casi-un-millon-personas-sin-casa-AH20274756 and https://tiempo.hn/de-un-1-millon-600-mil-viviendas-es-el-deficit-habitacional-honduras/

Honduran Institute of Anthropology and History (IHAH). *Publicaciones sobre etnoarquitectura y cultura Lenca*. Last access May 20, 2025. https://portalunico.iaip.gob.hn/410/

Hong, R. Y., & Chen, Q. (2015). Dispersion of inorganic nanoparticles in polymer matrices: challenges and solutions. *Organic-Inorganic Hybrid Nanomaterials*, 1-38 https://doi.org/DOI: 10.1007/12_2014_286

Hossain, M. R., Sultana, R., Patwary, M. M., Khunga, N., Sharma, P., & Shaker, S. J. (2022). Self-healing concrete for sustainable buildings. A review. *Environmental Chemistry Letters*, *20*(2), 1–9. DOI: 10.1007/s10311-021-01375-9

Huang, H., Lu, J., Jin, L., & Ren, H. (2024). The future of environmental engineering technology: A disruptive innovation perspective. *Engineering (Beijing)*, *41*, 153–160. DOI: 10.1016/j.eng.2024.06.009

Huang, L. M., Hu, C. W., Liu, H. C., Hsu, C. Y., Chen, C. H., & Ho, K. C. (2012). Photovoltaic electrochromic device for solar cell module and self-powered smart glass applications. *Solar Energy Materials and Solar Cells*, *99*, 154–159. DOI: 10.1016/j.solmat.2011.03.036

Huang, Y., Stonehouse, A., & Abeykoon, C. (2023). Encapsulation methods for phase change materials—A critical review. *International Journal of Heat and Mass Transfer*, *200*, 123458. DOI: 10.1016/j.ijheatmasstransfer.2022.123458

Hu, S., Zhang, Y., & Liu, X. (2021). Performance test and microstructure of modified PVC aggregate-hybrid fiber reinforced engineering cementitious composite (ECC). *Materials (Basel)*, *14*(8), 1856. Advance online publication. DOI: 10.3390/ma14081856

Hussein, Z., Ashour, T., Khalil, M., Bahnasawy, A., Ali, S., Hollands, J., & Korjenic, A. (2019). Rice straw and flax fiber particleboards as a product of agricultural waste: An evaluation of technical properties. *Applied Sciences (Basel, Switzerland), 9*(18), 3878. DOI: 10.3390/app9183878

Hýsek, Š., Neuberger, P., Sikora, A., Schönfelder, O., & Ditommaso, G. (2019). Waste utilization: Insulation panel from recycled polyurethane particles and wheat husks. *Materials (Basel), 12*(19), 3075. DOI: 10.3390/ma12193075

Idumah, C. I. (2023). Thermal expansivity of polymer nanocomposites and applications. *Polymer-Plastics Technology and Materials, 62*(9), 1178–1203. DOI: 10.1080/25740881.2023.2204952

Iftikhar, M., Asghar, A., Ramzan, N., Sajjadi, B., & Chen, W. Y. (2019). Biomass densification: Effect of cow dung on the physicochemical properties of wheat straw and rice husk based biomass pellets. *Biomass and Bioenergy, 122*, 1–16. DOI: 10.1016/j.biombioe.2019.01.005

Illankoon, W. A. M. A. N., Milanese, C., Karunarathna, A. K., Liyanage, K. D. H. E., Alahakoon, A. M. Y. W., Rathnasiri, P. G., Collivignarelli, M. C., & Sorlini, S. (2023). Evaluating sustainable options for valorization of rice by-products in Sri Lanka: An approach for a circular business model. *Agronomy (Basel), 13*(3), 803. DOI: 10.3390/agronomy13030803

Instituto de Desarrollo Económico del Principado de Asturias (IDEPA). (2016). *Aplicaciones industriales de la nanotecnología*. https://www.idepa.es/documents/20147/163848/AplicacionesIndustriales.pdf/6c110c65-76ef-fdfb-15dd-38c072d6e2ee

Instituto Mexicano del Cemento y del Concreto (IMCYC). (2016). Nanotecnología y concreto. *Revista IMCYC*, mayo, 18-25. https://www.imcyc.com/images/MAYO%202016.pdf

Inter-American Development Bank (IDB). (2023). *Resiliencia y sostenibilidad en los códigos de construcción de América Latina y el Caribe: análisis regional comparativo y oportunidades de fortalecimiento*. IDB. Last access May 2, 2025. https://publications.iadb.org/publications/spanish/document/Resiliencia-y-sostenibilidad-en-los-codigos-de-construccion-de-America-Latina-y-el-Caribe-analisis-regional-comparativo-y-oportunidades-de-fortalecimiento.pdf

International Organization for Standardization. (1991). *ISO 8302: Thermal insulation — Determination of steady-state thermal resistance and related properties — Guarded hot plate apparatus*. ISO.

International Organization for Standardization. (1998). *ISO 10534-2: Acoustics — Determination of sound absorption coefficient and impedance in impedance tubes — Part 2: Transfer-function method.* ISO.

International Organization for Standardization. (2006). *ISO 14040: Environmental management — Life cycle assessment — Principles and framework.* ISO.

International Organization for Standardization. (2006). *ISO 14044: Environmental management — Life cycle assessment — Requirements and guidelines.* ISO.

Isyaka, A., Ibrahim, A., & Bello, A. (2020). A review of the effect of waste tire rubber on the properties of ECC. *International Journal of Advanced and Applied Sciences*, *7*(8), 105–116. DOI: 10.21833/ijaas.2020.08.011

Iyiola, O. O. (2025). *Integrated Lifecycle Assessment and Carbon Accounting of Building Engineering Systems Under Circular Economy Frameworks.* ResearchGate.

Jami, T., Karade, S. R., & Singh, L. P. (2019). A review of the properties of hemp concrete for green building applications. *Journal of Cleaner Production*, *239*, 117852. DOI: 10.1016/j.jclepro.2019.117852

Jayakumari, B. Y., Swaminathan, E. N., & Partheeban, P. (2023). A review on characteristics studies on carbon nanotubes-based cement concrete. *Construction & Building Materials*, *367*, 130344. DOI: 10.1016/j.conbuildmat.2023.130344

Jaysawal, R. K., Chakraborty, S., Elangovan, D., & Padmanaban, S. (2022). Concept of net zero energy buildings (NZEB)-A literature review. *Cleaner Engineering and Technology*, *11*, 100582. DOI: 10.1016/j.clet.2022.100582

Jelle, B. P., Kalnæs, S. E., & Gao, T. (2015). Low-emissivity materials for building applications: A state-of-the-art review and future research perspectives. *Energy and Building*, *96*, 329–356. DOI: 10.1016/j.enbuild.2015.03.024

Jirawattanasomkul, T., Minakawa, H., Likitlersuang, S., Ueda, T., Dai, J. G., Wuttiwannasak, N., & Kongwang, N. (2021). Use of water hyacinth waste to produce fibre-reinforced polymer composites for concrete confinement: Mechanical performance and environmental assessment. *Journal of Cleaner Production*, *292*, 126041. DOI: 10.1016/j.jclepro.2021.126041

Jochem, E. (2009) Passive houses and buildings. Improving the Efficiency of R&D and the Market Diffusion of Energy Technologies, 105-141.

Jones, D., & Brischke, C. (2017). *Performance of bio-based building materials.* Woodhead Publishing. DOI: 10.1016/B978-0-08-100982-6.00001-X

Kaitouni, S. I., Charai, M., Es-Sakali, N., Mghazli, M. O., El Mankibi, M., Uk-Joo, S., Ahachad, M., & Brigui, J. (2024). Energy and hygrothermal performance investigation and enhancement of rammed earth buildings in hot climates: From material to field measurements. *Energy and Building*, *315*, 114325. DOI: 10.1016/j.enbuild.2024.114325

Kan, A., Zhang, Q., Chen, Z., Zhang, J., Wang, K., & Cao, D. (2023). Novel prediction of thermal conductivities for nano-aerogel and its composites as vacuum insulation panel core. *International Journal of Thermal Sciences*, *189*, 108277. DOI: 10.1016/j.ijthermalsci.2023.108277

Kanda, M., Puggal, S., Dhall, N., & Sharma, A. (2018). Recent developments in the fabrication, characterization, and properties enhancement of polymer nanocomposites: A critical review. *Materials Today: Proceedings*, *5*(14), 28243–28252. DOI: 10.1016/j.matpr.2018.10.069

Kang, D. J., Park, G. U., Im, H. G., Park, H. Y., & Jin, J. (2016). Silica nanoparticle-embedded urethane acrylate nanohybrid thermosets for photo-patternable transparent hard coating. *Polymer*, *105*, 19–24. DOI: 10.1016/j.polymer.2016.10.008

Kaushik, D., Singh, H., & Tassou, S. A. (2024). Vacuum insulation panels for high-temperature applications–Design principles, challenges and pathways. *Thermal Science and Engineering Progress*, *48*, 102415. DOI: 10.1016/j.tsep.2024.102415

Kewalramani, M. A., Mohamed, O. A., & Syed, Z. I. (2017). Engineered cementitious composites for modern civil engineering structures in hot arid coastal climatic conditions. *Procedia Engineering*, *180*, 767–774. DOI: 10.1016/j.proeng.2017.04.237

Khan, H., Zahid, Z. B., Malik, U. J., & Ahmad, J. (2025). Performance Evaluation of Innovative Connection Systems for Structural and Seismic Challenges in High-Rise Modular Construction: A State-of-the-Art Review. Journal of Building Engineering, Elsevier.DOI: 10.1016/j.jobe.2025.112808

Khan, S. A., Jassim, M., Ilcan, H., Sahin, O., Bayer, İ. R., Sahmaran, M., & Koc, M. (2023). 3D printing of circular materials: Comparative environmental analysis of materials and construction techniques. *Case Studies in Construction Materials*, *18*, e02059. DOI: 10.1016/j.cscm.2023.e02059

Khan, Z. I., Habib, U., Mohamad, Z. B., Rahmat, A. R. B., & Abdullah, N. A. S. B. (2022). Mechanical and thermal properties of sepiolite strengthened thermoplastic polymer nanocomposites: A comprehensive review. *Alexandria Engineering Journal*, *61*(2), 975–990. DOI: 10.1016/j.aej.2021.06.015

Khedher, B. I., & Al-Sulaimi, S. K. (2018). Optimization of hybrid fibres in engineered cementitious composites. *Construction & Building Materials*, *190*, 24–37. DOI: 10.1016/j.conbuildmat.2018.08.188

Kirme, S. K., & Kapse, V. S. (2024). A comprehensive review of residential building energy efficiency measures in India. *Energy and Building*, *319*, 114537. DOI: 10.1016/j.enbuild.2024.114537

Klaine, S. J., Alvarez, P. J., Batley, G. E., Fernandes, T. F., Handy, R. D., Lyon, D. Y., & Lead, J. R. (2008). Nanomaterials in the environment: Behavior, fate, bioavailability, and effects. *Environmental Toxicology and Chemistry*, *27*(9), 1825–1851. DOI: 10.1897/08-090.1 PMID: 19086204

Kliem, S., Kreutzbruck, M., & Bonten, C. (2020). Review on the biological degradation of polymers in various environments. *Materials (Basel)*, *13*(20), 4586. DOI: 10.3390/ma13204586

Koh, C. H. A., & Kraniotis, D. (2021). Hygrothermal performance, energy use and embodied emissions in straw bale buildings. *Energy and Building*, *245*, 111091. DOI: 10.1016/j.enbuild.2021.111091

Konsta-Gdoutos, M. S., Metaxa, Z. S., & Shah, S. P. (2010). Highly dispersed carbon nanotube reinforced cement-based materials. *Cement and Concrete Research*, *40*(7), 1052–1059. DOI: 10.1016/j.cemconres.2010.02.015

Krungthai, C. O. M. P. A. S. S. (2022). *Biomass product market: Opportunities from agricultural waste* [Presentation slides]. Krungthai Bank Public Company Limited. Retrieved from https://krungthai.com/Download/news/MediaFile_471slide_upcycling_food_wastes.pdf

Kumar, A. P., Depan, D., Tomer, N. S., & Singh, R. P. (2009). Nanoscale particles for polymer degradation and stabilization—Trends and future perspectives. *Progress in Polymer Science*, *34*(6), 479–515. DOI: 10.1016/j.progpolymsci.2009.01.002

Kumar, K., Ghosh, P. K., Kumar, A., & Singh, O. (2021). Enhanced Thermomechanical Properties of ZrO 2 Particle Reinforced Epoxy Nanocomposite. *Journal of Materials Engineering and Performance*, *30*(1), 145–153. DOI: 10.1007/s11665-020-05350-3

Kylili, A., & Fokaides, P. A. (2017). Methodologies for selection of thermal insulation materials for cost-effective, sustainable, and energy-efficient retrofitting. In *Cost-Effective Energy Efficient Building Retrofitting* (pp. 23–55). Woodhead Publishing. DOI: 10.1016/B978-0-08-101128-7.00002-2

Lai, Z., & Chen, Y. (2024). Enhancing the mechanical and environmental performance of solidified soil using construction waste and glass micro-powder. *Heliyon, 10*(22), e40187. Advance online publication. DOI: 10.1016/j.heliyon.2024.e40187 PMID: 39584091

Land Development Department. (2025, May 05). Plowing of stubble to improve soil and increase rice yield. https://www.ldd.go.th/menu_moc/POSTER/rice/rice.htm

Lavín, R. (2023). Nanotecnología y construcción. *Universidad Diego Portales.* https://www.udp.cl/roberto-lavin-nanotecnologia-y-construccion/

Lawanwadeekul, S., & Chindaprasirt, P. (2025). Sustainable construction materials from leveraging water hyacinth and wastepaper for superior sound absorption and thermal insulation. *Journal of Building Engineering, 103*, 112147. DOI: 10.1016/j.jobe.2025.112147

Lawanwadeekul, S., Jun-On, N., Kongthavorn, P., Sangkas, T., & Daothong, S. (2024). Chemical-free thermal-acoustic panels from agricultural waste for sustainable building materials. *Cleaner Materials, 12*, 100245. DOI: 10.1016/j.clema.2024.100245

Le, A., Rodrigo, N., Domingo, N., & Senaratne, S. (2023). Policy mapping for net-zero-carbon buildings: Insights from leading countries. *Buildings (Basel, Switzerland), 13*(11), 2766. DOI: 10.3390/buildings13112766

Lee, J., Mahendra, S., & Alvarez, P. J. J. (2010). Nanomaterials in the Construction Industry: Applications and Environmental Considerations. *Environmental Science & Technology, 44*(10), 3655–3661. DOI: 10.1021/es903264k PMID: 20397700

Lee, S. W., Lim, C. H., & Salleh, E. I. B. (2016). Reflective thermal insulation systems in building: A review on radiant barrier and reflective insulation. *Renewable & Sustainable Energy Reviews, 65*, 643–661. DOI: 10.1016/j.rser.2016.07.002

Leite Ribeiro, L. M., Piccinini Scolaro, T., & Ghisi, E. (2025). LEED certification in building energy efficiency: A review of its performance efficacy and global applicability. *Sustainability (Basel), 17*(5), 1876. DOI: 10.3390/su17051876

Liang, H., Miao, C., Zhou, C., & Wang, Q. (2025). Norfloxacin–derived carbon dots modifying mechanical properties, impermeability, and chloride binding of cement–based materials. *Construction & Building Materials, 409*, 134342. DOI: 10.1016/j.conbuildmat.2025.144191

Liang, Y., Wu, H., Huang, G., Yang, J., & Wang, H. (2017). Thermal performance and service life of vacuum insulation panels with aerogel composite cores. *Energy and Building, 154*, 606–617. DOI: 10.1016/j.enbuild.2017.08.085

Li, C. Z., Hu, M., Xiao, B., Chen, Z., Tam, V. W., & Zhao, Y. (2021). Mapping the knowledge domains of emerging advanced technologies in the management of prefabricated construction. *Sustainability (Basel)*, *13*(16), 8800. DOI: 10.3390/su13168800

Li, C., Wen, X., Cai, W., Wu, J., Shao, J. J., Yang, Y., Yu, H., Liu, D., & Wang, M. (2022). Energy performance of buildings with composite phase-change material wallboards in different climatic zones of China. *Energy and Building*, *272*, 112398. DOI: 10.1016/j.enbuild.2022.112398

Li, H., Luo, Z., Xu, X., Cang, Y., & Yang, L. (2021). Assessing the embodied carbon reduction potential of straw bale rural houses by hybrid life cycle assessment: A four-case study. *Journal of Cleaner Production*, *303*, 127002. DOI: 10.1016/j.jclepro.2021.127002

Li, L., Yi, Z., Zhang, S., & Shao, Z. (2023). Exploring the mechanism of digital transformation empowering green innovation in construction enterprises. *Developments in the Built Environment*, *15*, 100199. Advance online publication. DOI: 10.1016/j.dibe.2023.100199

Li, M., Dai, L., & Hu, Y. (2022). Machine learning for harnessing thermal energy: From materials discovery to system optimization. *ACS Energy Letters*, *7*(10), 3204–3226. DOI: 10.1021/acsenergylett.2c01836 PMID: 37325775

Ling, H. M., Yew, M. C., Yew, M. K., & Saw, L. H. (2024). Analyzing recent active and passive cool roofing technology in buildings, including challenges and optimization approaches. *Journal of Building Engineering*, *89*, 109326. DOI: 10.1016/j.jobe.2024.109326

Li, T., Sun, H., Lei, F., Li, D., Leng, J., Chen, L., Huang, Y., & Sun, D. (2019). High performance linear low density polyethylene nanocomposites reinforced by two-dimensional layered nanomaterials. *Polymer*, *172*, 142–151. DOI: 10.1016/j.polymer.2019.03.072

Liu, Z. (2025). IoT for Innovation in the Building Process: IT for Design and Life Cycle Management of the Building Envelope [Doctoral Thesis]. University of Ferrara.

Liu, Z., Zhao, Y. G., Ma, L., & Lin, S. (2024, June) Review on high-strength recycled aggregate concrete: Mix design, properties, models and structural behaviour. In Structures (Vol. 64, 106598). Elsevier.

Liu, C., Lyu, Y., Li, C., Li, J., Zhuo, K., & Su, H. (2021). Thermal performance testing of triple-glazing water flow window in cooling operation. *Solar Energy*, *218*, 108–116. DOI: 10.1016/j.solener.2021.02.048

Liu, D., Li, X., & Chen, Y. (2023). Mechanical performance of high-strength engineering cementitious composites (ECC) with hybriding PE and steel fibers. *Case Studies in Construction Materials, 18*, e01961. Advance online publication. DOI: 10.1016/j.cscm.2023.e01961

Liu, N., & Morawska, L. (2020). Modeling the urban heat island mitigation effect of cool coatings in realistic urban morphology. *Journal of Cleaner Production, 264*, 121560. DOI: 10.1016/j.jclepro.2020.121560

Liu, P., Wu, W., Du, B., Tian, G., & Gong, Y. (2022). Study on the heat and moisture transfer characteristics of aerogel-enhanced foam concrete precast wall panels and the influence of building energy consumption. *Energy and Building, 256*, 111707. DOI: 10.1016/j.enbuild.2021.111707

Liu, W., Zhang, H., Wang, Q., Hua, T., & Xue, H. (2021). A review and scientometric analysis of global research on prefabricated buildings. *Advances in Civil Engineering, 2021*(1), 8869315. DOI: 10.1155/2021/8869315

Liu, Y. (2015). Sound transmission through triple-panel structures lined with poroelastic materials. *Journal of Sound and Vibration, 339*, 376–395. DOI: 10.1016/j.jsv.2014.11.014

Liu, Y., Zhang, J., Xu, J., Wang, Y., Li, B., & Zhang, S. (2023). Carbon emission-based life cycle assessment of rural residential buildings constructed with engineering bamboo: A case study in China. *Journal of Building Engineering, 76*, 107182. DOI: 10.1016/j.jobe.2023.107182

Liu, Z., Lyu, J., Fang, D., & Zhang, X. (2019). Nanofibrous kevlar aerogel threads for thermal insulation in harsh environments. *ACS Nano, 13*(5), 5703–5711. DOI: 10.1021/acsnano.9b01094 PMID: 31042355

Liu, Z., Zhang, Y., Yuan, X., Liu, Y., Xu, J., Zhang, S., & He, B. J. (2021). A comprehensive study of feasibility and applicability of building integrated photovoltaic (BIPV) systems in regions with high solar irradiance. *Journal of Cleaner Production, 307*, 127240. DOI: 10.1016/j.jclepro.2021.127240

Li, V. C., Wu, C., Wang, S., Ogawa, A., & Saito, T. (2020). On the emergence of 3D printable engineered, strain hardening cementitious composites (ECC/SHCC). *Cement and Concrete Research, 132*, 106038. Advance online publication. DOI: 10.1016/j.cemconres.2020.106038

Li, V. C., Yang, E. H., & Li, M. (2017). Self-healing of microcracks in engineered cementitious composites under sulfate and chloride environment. *Construction & Building Materials, 153*, 948–956. DOI: 10.1016/j.conbuildmat.2017.07.126

Li, Z., Yang, C., Yan, K., Xia, M., Yan, Z., Wang, D., & Wang, W. (2024). Rational design of a polypropylene composite foam with open-cell structure via graphite conductive network for sound absorption. *Soft Matter*, *20*(5), 1089–1099. DOI: 10.1039/D3SM01432K PMID: 38221881

López, M., & Pérez, J. (2024). Propiedades de concretos y morteros modificados con nanomateriales: Estado del arte. *Revista de Ingeniería Civil*, 30(2), 45-60. https://www.researchgate.net/publication/329301218_Propiedades_de_concretos_y_morteros_modificados_con_nanomateriales_estado_del_arte

Luo, L., Jin, X., Shen, G. Q., Wang, Y., Liang, X., Li, X., & Li, C. Z. (2020). Supply chain management for prefabricated building projects in Hong Kong. *Journal of Management Engineering*, *36*(2), 05020001. DOI: 10.1061/(ASCE)ME.1943-5479.0000739

Macarthur, E. L. L. E. N., & Heading, H. E. A. D. I. N. G. (2019). How the circular economy tackles climate change. *Ellen MacArthur Found*, *1*, 1–71.

Maderuelo-Sanz, R., García-Cobos, F. J., Sánchez-Delgado, F. J., Mota-López, M. I., Meneses-Rodríguez, J. M., Romero-Casado, A., Acedo-Fuentes, P., & López-Ramos, L. (2022). Mechanical, thermal and acoustical evaluation of biocomposites made of agricultural waste for ceiling tiles. *Applied Acoustics*, *191*, 108689. DOI: 10.1016/j.apacoust.2022.108689

Ma, H., Yi, C., & Wu, C. (2021). Review and outlook on durability of engineered cementitious composite (ECC). *Construction & Building Materials*, *287*, 122719. Advance online publication. DOI: 10.1016/j.conbuildmat.2021.122719

Mahame, C., Kikwasi, G. J., & Baruti, M. M. (2024). Barriers to the effective selection of sustainable materials for residential building projects: A qualitative study. *Sustainability (Basel)*, *16*(21), 9526. DOI: 10.3390/su16219526

Maiti, B., Abramov, A., Finn, M. G., & Díaz, D. D. (2021). Biopolymers as sustainable metal bio-adhesives. *Journal of Applied Polymer Science*, *138*(5), 49783. DOI: 10.1002/app.49783

Malanho, S., Veiga, R., & Farinha, C. B. (2021). Global performance of sustainable thermal insulating systems with cork for building facades. *Buildings (Basel, Switzerland)*, *11*(3), 83. DOI: 10.3390/buildings11030083

Manandhar, R., Kim, J. H., & Kim, J. T. (2019). Environmental, social and economic sustainability of bamboo and bamboo-based construction materials in buildings. *Journal of Asian Architecture and Building Engineering*, *18*(2), 49–59. DOI: 10.1080/13467581.2019.1595629

Marconi, M., Barbanera, M., Calabrò, G., & Baffo, I. (2020). Reuse of leather scraps for insulation panels: Technical and environmental feasibility evaluation. *Procedia CIRP*, *90*, 55–60. DOI: 10.1016/j.procir.2020.01.053

Martín-Chivelet, N., Kapsis, K., Wilson, H. R., Delisle, V., Yang, R., Olivieri, L., Polo, J., Eisenlohr, J., Roy, B., Maturi, L., Otnes, G., Dallapiccola, M., & Upalakshi Wijeratne, W. M. P. (2022). Building-integrated photovoltaic (BIPV) products and systems: A review of energy-related behavior. *Energy and Building*, *262*, 111998. DOI: 10.1016/j.enbuild.2022.111998

Martínez, L., & Rodríguez, P. (2024). Materiales inteligentes en construcciones modernas. *Revista de Arquitectura y Tecnología*, 8(3), 55-70. https://es.scribd.com/document/479501199/MATERIALES-INTELIGENTES-EN-CONSTRUCCIONES-MODERNAS

Martínez, B., Bernat-Maso, E., & Gil, L. (2023). Applications and properties of hemp stalk-based insulating biomaterials for buildings. *Materials (Basel)*, *16*(8), 3245. DOI: 10.3390/ma16083245 PMID: 37110081

Martwong, E., Minanandana, T., Puksuwan, K., Junthip, J., & Sukhawipat, N. (2024). Sound absorption and thermal insulation materials from waste palm oil for housing application: Green polyurethane/water hyacinth fiber sheet composite. *Construction & Building Materials*, *438*, 137007. DOI: 10.1016/j.conbuildmat.2024.137007

Mendes, L. C., Mariano, D. M., Freitas, D. F. S., Albitres, G. A. V., Tavares, M. I. B., & Garcia, E. E. (2024). Ecofriendly composites based on poly (lactic acid) with nano-zirconium phosphate and nano-zinc oxide/zirconium phosphate: Physico-chemical and aging characteristics. *Journal of Thermal Analysis and Calorimetry*, *149*(17), 9297–9306. DOI: 10.1007/s10973-024-13541-6

Méndez, M., & Pineda, L. (2017). Estabilización de adobes con fibra de bambú para la construcción de viviendas de interés social en el barrio “El Pantanal”, Granada. *Revista de Arquitectura*, 13(1), 18-29. https://camjol.info/index.php/arquitectura/article/download/8981/10145/31194

Menéndez, R. (2023). *El impacto de la nanotecnología y la revolución del grafeno*. CSIC. https://digital.csic.es/bitstream/10261/351671/1/Materiales_del_Futuro_R_Menendez_AACI-vol.-3-2023.pdf

Meng, Q., Hu, L., Li, M., & Qi, X. (2024). Assessing the environmental impact of building life cycle: A carbon reduction strategy through innovative design, intelligent construction, and secondary utilization. *Developments in the Built Environment*, *16*, 100230. Advance online publication. DOI: 10.1016/j.dibe.2023.100230

Mensah, R. A., Xiao, J., Asamoah, R. B., & Amponsah, K. (2021). Biochar-added cementitious materials—A review on mechanical, thermal, and environmental properties. *Sustainability (Basel), 13*(16), 9073. Advance online publication. DOI: 10.3390/su13169336

Merli, F., Belloni, E., & Buratti, C. (2021). Eco-sustainable wood waste panels for building applications: Influence of different species and assembling techniques on thermal, acoustic, and environmental performance. *Buildings (Basel, Switzerland), 11*(8), 361. DOI: 10.3390/buildings11080361

Mihai, M., Tanasiev, V., Dinca, C., Badea, A., & Vidu, R. (2017). Passive house analysis in terms of energy performance. *Energy and Building*, *144*, 74–86. DOI: 10.1016/j.enbuild.2017.03.025

Mihalakakou, G., Souliotis, M., Papadaki, M., Menounou, P., Dimopoulos, P., Kolokotsa, D., Paravantis, J. A., Tsangrassoulis, A., Panaras, G., Giannakopoulos, E., & Papaefthimiou, S. (2023). Green roofs as a nature-based solution for improving urban sustainability: Progress and perspectives. *Renewable & Sustainable Energy Reviews*, *180*, 113306. DOI: 10.1016/j.rser.2023.113306

Mirindi, D. (2024). A review of particleboard development and performance using non-toxic and biodegradable adhesives. *International Journal of Engineering Trends and Technology*, *72*(5), 252–260. DOI: 10.14445/22315381/IJETT-V72I5P126

Mishra, T., Mandal, P., Rout, A. K., & Sahoo, D. (2022). A state-of-the-art review on potential applications of natural fiber-reinforced polymer composite filled with inorganic nanoparticle. *Composites Part C: Open Access*, *9*, 100298. DOI: 10.1016/j.jcomc.2022.100298

Mohammed, B. S., Achara, B. E., & Liew, M. S. (2018). The influence of high temperature on microstructural damage and residual properties of nano-silica-modified (NS-modified) self-consolidating engineering cementitious composites (SC-ECC) using response surface methodology (RSM). *Construction & Building Materials*, *192*, 450–466. DOI: 10.1016/j.conbuildmat.2018.10.114

Moini, N., Khaghanipour, M., Faridani, F., & Jahandideh, A. (2022). Green adhesives—Past, present, and future outlook. In *Green Sustainable Process for Chemical and Environmental Engineering and Science* (pp. 341–372). Elsevier. DOI: 10.1016/B978-0-323-99643-3.00006-1

Molod, M. A., Spyridis, P., & Barthold, F. J. (2022). Applications of shape memory alloys in structural engineering with a focus on concrete construction—A comprehensive review. *Construction & Building Materials*, *337*, 127565. DOI: 10.1016/j.conbuildmat.2022.127565

Mondal, M. K., Bose, B. P., & Bansal, P. (2019). Recycling waste thermoplastic for energy efficient construction materials: An experimental investigation. *Journal of Environmental Management*, *240*, 119–125. DOI: 10.1016/j.jenvman.2019.03.016 PMID: 30928789

Mondem, N., & Balunaini, U. (2025). Optimal Blends of Artificial and Natural Aggregates for Enhanced Pavement Performance. In *Conference on GeoPractices towards Sustainable Infrastructure* (pp. 41-51). Springer, Singapore. DOI: 10.1007/978-981-96-3220-6_4

Morris, F., Allen, S., & Hawkins, W. (2021). On the embodied carbon of structural timber versus steel, and the influence of LCA methodology. *Building and Environment*, *206*, 108285. DOI: 10.1016/j.buildenv.2021.108285

Moussa, R. R., Ismaeel, W. S., & Solban, M. M. (2022). Energy generation in public buildings using piezoelectric flooring tiles; A case study of a metro station. *Sustainable Cities and Society*, *77*, 103555. DOI: 10.1016/j.scs.2021.103555

Muhit, I. B., Omairey, E. L., & Pashakolaie, V. G. (2024). A holistic sustainability overview of hemp as building and highway construction materials. *Building and Environment*, *256*, 111470. DOI: 10.1016/j.buildenv.2024.111470

Mulder, W. J., Gosselink, R. J. A., Vingerhoeds, M. H., Harmsen, P. F. H., & Eastham, D. (2011). Lignin based controlled release coatings. *Industrial Crops and Products*, *34*(1), 915–920. DOI: 10.1016/j.indcrop.2011.02.011

Mulla, M. Z., Rahman, M. R. T., Marcos, B., Tiwari, B., & Pathania, S. (2021). Poly Lactic Acid (PLA) Nanocomposites: Effect of inorganic nanoparticles reinforcement on its performance and food packaging applications. *Molecules (Basel, Switzerland)*, *26*(7), 1967. DOI: 10.3390/molecules26071967 PMID: 33807351

Mundo Nano. (2023). *Nanopartículas de TiO2 fotocatalíticas incorporadas en materiales de construcción*. https://www.mundonano.unam.mx/ojs/index.php/nano/article/view/69737

Municipal del Distrito Central, A., & de Ordenamiento Territorial, D. (2020). *Reglamento verde para la construcción.* [Acuerdo municipal n.° 024, Acta n.° 016 del 14 de diciembre de 2020]. *La Gaceta: Diario Oficial de la República de Honduras*, 35 (493), 1-43. Last access May 22, 2025. https://www.tsc.gob.hn/web/leyes/Acuerdo-024-AMDC.pdf

Muñoz Torres, L. C. (2015). *Análisis tipológico y bioconstructivo de las viviendas vernáculas en el municipio San Antonio de Oriente, Honduras* [thesis, Zamorano University, Escuela Agrícola Panamericana]. https://bdigital.zamorano.edu/handle/11036/4550

Muralidharan, R., Sathyan, D., & Anand, K. B. (2021). Review of the effects of supplementary cementitious materials and chemical additives on the physical, mechanical and durability properties of hydraulic concrete. *Materials (Basel), 14*(23), 7270. Advance online publication. DOI: 10.3390/ma14237270

Mushtaq, A., Ali, S., Chaudhry, A.H., Sial, N., Aslam, M. & Batool, H. (2025) Geopolymers as supplementary cementitious materials to reduce carbon dioxide emissions. Nature Environment & Pollution Technology, 24.

Naeem, S., Gilani, S. Q. Z., Baheti, V., Wiener, J., Militky, J., Javed, S., ... & ul Hassan, S. Z. (2018). Electrical Conductivity of PLA Films Reinforced With Carbon Nano Particles from Waste Acrylic Fibers. In *Advances in Natural Fibre Composites: Raw Materials, Processing and Analysis* (pp. 205-222). Springer International Publishing DOI: 10.1007/978-3-319-64641-1_18

Naffakh, M., Díez-Pascual, A. M., Marco, C., Ellis, G. J., & Gomez-Fatou, M. A. (2013). Opportunities and challenges in the use of inorganic fullerene-like nanoparticles to produce advanced polymer nanocomposites. *Progress in Polymer Science, 38*(8), 1163–1231. DOI: 10.1016/j.progpolymsci.2013.04.001

Naghibalsadati, F., Karimi, N., Mim, S. J., & Ng, K. T. W. (2025). Thematic evolution of Life Cycle Assessment in construction and demolition waste management: Before and after ISO 14040 and the Paris Agreement. Journal of Building Engineering, Elsevier.

Nandy, A., Houl, Y., Zhao, W., & D'Souza, N. A. (2023). Thermal heat transfer and energy modeling through incorporation of phase change materials (PCMs) into polyurethane foam. *Renewable & Sustainable Energy Reviews, 182*, 113410. DOI: 10.1016/j.rser.2023.113410

National Congress of Honduras. (2020). *Ley Marco de Vivienda y Asentamientos Humanos* (Decree No. 173-2019). Last access May 7, 2025. https://dinero.hn/congreso-nacional-aprobo-ley-marco-de-vivienda-y-fondo-para-construccion/

Ndukwe, C. V., Liu, J., & Chan, T. K. (2020, November) Impact of COVID-19 on the China-Australia construction supply chain. In International Symposium on Advancement of Construction Management and Real Estate (pp. 1275-1291). Singapore: Springer Singapore.

Neila González, F. J. (2004). *Arquitectura Bioclimática en un entorno sostenible*. Munilla-Lería Publisher.

Netto, L. G., Agostinho, F., Almeida, C. M., Liu, G., & Giannetti, B. F. (2024). Environmental performance of bamboo-based office paper production: A comparative study with eucalyptus. *Industrial Crops and Products*, *211*, 118234. DOI: 10.1016/j.indcrop.2024.118234

Nguyen, M. T., Fernandez, C. A., Haider, M. M., Chu, K. H., Jian, G., Nassiri, S., Zhang, D., Rousseau, R., & Glezakou, V. A. (2023). Toward self-healing concrete infrastructure: Review of experiments and simulations across scales. *Chemical Reviews*, *123*(18), 10838–10876. DOI: 10.1021/acs.chemrev.2c00709 PMID: 37286529

Nguyen-Van, V., Li, S., Liu, J., Nguyen, K., & Tran, P. (2023). Modelling of 3D concrete printing process: A perspective on material and structural simulations. *Additive Manufacturing*, *61*, 103333. DOI: 10.1016/j.addma.2022.103333

Nicolini, E. (2022). Built environment and wellbeing—Standards, multi-criteria evaluation methods, certifications. *Sustainability (Basel)*, *14*(8), 4754. DOI: 10.3390/su14084754

Nikam, P. C., Rao, A. R., & Shertukde, V. V. (2021). Enhancement of thermomechanical and chemical resistance properties of polyurethane composite reinforced with hydrophobic nano-silica and scrape PET derived bis (2-hydroxyethyl terephthalate). *Materials Today. Communications*, *29*, 102788. DOI: 10.1016/j.mtcomm.2021.102788

Novais, R. M., Senff, L., Carvalheiras, J., Lacasta, A. M., Cantalapiedra, I. R., & Labrincha, J. A. (2021). Simple and effective route to tailor the thermal, acoustic and hygrothermal properties of cork-containing waste derived inorganic polymer composites. *Journal of Building Engineering*, *42*, 102501. DOI: 10.1016/j.jobe.2021.102501

Nsaif, M. A., Jalil, J. M., & Baccar, M. (2024). An experimental and numerical investigation of the thermal performance of phase change materials in different triple-glazed window configurations. *International Journal of Thermofluids*, *24*, 100889. DOI: 10.1016/j.ijft.2024.100889

Nur-E-Alam, M., Vasiliev, M., Yap, B. K., Islam, M. A., Fouad, Y., & Kiong, T. S. (2024). Design, fabrication, and physical properties analysis of laminated Low-E coated glass for retrofit window solutions. *Energy and Building*, *318*, 114427. DOI: 10.1016/j.enbuild.2024.114427

Ogbonna, V. E., Popoola, A. P. I., Popoola, O. M., & Adeosun, S. O. (2022). A review on the recent advances on improving the properties of epoxy nanocomposites for thermal, mechanical, and tribological applications: Challenges and recommendations. *Polymer-Plastics Technology and Materials*, *61*(2), 176–195. DOI: 10.1080/25740881.2021.1967391

Ojo, A. O. (2023). An overview of lignocellulose and its biotechnological importance in high-value product production. *Fermentation (Basel, Switzerland)*, *9*(11), 990. DOI: 10.3390/fermentation9110990

Oliver, P. (2006). *Built to meet needs: cultural issues in vernacular architecture*. Architectural Press. Last access May 2, 2025, https://www.routledge.com/Built-to-Meet-Needs-Cultural-Issues-in-Vernacular-Architecture/Oliver/p/book/9780750666572

Ormondroyd, G. A. (2015). Adhesives for wood composites. In *Wood composites* (pp. 47–66). Woodhead Publishing. DOI: 10.1016/B978-1-78242-454-3.00003-2

Ouakarrouch, M., Bousshine, S., Bybi, A., Laaroussi, N., & Garoum, M. (2022). Acoustic and thermal performances assessment of sustainable insulation panels made from cardboard waste and natural fibers. *Applied Acoustics*, *199*, 109007. DOI: 10.1016/j.apacoust.2022.109007

Owen, M. M., Wong, L. S., Achukwu, E. O., Saharudin, M. S., & Akil, H. M. (2024). Enhanced mechanical, thermal, and morphological properties of waste PET plastics reinforced with coated biodegradable kenaf fibers for infrastructure applications. *Construction & Building Materials*, *442*, 137659. DOI: 10.1016/j.conbuildmat.2024.137659

Pacheco-Torgal, F. (2014). *Nanotechnology in eco-efficient construction: Materials, processes and applications*. Elsevier.

Pacheco-Torgal, F., & Jalali, S. (2011). Nanotechnology: Advantages and drawbacks in the field of construction and building materials. *Construction & Building Materials*, *25*(2), 582–590. DOI: 10.1016/j.conbuildmat.2010.07.009

Pajek, L., Hudobivnik, B., Kunič, R., & Košir, M. (2017). Improving thermal response of lightweight timber building envelopes during cooling season in three European locations. *Journal of Cleaner Production*, *156*, 939–952. DOI: 10.1016/j.jclepro.2017.04.098

Palma Bravo, J. C. (2024). *Avances en la nanotecnología para la construcción sostenible*. ResearchGate. https://www.researchgate.net/publication/380729400_Avances_en_la_Nanotecnologia_para_la_Construccion_Sostenible

Papadakis, V. G. (2000). Effect of fly ash on Portland cement systems: Part II. High-calcium fly ash. *Cement and Concrete Research*, *30*(10), 1647–1654. DOI: 10.1016/S0008-8846(00)00388-4

Parhizi, Z., Dearnaley, J., Kauter, K., Mikkelsen, D., & Pal, P. (2025). Spore-Tacular Future Materials: Innovations and Applications of Mycelium-Based Composites. *Preprints*, 2025.

Parmesan, C., Anshari, G., Buotte, P., Campbell, D., Castellanos, E., & Cowie, A. (2022). Nature-based solutions for climate change: Mitigation and adaptation. *Climatic Change*, •••, 303–309.

Pathak, R., Punetha, M., Bhatt, S., Pillai, S. A., Dhapola, P. S., & Punetha, V. D. (2024). Organic and inorganic nanofillers for polymer nanocomposites: Trends, opportunities, and challenges. *Advances in Functionalized Polymer Nanocomposites*, 1-34 DOI: 10.1016/B978-0-443-18860-2.00003-7

Paul, D. R., & Robeson, L. M. (2008). Polymer nanotechnology: Nanocomposites. *Polymer*, *49*(15), 3187–3204. DOI: 10.1016/j.polymer.2008.04.017

Petousis, M., Moutsopoulou, A., Korlos, A., Papadakis, V., Mountakis, N., Tsikritzis, D., Ntintakis, I., & Vidakis, N. (2023). The effect of nano zirconium dioxide (ZrO2)-optimized content in polyamide 12 (PA12) and polylactic acid (PLA) matrices on their thermomechanical response in 3D printing. *Nanomaterials (Basel, Switzerland)*, *13*(13), 1906. DOI: 10.3390/nano13131906 PMID: 37446421

Philip, S., & Rakendu, R. (2020). Thermal insulation materials based on water hyacinth for application in sustainable buildings. *Materials Today: Proceedings*, *33*, 3803–3809. DOI: 10.1016/j.matpr.2020.06.219

Pisello, A. L. (2017). State of the art on the development of cool coatings for buildings and cities. *Solar Energy*, *144*, 660–680. DOI: 10.1016/j.solener.2017.01.068

Pizzi, A., Papadopoulos, A. N., & Policardi, F. (2020). Wood composites and their polymer binders. *Polymers*, *12*(5), 1115. DOI: 10.3390/polym12051115

Poopalam, K. D., Ismail, T. N. M. T., Hanzah, N. A., Alias, A. H., Wahab, N. A., Ibrahim, Z., Subramaniam, V., Armylisas, A. H. N., & Idris, Z. (2024). Utilization of oil palm biomass and polyurethanes as sustainable construction materials: A review. *Renewable and Sustainable Materials Reviews*, *18*, 100123. Advance online publication. DOI: 10.1016/j.dibe.2024.100380

Porter, W. A. (2020) Insulation: facilities. In Managing Human and Social Systems (pp. 227-249). CRC Press. DOI: 10.1201/9781003053514-26

Presidency of the Republic of Honduras. (2022). *Creación del Programa de Vivienda y Asentamientos Humanos (PVAH)* (PCM-005-2022). Last access May 27, 2025. https://pvah.gob.hn/

Proietti, S., Sdringola, P., Desideri, U., Zepparelli, F., Masciarelli, F., & Castellani, F. (2013). Life cycle assessment of a passive house in a seismic temperate zone. *Energy and Building*, *64*, 463–472. DOI: 10.1016/j.enbuild.2013.05.013

Qaidi, S. M., Atrushi, D. S., Mohammed, A. S., Ahmed, H. U., Faraj, R. H., Emad, W., Tayeh, B. A., & Najm, H. M. (2022). Ultra-high-performance geopolymer concrete: A review. *Construction & Building Materials*, *346*, 128495. DOI: 10.1016/j.conbuildmat.2022.128495

Qian, C., Zheng, T., Zhang, X., & Su, Y. (2021). Application of microbial self-healing concrete: Case study. *Construction & Building Materials*, *290*, 123226. DOI: 10.1016/j.conbuildmat.2021.123226

Qiu, J., Aw-Yong, W. L., & Yang, E. H. (2018). Effect of self-healing on fatigue of engineered cementitious composites (ECCs). *Cement and Concrete Composites*, *94*, 145–152. DOI: 10.1016/j.cemconcomp.2018.09.007

Rafiei, P., Nili, M., & Amini, M. (2021). Temperature impact on engineered cementitious composite containing basalt fibers. *Applied Sciences (Basel, Switzerland)*, *11*(15), 6848. Advance online publication. DOI: 10.3390/app11156848

Rahim, M., Douzane, O., Le, A. T., Promis, G., & Langlet, T. (2016). Characterization and comparison of hygric properties of rape straw concrete and hemp concrete. *Construction & Building Materials*, *102*, 679–687. DOI: 10.1016/j.conbuildmat.2015.11.021

Ramesh, M., Rajeshkumar, L., Sasikala, G., Balaji, D., Saravanakumar, A., Bhuvaneswari, V., & Bhoopathi, R. (2022). A critical review on wood-based polymer composites: Processing, properties, and prospects. *Polymers*, *14*(3), 589. DOI: 10.3390/polym14030589

Razali, N., Aziz, A. H. A., & Hassan, R. (2024). Governments' initiatives towards Islamic green finance. In *Islamic Green Finance* (pp. 135-143). Routledge.

Raza, S., Shafei, B., Saiidi, M. S., Motavalli, M., & Shahverdi, M. (2022). Shape memory alloy reinforcement for strengthening and self-centering of concrete structures—State of the art. *Construction & Building Materials*, *324*, 126628. DOI: 10.1016/j.conbuildmat.2022.126628

Rezaei, S. D., Shannigrahi, S., & Ramakrishna, S. (2017). A review of conventional, advanced, and smart glazing technologies and materials for improving indoor environment. *Solar Energy Materials and Solar Cells*, *159*, 26–51. DOI: 10.1016/j.solmat.2016.08.026

Ricciardi, P., Belloni, E., Merli, F., & Buratti, C. (2021). Sustainable panels made with industrial and agricultural waste: Thermal and environmental critical analysis of the experimental results. *Applied Sciences (Basel, Switzerland)*, *11*(2), 494. DOI: 10.3390/app11020494

Riffat, S., Ahmad, M. I., & Shakir, A. (2024). Zero-energy and low carbon buildings. In *Sustainable Energy Technologies and Low Carbon Buildings* (pp. 219–258). Springer Nature Switzerland.

Rincón, J., Manrique, A., & Galvis, F. (2020). Prototipo de vivienda social bioclimática y sostenible en Quibdó, Chocó, Colombia. *Revista de Arquitectura (Bogotá)*, 22(1), 86-98. https://repository.ugc.edu.co/server/api/core/bitstreams/94473d6f-3f53-4428-bc40-952ac9dc1f98/content

Rodriguez, N., Katooziani, A., & Jeelani, I. (2024). Barriers to energy-efficient design and construction practices: A comprehensive analysis. *Journal of Building Engineering*, *82*, 108349. DOI: 10.1016/j.jobe.2023.108349

Rodríguez-Romero, R., Compán, V., Sáez, A., & García-Macías, E. (2023). *Hierarchical meta-modelling of stone injected with CNT/cement mortar*. arXiv. https://arxiv.org/abs/2310.05759

Roque, A. J., Paleologos, E. K., O'Kelly, B. C., Tang, A. M., Reddy, K. R., Vitone, C., Mohamed, A.-M. O., Koda, E., Goli, V. S. N. S., Vieira, C. S., Fei, X., Sollecito, F., Vaverková, M. D., Plötze, M., Petti, R., Podlasek, A., Puzrin, A. M., Cotecchia, F., Osiński, P., & Singh, D. N. (2021). Sustainable environmental geotechnics practices for a green economy. *Environmental Geotechnics*, *9*(2), 68–84. DOI: 10.1680/jenge.21.00091

Rosati, A., Fedel, M., & Rossi, S. (2021). NIR reflective pigments for cool roof applications: A comprehensive review. *Journal of Cleaner Production*, *313*, 127826. DOI: 10.1016/j.jclepro.2021.127826

Rudofsky, B. (1964). *Architecture Without Architects: A Short Introduction to Non-Pedigreed Architecture*. Museum of Modern Art. Last access May 2, 2025. https://www.blinkist.com/en/books/architecture-without-architects-en

Sabapathy, L., Mahalingam, B., & Ravichandran, P. T. (2020). Acid and sulphate attacks on a rubberized engineered cementitious composite containing graphene oxide. *Materials (Basel), 13*(14), 3125. Advance online publication. DOI: 10.3390/ma13143125

Sadineni, S. B., Madala, S., & Boehm, R. F. (2011). Passive building energy savings: A review of building envelope components. *Renewable & Sustainable Energy Reviews, 15*(8), 3617–3631. DOI: 10.1016/j.rser.2011.07.014

Sáez-de-Guinoa, A., Zambrana-Vasquez, D., Fernández, V., & Bartolomé, C. (2022). Circular economy in the European construction sector: A review of strategies for implementation in building renovation. *Energies, 15*(13), 4747. DOI: 10.3390/en15134747

Saffari, M., Roe, C., & Finn, D. P. (2022). Improving the building energy flexibility using PCM-enhanced envelopes. *Applied Thermal Engineering, 217*, 119092. DOI: 10.1016/j.applthermaleng.2022.119092

Şahmaran, M., Lachemi, M., Hossain, K. M. A., Ranade, R., & Li, V. C. (2009). Influence of aggregate type and size on ductility and mechanical properties of engineered cementitious composites. *ACI Materials Journal, 106*(3), 308–316. DOI: 10.14359/56556

Şahmaran, M., & Li, V. C. (2010). Engineered cementitious composites. *Transportation Research Record: Journal of the Transportation Research Board, 2164*(1), 1–8. DOI: 10.3141/2164-01

Said, H. S., Ibrahim, A., & Elrahman, M. A. (2018). Use of recycled glass powder to improve the performance properties of high volume fly ash-engineered cementitious composites. *Construction & Building Materials, 163*, 53–62. DOI: 10.1016/j.conbuildmat.2017.12.067

Sain, A., Gaur, A., Khichad, J. S., & Somani, P. (2024, June). Treatment of bamboo for sustainable construction practice: A comprehensive review. [). IOP Publishing.]. *IOP Conference Series. Earth and Environmental Science, 1326*(1), 012049. DOI: 10.1088/1755-1315/1326/1/012049

Salah, A., Nada, S., & Mahmoud, H. (2024). Applicability of developing an affordable eco-friendly switchable insulation for sustainable building envelopes in a hot climate: Comprehensive review. *Energy and Building, 323*, 114757. DOI: 10.1016/j.enbuild.2024.114757

Salas-Ruiz, A., del Mar Barbero-Barrera, M., & Ruiz-Téllez, T. (2019). Micro-structural and thermo-physical characterization of a Water Hyacinth petiole for thermal insulation particle board manufacture. *Materials (Basel), 12*(4), 560. DOI: 10.3390/ma12040560

Salinas, I. M. (1991). *Arquitectura de los grupos étnicos de Honduras*. Guaymuras Publishing.

Saloni, S., Singh, A., Jatin, , Kumar, J., & Parveen, . (2019). Influence of fly ash, cement and ground river sand on compressive strength and chloride penetration of mortar. *Materials Today: Proceedings, 33*, 1690–1694. DOI: 10.1016/j.matpr.2020.07.269

Samadianfard, S., & Toufigh, V. (2020). Energy use and thermal performance of rammed-earth materials. *Journal of Materials in Civil Engineering, 32*(10), 04020276. DOI: 10.1061/(ASCE)MT.1943-5533.0003364

Sambucci, M., Savoni, F., & Valente, M. (2023). Aerogel technology for thermal insulation of cryogenic tanks—Numerical analysis for comparison with traditional insulating materials. *Gels (Basel, Switzerland), 9*(4), 307. DOI: 10.3390/gels9040307 PMID: 37102919

Sambucci, M., Sibai, A., & Valente, M. (2021). Recent advances in geopolymer technology. A potential eco-friendly solution in the construction materials industry: A review. *Journal of Composites Science, 5*(4), 109. DOI: 10.3390/jcs5040109

Samieifakhr, M., & Shojaei, A. (2024). Improved crystallization behavior and enhanced impact strength of melt processed poly (ethylene terephthalate)/UiO-66 nanocomposites. *Polymer, 290*, 126593. DOI: 10.1016/j.polymer.2023.126593

Sánchez, F., & Sobolev, K. (2010). Nanotechnology in concrete – A review. *Construction & Building Materials, 24*(11), 2060–2071. DOI: 10.1016/j.conbuildmat.2010.03.014

Sandanayake, M., Kraus, R., Haigh, R., Yaghoubi, E., & Vrcelj, Z. (2025). Tex-Crete—Carbon and Cost Assessment of Concrete with Textile and Cardboard Fibres—Case Studies Towards Circular Economy. *Applied Sciences (Basel, Switzerland), 15*(13), 6962. DOI: 10.3390/app15136962

Sangmesh, B., Patil, N., Jaiswal, K. K., Gowrishankar, T. P., Selvakumar, K. K., Jyothi, M. S., & Kumar, S. (2023). Development of sustainable alternative materials for the construction of green buildings using agricultural residues: A review. *Construction & Building Materials, 368*, 130457. DOI: 10.1016/j.conbuildmat.2023.130457

Santamouris, M. (2013). Using cool pavements as a mitigation strategy to fight urban heat island—A review of the actual developments. *Renewable & Sustainable Energy Reviews*, *26*, 224–240. DOI: 10.1016/j.rser.2013.05.047

Santos, A. K., Ferreira, V. M., & Dias, A. C. (2024). Promoting decarbonisation in the construction of new buildings: A strategy to calculate the embodied carbon footprint. *Journal of Building Engineering*, *103*, 112037. DOI: 10.1016/j.jobe.2025.112037

Sarapin, P. Netpae, T., Phothi, R., Sangon, A. . ., & Chaothum, C. . . (2024). The Relationship Between Hotspots From Open Field Burning of Rice Straw and PM2.5 Change in Nakhon Sawan Province PSRU Journal of Science and Technology, 8(3), 54–68. retrieved from https://ph01.tci-thaijo.org/index.php/Scipsru/article/view/253255 (Original work published December 20, 2023)

Saxena, M., Morchhale, R. K., Asokan, P., & Prasad, B. K. (2008). Plant fiber—Industrial waste reinforced polymer composites as a potential wood substitute material. *Journal of Composite Materials*, *42*(4), 367–384. DOI: 10.1177/0021998307087014

Sbar, N. L., Podbelski, L., Yang, H. M., & Pease, B. (2012). Electrochromic dynamic windows for office buildings. *International Journal of Sustainable Built Environment*, *1*(1), 125–139. DOI: 10.1016/j.ijsbe.2012.09.001

Schmidt, H., Jonschker, G., Goedicke, S., & Mennig, M. (2000). The sol-gel process as a basic technology for nanoparticle-dispersed inorganic-organic composites. *Journal of Sol-Gel Science and Technology*, *19*(1-3), 39–51. DOI: 10.1023/A:1008706003996

Sedaghat, A., Salem, H., Hussam, W. K., Mahdizadeh, A., Al-Khiami, M. I., Malayer, M. A., Soleimani, S. M., Sabati, M., Narayanan, R., Rasul, M., & Khan, M. M. K. (2023). Exploring energy-efficient building solutions in hot regions: A study on bio-phase change materials and cool roof coatings. *Journal of Building Engineering*, *76*, 107258. DOI: 10.1016/j.jobe.2023.107258

Serrano, H., Barba, E., & Del Real, C. (2014). Prototipo de vivienda de adobe con energías renovables. *Ra Ximhai*, *10*(4), 187–200. https://dialnet.unirioja.es/descarga/articulo/5035106.pdf

Shanmugasundaram, N., & Praveenkumar, S. (2021). Influence of supplementary cementitious materials, curing conditions and mixing ratios on fresh and mechanical properties of engineered cementitious composites—A review. *Construction & Building Materials*, *309*, 125038. Advance online publication. DOI: 10.1016/j.conbuildmat.2021.125038

Shanmugasundaram, N., & Praveenkumar, S. (2023). Influence of manufactured sand gradation and water cement ratios on compressive strength of engineered cementitious composites. *Materials Today: Proceedings*. Advance online publication. DOI: 10.1016/j.matpr.2023.05.024

Shanmugasundaram, N., & Praveenkumar, S. (2024). Prediction on mechanical properties of engineered cementitious composites: An experimental and machine learning approach. *Structural Concrete*. Advance online publication. DOI: 10.1002/suco.202400231

Shanmugasundaram, N., Praveenkumar, S., & Kumar, K. V. (2025). Development of sustainable engineered cementitious composites using limestone calcined clay cement (LC3): A review. *Civil Engineering (Shiraz)*. Advance online publication. DOI: 10.1007/s40996-025-01966-4

Shanmugasundaram, N., Praveenkumar, S., & Kumar, K. V. (2025). Effects of basalt fiber content on the properties of ambient-cured engineered geopolymer composites. *Fibers and Polymers*, *26*(7), 3147–3168. Advance online publication. DOI: 10.1007/s12221-025-01019-7

Shanmugasundaram, N., Praveenkumar, S., & Kumar, K. V. (2025). Structural behavior of engineered cementitious composite substrate slab overlays for bridge deck and pavement applications. *Construction & Building Materials*, *459*, 139736. Advance online publication. DOI: 10.1016/j.conbuildmat.2024.139736

Sharma, R. K., Ganesan, P., Tyagi, V. V., Metselaar, H. S. C., & Sandaran, S. C. (2015). Developments in organic solid–liquid phase change materials and their applications in thermal energy storage. *Energy Conversion and Management*, *95*, 193–228. DOI: 10.1016/j.enconman.2015.01.084

Shekarchian, M., Moghavvemi, M., Rismanchi, B., Mahlia, T. M. I., & Olofsson, T. (2012). The cost benefit analysis and potential emission reduction evaluation of applying wall insulation for buildings in Malaysia. *Renewable & Sustainable Energy Reviews*, *16*(7), 4708–4718. DOI: 10.1016/j.rser.2012.04.045

Sheng, J., Ji, D., Yu, F., Cui, L., Zeng, Q., Ai, N., & Ji, J. (2014). Influence of chemical treatment on rice straw pyrolysis by TG-FTIR. *IERI Procedia*, *8*, 30–34. DOI: 10.1016/j.ieri.2014.09.006

Sherir, M. A. A., Hossain, K. M. A., & Lachemi, M. (2018). Fresh state, mechanical & durability properties of strain hardening cementitious composite produced with locally available aggregates and high volume of fly ash. *Construction & Building Materials*, *189*, 253–264. DOI: 10.1016/j.conbuildmat.2018.08.204

Shi, D., Gao, Y., Zeng, P., Li, B., Shen, P., & Zhuang, C. (2022). Climate adaptive optimization of green roofs and natural night ventilation for lifespan energy performance improvement in office buildings. *Building and Environment*, *223*, 109505. DOI: 10.1016/j.buildenv.2022.109505

Shi, L., Li, Z., & Wu, Z. (2018). Artificial neural network based mechanical and electrical property prediction of engineered cementitious composites. *Construction & Building Materials*, *174*, 667–674. DOI: 10.1016/j.conbuildmat.2018.04.127

Shoeibi, S., Jamil, F., Parsa, S. M., Mehdi, S., Kargarsharifabad, H., Mirjalily, S. A. A., Guo, W., Ngo, H. H., Ni, B. J., & Khiadani, M. (2024). Recent advancements in applications of encapsulated phase change materials for solar energy systems: A state-of-the-art review. *Journal of Energy Storage*, *94*, 112401. DOI: 10.1016/j.est.2024.112401

Siad, H., Lachemi, M., Sahmaran, M., & Hossain, K. M. A. (2015). Influence of limestone powder on mechanical, physical and self-healing behavior of engineered cementitious composites. *Construction & Building Materials*, *99*, 1–10. DOI: 10.1016/j.conbuildmat.2015.09.007

Silva, A., & Gómez, R. (2024). Uso de nanomateriales en la producción del concreto. *Revista de Ingeniería de Materiales*, 12(1), 15-28. https://revistas.poli.br/index.php/repa/article/view/1676/762

Singh, L. P., Dhaka, R. K., Ali, D., Tyagi, I., Sharma, U., & Banavath, S. N. (2021). Remediation of noxious pollutants using nano-titania-based photocatalytic construction materials: A review. *Environmental Science and Pollution Research International*, *28*(26), 34087–34107. DOI: 10.1007/s11356-021-14189-7 PMID: 33963998

Singh, M., Saini, B., & Chalak, H. D. (2019). Performance and composition analysis of engineered cementitious composite (ECC)—A review. *Journal of Building Engineering*, *26*, 100851. Advance online publication. DOI: 10.1016/j.jobe.2019.100851

Singh, M., Saini, B., & Chalak, H. D. (2020). Long term evaluation of engineered cementitious composite containing stone slurry powder. *Construction & Building Materials*, *264*, 120183. Advance online publication. DOI: 10.1016/j.conbuildmat.2020.120183

Singh, S., Nateriya, R., & Raheem, J. (2024). Self-Healing Concrete: A Comprehensive Review. In *International Conference on Recent Advances in Structural Engineering and Construction Management* (pp. 285–297). Springer.

Skullestad, J. L., Bohne, R. A., & Lohne, J. (2016). High-rise timber buildings as a climate change mitigation measure–A comparative LCA of structural system alternatives. *Energy Procedia*, *96*, 112–123. DOI: 10.1016/j.egypro.2016.09.112

Soares, M., Faria, L., Miranda, T., Pereira, E., Vilarinho, C., & Carvalho, J. (2025). The potential of agri-food waste to solve construction's environmental problems: A review. *Cleaner and Circular Bioeconomy*, *10*, 100138. DOI: 10.1016/j.clcb.2025.100138

Soe, K. T., Zhang, Y. X., & Zhang, L. C. (2013). Material properties of a new hybrid fibre-reinforced engineered cementitious composite. *Construction & Building Materials*, *43*, 399–407. DOI: 10.1016/j.conbuildmat.2013.02.021

Somasundaram, S., Thangavelu, S. R., & Chong, A. (2020). Improving building efficiency using low-e coating based retrofit double glazing with solar films. *Applied Thermal Engineering*, *171*, 115064. DOI: 10.1016/j.applthermaleng.2020.115064

Song, Q., Guo, M. Z., Wang, L., & Ling, T. C. (2021). Use of steel slag as sustainable construction materials: A review of accelerated carbonation treatment. *Resources, Conservation and Recycling*, *173*, 105740. DOI: 10.1016/j.resconrec.2021.105740

Sudhakar, Y. N., Selvakumar, M., & Bhat, D. K. (2021). Investigations on thermo-mechanical properties of organically modified polymer clay nanocomposites for packaging application. *Polymers & Polymer Composites*, *29*(8), 1191–1199. DOI: 10.1177/0967391120960645

Sun, C., Gu, J., Dong, Q., Qu, D., Chang, W., & Yin, X. (2023). Are straw bales better insulation materials for constructions? A review. *Developments in the Built Environment*, *15*, 100209. DOI: 10.1016/j.dibe.2023.100209

Sun, C., Yan, C., Li, J., Ma, J., & Li, Y. (2025). Ultra-high-performance concrete from 100% solid waste: Mechanical strength, microstructure and sustainability analysis. *Case Studies in Construction Materials*, *24*, 100000. DOI: 10.1016/j.cscm.2025.e05482

Sun, X., Brown, M. A., Cox, M., & Jackson, R. (2016). Mandating better buildings: A global review of building codes and prospects for improvement in the United States. *Wiley Interdisciplinary Reviews. Energy and Environment*, *5*(2), 188–215. DOI: 10.1002/wene.168

Supraja, A., Vamsi, T., Rao, S. K., Bhavitha, P., & Kumar, K. K. (2024, June). Green porous concrete for rainwater harvesting and urban pavements. []). IOP Publishing.]. *Journal of Physics: Conference Series*, *2779*(1), 012071. DOI: 10.1088/1742-6596/2779/1/012071

Suresh, C., Hotta, T. K., & Saha, S. K. (2022). Phase change material incorporation techniques in building envelopes for enhancing the building thermal comfort—A review. *Energy and Building*, *268*, 112225. DOI: 10.1016/j.enbuild.2022.112225

Suriya Kumar, T., Adebanjo, A. U., & Shafiq, N. (2025). *Mechanical and sustainability assessment of the influence of oily sludge ash on fly ash-based geopolymer composites. Innovative Infrastructure Solutions*. Springer.

Tabrizikahou, A., Kuczma, M., Łasecka-Plura, M., Farsangi, E. N., Noori, M., Gardoni, P., & Li, S. (2022). Application and modelling of shape-memory alloys for structural vibration control: State-of-the-art review. *Construction & Building Materials*, *342*, 127975. DOI: 10.1016/j.conbuildmat.2022.127975

Tajadod, O. E., Ravanshadnia, M., & Ghanbari, M. (2025). Integrating circular economy and life cycle assessment strategies in climate-resilient buildings: An artificial intelligence approach to enhance thermal comfort and minimize CO_2. *Energy*, *294*, 130468. DOI: 10.1016/j.energy.2025.135064

Talwar, P., Verma, N., Khatri, H., Ahire, P. D., Chaudhary, G., Lindenberger, C., & Vivekanand, V. (2023). A systematic review of photovoltaic-green roof systems in different climatic conditions focusing on sustainable cities and societies. *Sustainable Cities and Society*, *98*, 104813. DOI: 10.1016/j.scs.2023.104813

Tanasombat, M., Okabayashi, Y., Sakurai, K., Thaiutsa, B., Thammincha, S., & Suekeaw, P. (2005). Silvicultural performance of paper mulberry in Thailand. *Tropics*, *14*(2), 149–162. DOI: 10.3759/tropics.14.149

Tang, L., Wu, T., & Li, Q. (2025). Construction and optimization strategy for collaborative governance of construction waste resource utilization. *KSCE Journal of Civil Engineering*, *29*(3), 100034. DOI: 10.1016/j.kscej.2024.100034

Taşer, A., Koyunbaba, B. K., & Kazanasmaz, T. (2023). Thermal, daylight, and energy potential of building-integrated photovoltaic (BIPV) systems: A comprehensive review of effects and developments. *Solar Energy*, *251*, 171–196. DOI: 10.1016/j.solener.2022.12.039

Terzis, D. (2025, December 2). Monitoring innovation metrics in construction and civil engineering: Trends, drivers and laggards. *Developments in the Built Environment*, *9*, 100064. Advance online publication. DOI: 10.1016/j.dibe.2021.100064

Thai Watsadu. (n.d.). *Thai Watsadu: Building materials and home improvement*. Retrieved December 25, 2024, from https://www.thaiwatsadu.com

Thomson, A., & Walker, P. (2014). Durability characteristics of straw bales in building envelopes. *Construction & Building Materials*, *68*, 135–141. DOI: 10.1016/j.conbuildmat.2014.06.041

Tien, P. W., Wei, S., Darkwa, J., Wood, C., & Calautit, J. K. (2022). Machine learning and deep learning methods for enhancing building energy efficiency and indoor environmental quality–a review. *Energy and AI*, *10*, 100198. DOI: 10.1016/j.egyai.2022.100198

Tiza, M. T., Ogunleye, E., Jiya, V. H., Onuzulike, C., Akande, E. O., & Terlumun, S. (2023). Integrating sustainability into civil engineering and the construction industry. *Journal of Cement Based Composites*, *4*(1), 1–11. DOI: 10.36937/cebacom.2023.5756

Tlaiji, G., Ouldboukhitine, S., Pennec, F., & Biwole, P. (2022). Thermal and mechanical behavior of straw-based construction: A review. *Construction & Building Materials*, *316*, 125915. DOI: 10.1016/j.conbuildmat.2021.125915

Tseng, K. H., Chung, M. Y., Chen, L. H., & Chou, L. A. (2022). A study of green roof and impact on the temperature of buildings using integrated IoT system. *Scientific Reports*, *12*(1), 16140. DOI: 10.1038/s41598-022-20552-6 PMID: 36167970

Tsotsis, G. (2018). *Aplicaciones de la nanotecnología en los materiales de la construcción* [Trabajo de fin de máster, Universitat Politècnica de Catalunya]. https://upcommons.upc.edu/bitstream/handle/2117/133970/Mem%C3%B2ria_TsotsisGeorgios.pdf?isAllowed=y&sequence=1

UNIAPRAVI. (2020). *Datos del déficit habitacional*. Last access May 2, 2025. https://www.uniapravi.org/objetos/publicacion/MTE=/04092024111559.pdf

United Nations Environment Programme & Global Alliance for Buildings and Construction. (2025). *Global Status Report for Buildings and Construction 2024/2025: Not just another brick in the wall*. Last access May 2, 2025. https://www.unep.org/resources/report/global-status-report-buildings-and-construction-20242025

Urban and Cities Platform - ECLAC. *Derecho a la vivienda en Honduras*. Last access May 2, 2025. https://plataformaurbana.cepal.org/es/derecho-la-vivienda-en-honduras

Utracki, L. A., Sepehr, M., & Boccaleri, E. (2007). Synthetic, layered nanoparticles for polymeric nanocomposites (PNCs). *Polymers for Advanced Technologies*, *18*(1), 1–37. DOI: 10.1002/pat.852

Vaia, R. A., & Maguire, J. F. (2007). Polymer nanocomposites with prescribed morphology: Going beyond nanoparticle-filled polymers. *Chemistry of Materials*, *19*(11), 2736–2751. DOI: 10.1021/cm062693+

Vásquez, V., Puyen, Z., & Valdivia, P. (2020). *Módulo de vivienda rural bioclimatizado (VRB): mejora del confort térmico en la zona altoandina de Puno, Perú.* INUDI Publishing. Acccesed May 27, 2025. https://editorial.inudi.edu.pe/index.php/editorialinudi/catalog/book/30

Vassiliades, C., Agathokleous, R., Barone, G., Forzano, C., Giuzio, G. F., Palombo, A., Buonomano, A., & Kalogirou, S. (2022). Building integration of active solar energy systems: A review of geometrical and architectural characteristics. *Renewable & Sustainable Energy Reviews, 164*, 112482. DOI: 10.1016/j.rser.2022.112482

Vega-Baudrit, J. R., Cordero, C., Murrell Blanco, M., & Campos, A. (2024). *Nanotecnología en la arquitectura: Una visión integral en el uso de nanomateriales.* ResearchGate. https://www.researchgate.net/publication/382936049_Nanotecnologia_en_la_arquitectura_Una_vision_integral_en_el_uso_de_nanomateriales

Verma, D., Goh, K. L., & Vimal, V. (2022). Interfacial studies of natural fiber-reinforced particulate thermoplastic composites and their mechanical properties. *Journal of Natural Fibers, 19*(6), 2299–2326. DOI: 10.1080/15440478.2020.1808147

Visakh, P. M. (2015). Polyethylene terephthalate: blends, composites, and nanocomposites–state of art, new challenges, and opportunities. *Poly (Ethylene Terephthalate) Based Blends, Composites and Nanocomposites*, 1-14 https://doi.org/ DOI: 10.1016/B978-0-323-31306-3.00001-4

Wahyuni, N. L. E., Soeswanto, B., Aulia, H., & Fadhilah, R. A. (2019, September). The effects of rice straw weight fraction and particle size on thermal conductivity and mechanical properties of polypropylene composite. [). IOP Publishing.]. *Journal of Physics: Conference Series, 1295*(1), 012067. DOI: 10.1088/1742-6596/1295/1/012067

Wai, C. Y., Chau, H. W., Paresi, P., & Muttil, N. (2025). Experimental analysis of cool roof coatings as an urban heat mitigation strategy to enhance thermal performance. *Buildings (Basel, Switzerland), 15*(5), 685. DOI: 10.3390/buildings15050685

Wang, C., Zhang, Y., Hu, X., Jia, X., Li, K., Wang, C., & Wang, Y. (2024). Sustainability-oriented construction materials for traditional residential buildings: From material characteristics to environmental suitability. *Case Studies in Construction Materials, 2024*, e03820. Advance online publication. DOI: 10.1016/j.cscm.2024.e03820

Wang, D., Zhou, X., Meng, Y., & Chen, Z. (2017). Durability of concrete containing fly ash and silica fume against combined freezing-thawing and sulfate attack. *Construction & Building Materials, 147*, 398–406. DOI: 10.1016/j.conbuildmat.2017.04.172

Wang, Q., Zhang, J., & Ho, J. C. M. (2020). Zeolite to improve strength-shrinkage performance of high-strength engineered cementitious composite. *Construction & Building Materials*, *234*, 117335. Advance online publication. DOI: 10.1016/j.conbuildmat.2019.117335

Wang, R., Sui, J., & Wang, X. (2022). Natural piezoelectric biomaterials: A biocompatible and sustainable building block for biomedical devices. *ACS Nano*, *16*(11), 17708–17728. DOI: 10.1021/acsnano.2c08164 PMID: 36354375

Wang, W., Guo, H., Li, X., Tang, S., Xia, J., & Lv, Z. (2022). Deep learning for assessment of environmental satisfaction using BIM big data in energy-efficient building digital twins. *Sustainable Energy Technologies and Assessments*, *50*, 101897. DOI: 10.1016/j.seta.2021.101897

Wang, X., Li, W., Luo, Z., Wang, K., & Shah, S. P. (2022). A critical review on phase change materials (PCM) for sustainable and energy efficient building: Design, characteristic, performance and application. *Energy and Building*, *260*, 111923. DOI: 10.1016/j.enbuild.2022.111923

Wang, Z., Liu, J., Li, D., Yang, K., Chen, M., & Wang, C. (2024). Experimental and numerical study on load-bearing performance in triple-glazed insulating glass units. *Construction & Building Materials*, *418*, 135385. DOI: 10.1016/j.conbuildmat.2024.135385

Wąs, K., Radoń, J., & Sadłowska-Sałęga, A. (2020). Maintenance of passive house standard in the light of long-term study on energy use in a prefabricated lightweight passive house in Central Europe. *Energies*, *13*(11), 2801. DOI: 10.3390/en13112801

Wells, L., Rismanchi, B., & Aye, L. (2018). A review of net zero energy buildings with reflections on the Australian context. *Energy and Building*, *158*, 616–628. DOI: 10.1016/j.enbuild.2017.10.055

White, S. R., Sottos, N. R., Geubelle, P. H., Moore, J. S., Kessler, M. R., Sriram, S. R., & Viswanathan, S. (2001). Autonomic healing of polymer composites. *Nature*, *409*(6822), 794–797. DOI: 10.1038/35057232 PMID: 11236987

Woltman, G., Noel, M., & Fam, A. (2017). Experimental and numerical investigations of thermal properties of insulated concrete sandwich panels with fiberglass shear connectors. *Energy and Building*, *145*, 22–31. DOI: 10.1016/j.enbuild.2017.04.007

World Health Organization (WHO). (2022). *WHO global air quality guidelines: Particulate matter (PM2.5 and PM10), ozone, nitrogen dioxide, sulfur dioxide and carbon monoxide*. Geneva: WHO Press. https://www.who.int/publications/i/item/9789240034228

Worrell, E., Allwood, J., & Gutowski, T. (2016). The role of material efficiency in environmental stewardship. *Annual Review of Environment and Resources*, *41*(1), 575–598. DOI: 10.1146/annurev-environ-110615-085737

Wu, H., Zhang, H., Zhang, G., Liu, J., Liu, Z., & Du, F. (2023). Study on preparation and performance of advanced aerogel foamed concrete with ultra-light aerogel. *Construction & Building Materials*, *366*, 130166. DOI: 10.1016/j.conbuildmat.2022.130166

Xiang, J., Qiu, J., Song, Y., Zhao, Y., & Fei, X. (2024). A novel conceptual design for self-healing of cracked cementitious composites incorporating two bacteria-based capsules. *Composites. Part B, Engineering*, *287*, 111872. DOI: 10.1016/j.compositesb.2024.111872

Xiao, J., Wang, C., Ding, T., & Akbarnezhad, A. (2018). A recycled aggregate concrete high-rise building: Structural performance and embodied carbon footprint. *Journal of Cleaner Production*, *199*, 868–881. DOI: 10.1016/j.jclepro.2018.07.210

Xiao, J., Zhang, H., Tang, Y., Deng, Q., Wang, D., & Poon, C. S. (2022). Fully utilizing carbonated recycled aggregates in concrete: Strength, drying shrinkage and carbon emissions analysis. *Journal of Cleaner Production*, *377*, 134520. DOI: 10.1016/j.jclepro.2022.134520

Xu, G., & Shi, X. (2018). Characteristics and applications of fly ash as a sustainable construction material: A state-of-the-art review. *Resources, Conservation and Recycling*, *136*, 95–109. DOI: 10.1016/j.resconrec.2018.04.010

Xu, G., Xia, H., Chen, P., She, W., Zhang, H., Ma, J., Ruan, Q., Zhang, W., & Sun, Z. (2022). Thermochromic hydrogels with dynamic solar modulation and regulatable critical response temperature for energy-saving smart windows. *Advanced Functional Materials*, *32*(5), 2109597. DOI: 10.1002/adfm.202109597

Xu, J., Dai, J., Wu, X., Wu, S., Zhang, Y., Wang, F., Gao, A., & Tan, Y. (2023). Urban rainwater utilization: A review of management modes and harvesting systems. *Frontiers in Environmental Science*, *11*, 1025665. DOI: 10.3389/fenvs.2023.1025665

Xu, S., Li, H., & Li, Q. (2021). Thermal stability study of fiber-reinforced cementitious material with high ductility under high-temperature casting. *Construction & Building Materials*, *282*, 122700. Advance online publication. DOI: 10.1016/j.conbuildmat.2021.122700

Xu, X., Liu, M., Gao, H., Xiao, P., & Zhang, Y. (2025). Rainwater harvest with desert-sand bricks: An adaptive water resources management strategy for climate change. *Water Resources Management*, *39*(2), 809–819. DOI: 10.1007/s11269-024-03996-5

Yang, J., Yang, Y., Wu, W. M., Zhao, J., & Jiang, L. (2014). Evidence of polyethylene biodegradation by bacterial strains from the guts of plastic-eating waxworms. *Environmental Science & Technology*, *48*(23), 13776–13784. DOI: 10.1021/es504038a

Yang, M., Chen, L., Lai, J., Osman, A. I., Farghali, M., Rooney, D. W., & Yap, P. S. (2024). Advancing environmental sustainability in construction through innovative low-carbon, high-performance cement-based composites: A review. *Materials Today Sustainability*, *26*, 100712. Advance online publication. DOI: 10.1016/j.mtsust.2024.100712

Yang, S., Zhang, Y., Zhao, Y., Torres, J. F., & Wang, X. (2023). PCM-based ceiling panels for passive cooling in buildings: A CFD modelling. *Energy and Building*, *285*, 112898. DOI: 10.1016/j.enbuild.2023.112898

Yang, X., Liu, J., Li, H., & Xu, L. (2021). Behavior of biochar-modified cementitious composites exposed to high temperatures. *Materials (Basel)*, *14*(18), 5414. Advance online publication. DOI: 10.3390/ma14185414

Yang, Y., Dang, B., Wang, C., Chen, Y., Chen, K., Chen, X., Li, Y., & Sun, Q. (2024). Ultrastrong lightweight nanocellulose-based composite aerogels with robust superhydrophobicity and durable thermal insulation under extremely environment. *Carbohydrate Polymers*, *323*, 121392. DOI: 10.1016/j.carbpol.2023.121392 PMID: 37940285

Yaraş, A., Bayram, M., Ustaoğlu, A., Erdoğmuş, E., Hekimoğlu, G., Sarı, A., Gencel, O., Tyagi, V. V., & Ozbakkaloglu, T. (2024). Advancing thermal control in buildings with innovative cementitious mortar and recycled expanded glass/n-octadecane phase change material composites. *Renewable & Sustainable Energy Reviews*, *202*, 114680. DOI: 10.1016/j.rser.2024.114680

Ye, B., Zhang, Y., & Yu, J. (2019). Effect of water to binder ratio and sand to binder ratio on shrinkage and mechanical properties of high-strength engineered cementitious composite. *Construction & Building Materials*, *226*, 899–909. DOI: 10.1016/j.conbuildmat.2019.07.303

Yeon, J. O., Kim, K. W., Yang, K. S., Kim, J. M., & Kim, M. J. (2014). Physical properties of cellulose sound absorbers produced using recycled paper. *Construction & Building Materials*, *70*, 494–500. DOI: 10.1016/j.conbuildmat.2014.07.088

Ye, S., Li, V. C., & Li, H. (2015). Nano-modification to improve the ductility of cementitious composites. *Cement and Concrete Research*, *76*, 170–179. DOI: 10.1016/j.cemconres.2015.05.026

Yoo, D.-Y., & Banthia, N. (2022). High-performance strain-hardening cementitious composites with tensile strain capacity exceeding 4%: A review. *Cement and Concrete Composites*, *125*, 104325. Advance online publication. DOI: 10.1016/j.cemconcomp.2021.104325

Younsi, A., Turcry, P., Aït-Mokhtar, A., & Staquet, S. (2021). Hydration-drying interactions in a high-volume ground granulated blast-furnace slag mortar. *Construction & Building Materials*, *279*, 122427. Advance online publication. DOI: 10.1016/j.conbuildmat.2021.122427

Yu, J., Sui, H., Wang, S., Ma, Y., Xu, X., & Liu, Y. (2025). Influence of graphene oxide dispersion strategy on recycled concrete powder for cementitious systems. SSRN Electronic Journal. DOI: 10.1016/j.conbuildmat.2025.144771

Yücel, H. E., Dutkiewicz, M., & Yıldızhan, F. (2022). Application of ECC as a repair/retrofit and pavement/bridge deck material for sustainable structures: A review. *Materials (Basel)*, *15*(24), 8752. Advance online publication. DOI: 10.3390/ma15248752

Yu, H., Zahidi, I., Chow, M. F., Liang, D., & Madsen, D. Ø. (2024). Reimagining resources policy: Synergizing mining waste utilization for sustainable construction practices. *Journal of Cleaner Production*, *464*, 142795. Advance online publication. DOI: 10.1016/j.jclepro.2024.142795

Yu, K., Wang, Y., & Yu, J. (2018). Direct tensile properties of engineered cementitious composites: A review. *Construction & Building Materials*, *165*, 346–362. DOI: 10.1016/j.conbuildmat.2017.12.124

Yu, Y., Fang, G. H., Kurda, R., Sabuj, A. R., & Zhao, X. Y. (2024). An agile, intelligent and scalable framework for mix design optimization of green concrete incorporating recycled aggregates from precast rejects. *Case Studies in Construction Materials*, *20*, e03156. DOI: 10.1016/j.cscm.2024.e03156

Yu, Z., Nurdiawati, A., Kanwal, Q., Al-Humaiqani, M. M., & Al-Ghamdi, S. G. (2024). Assessing and mitigating environmental impacts of construction materials: Insights from environmental product declarations. *Journal of Building Engineering*, *98*, 110929. Advance online publication. DOI: 10.1016/j.jobe.2024.110929

Zandifaez, P., Shamsabadi, E. A., Nezhad, A. A., Zhou, H., & Dias-da-Costa, D. (2023). AI-assisted optimisation of green concrete mixes incorporating recycled concrete aggregates. *Construction & Building Materials*, *391*, 131851. DOI: 10.1016/j.conbuildmat.2023.131851

Zeng, Z., Brown, J., & Augenbroe, G. (2019, September). A comparative study on the energy performance of movable insulation and triple-glazed windows. In *Building Simulation 2019* (Vol. 16, pp. 4863–4870). IBPSA.

Zhang, B., Zhu, H., Feng, P., & Zhang, P. (2022). A review on shrinkage-reducing methods and mechanisms of alkali-activated/geopolymer systems: Effects of chemical additives. *Journal of Building Engineering*, *49*, 104056. DOI: 10.1016/j.jobe.2022.104056

Zhang, D., Jaworska, B., Zhu, Y., Yu, J., & Li, V. C. (2020). Engineered cementitious composites (ECC) with limestone calcined clay cement (LC3). *Cement and Concrete Composites*, *114*, 103766. Advance online publication. DOI: 10.1016/j.cemconcomp.2020.103766

Zhang, D., Yu, J., Wu, H., Jaworska, B., & Li, V. C. (2020). Discontinuous microfibers as intrinsic reinforcement for ductile engineered cementitious composites (ECC). *Composites. Part B, Engineering*, *184*, 107741. Advance online publication. DOI: 10.1016/j.compositesb.2020.107741

Zhang, H. Y., Kodur, V., Wu, B., Yan, J., & Yuan, Z. S. (2018). Effect of temperature on bond characteristics of geopolymer concrete. *Construction & Building Materials*, *163*, 277–285. DOI: 10.1016/j.conbuildmat.2017.12.043

Zhang, L., Zheng, M., Zhao, D., & Feng, Y. (2024). A review of novel self-healing concrete technologies. *Journal of Building Engineering*, *89*, 109331. DOI: 10.1016/j.jobe.2024.109331

Zhang, M., Yan, T., Wang, W., Jia, X., Wang, J., & Klemeš, J. J. (2022). Energy-saving design and control strategy towards modern sustainable greenhouse: A review. *Renewable & Sustainable Energy Reviews*, *164*, 112602. DOI: 10.1016/j.rser.2022.112602

Zhang, M., Zhou, J., Zhang, Q., Zhang, C., & Song, W. (2025). Construction of flame-retardant and water-resistant polybutylene terephthalate composites by reactive extrusion. *Frontiers in Materials*, *11*, 1536536. DOI: 10.3389/fmats.2025.1536536

Zhang, P., Su, J., Guo, J., & Hu, S. (2023). Influence of carbon nanotube on properties of concrete: A review. *Construction & Building Materials*, *369*, 130388. DOI: 10.1016/j.conbuildmat.2023.130388

Zhang, Q., Zhao, X., Lu, H., Ni, T., & Li, Y. (2017). Waste energy recovery and energy efficiency improvement in China's iron and steel industry. *Applied Energy*, *191*, 502–520. DOI: 10.1016/j.apenergy.2017.01.072

Zhang, R., Li, R., Xu, P., Zhong, W., Zhang, Y., Luo, Z., & Xiang, B. (2023). Thermochromic smart window utilizing passive radiative cooling for self-adaptive thermoregulation. *Chemical Engineering Journal*, *471*, 144527. DOI: 10.1016/j.cej.2023.144527

Zhang, R., Tang, Y., Liu, G., Wang, Z., & Zhang, Y. (2024). Unveiling the impact of innovation on competitiveness among construction projects: Moderating and mediating role of environmental regulation and sustainability. *Ain Shams Engineering Journal*, *15*(4), 102558. Advance online publication. DOI: 10.1016/j.asej.2023.102558

Zhang, S., Ocłoń, P., Klemeš, J. J., Michorczyk, P., Pielichowska, K., & Pielichowski, K. (2022). Renewable energy systems for building heating, cooling and electricity production with thermal energy storage. *Renewable & Sustainable Energy Reviews*, *165*, 112560. DOI: 10.1016/j.rser.2022.112560

Zhang, X., Xu, J., Zhang, X., & Li, Y. (2021). Life cycle carbon emission reduction potential of a new steel-bamboo composite frame structure for residential houses. *Journal of Building Engineering*, *39*, 102295. DOI: 10.1016/j.jobe.2021.102295

Zhao, W., Yao, M., Liu, L., & Wang, H. (2024). Preparation and characterization of PBT@PP-CNC@cellulose wood pulp paper double-layers fuel filtration materials. *Journal of Materials Science*, *59*(6), 4123–4138. DOI: 10.1007/s10853-023-09100-3

Zhao, Y., Lou, Z., Wang, Q., Yuan, T., Chen, M., Han, H., Wu, X., Xu, L., & Li, Y. (2023). Fabrication of a bamboo-based glulam based on reconstitution unit innovation: Mechanical property investigation and carbon footprint evaluation. *Industrial Crops and Products*, *202*, 117046. DOI: 10.1016/j.indcrop.2023.117046

Zheng, P. D., Li, Y., & Li, Z. (2021). Compressive behavior of steel stirrups-confined square engineered cementitious composite (ECC) columns. *Advances in Concrete Construction*, *11*(3), 193–206. DOI: 10.12989/acc.2021.11.3.193

Zhong, Q., Li, H., Liang, S., Jetashree, , Wu, X., Qi, J., & Wang, S. (2022). Changes of production and consumption structures in coastal regions lead to mercury emission control in China. *Journal of Industrial Ecology*, *26*(5), 1760–1770. DOI: 10.1111/jiec.13314

Zhou, D., Zhao, C. Y., & Tian, Y. (2012). Review on thermal energy storage with phase change materials (PCMs) in building applications. *Applied Energy*, *92*, 593–605. DOI: 10.1016/j.apenergy.2011.08.025

Zhou, Y., Zhang, J., Li, Y., & Wu, Z. (2021). Feasibility of incorporating recycled fine aggregate in high performance green lightweight engineered cementitious composites. *Journal of Cleaner Production*, *280*, 124445. Advance online publication. DOI: 10.1016/j.jclepro.2020.124445

Zhu, H., Zhang, D., Wu, C., & Li, V. C. (2020). Mechanical and self-healing behavior of low carbon engineered cementitious composites reinforced with PP-fibers. *Construction & Building Materials*, *259*, 119805. Advance online publication. DOI: 10.1016/j.conbuildmat.2020.119805

Zhu, J., Zhang, C., Zhang, M., & Yuan, J. (2024). Modeling and simulation of technological innovation diffusion of high-performance construction materials in transportation engineering. *Case Studies in Construction Materials*, *2024*, e03838. Advance online publication. DOI: 10.1016/j.cscm.2024.e03838

Zhu, P., Gao, L., & Wang, Z. (2019). Recent progress in self-healing materials with micro/nanocapsules. *Advanced Functional Materials*, *29*(47), 1806810.

Zilberberg, E., Trapper, P., Meir, I. A., & Isaac, S. (2021). The impact of thermal mass and insulation of building structure on energy efficiency. *Energy and Building*, *241*, 110954. DOI: 10.1016/j.enbuild.2021.110954

Zucaro, A., & Agostinho, F. (2025). Urban sustainability: Challenges and opportunities for resilient and resource-efficient cities. *Frontiers in Sustainable Cities*, *7*, 1556974. DOI: 10.3389/frsc.2025.1556974

About the Contributors

Roberto Alonso González-Lezcano Full Professor at the Department of Architecture and Design, area of Building Systems, within the Institute of Technology of Universidad CEU San Pablo. Coordinator of the subjects Building Systems as well as Electrical and Lighting systems. Professor Accredited by ANECA in the figures of Tenured Professor. Two six-year research period by CNEAI (last period 2016-2021). Coordinator of the post-graduate degree of Energy Efficiency and Sustainability in Buildings and Coordinator of the Laboratory of Building Systems within Universidad CEU San Pablo. Coordinator of line 2: Innovation, Construction and Technology, in the Doctoral Program of Composition, History and Techniques in Architecture and Urbanism. Extraordinary Doctorate Award. San Pablo CEU University. 57 impact articles in the last 5 years. H Index:17. Editor of 8 JCR-indexed journals (140 edited articles, indexed in the JCR). Principal Investigator of the ARIE Group: Architecture, Efficient Installations. CEU San Pablo University.

Mehmet Serkan Kirgiz is a pro-active, full-time, tenured-track professor at the Academia of ACMCEN, Oregon, USA. He has also been a full professor at various universities in Turkey, UK, and USA. His professional experience includes research and development with Scientific and Technological Research Council of Turkey (TUBITAK), teaching graduate- and post-graduate level students in international institutions and universities collaboratively, and advising students in their theses since 1998. He has authored 260 peer-reviewed articles, 6 books, and 37 book chapters published in highly reputable SCI journals and Scopus-based publications. He also holds an international patent on cement manufacturing with marble powder and brick powder as farina.

***

Md Faruk Abdulla is an educator at Parul University, serving as Project and IEP Coordinator. With expertise in Full Stack Development, AI, and Cyber Security, he actively bridges academics and industry through workshops, bootcamps, and project guidance.

Dibyendu Adak serves as an Assistant Professor in the Civil Engineering Department at the National Institute of Technology (NIT) Meghalaya. He earned his PhD from Jadavpur University, Kolkata, with a specialization in Structural Engineering. His research is on advanced materials in civil engineering, specifically alkali-activated concrete, geopolymer concrete, and microbially induced sustainable construction materials. Dr Adak has made substantial contributions to the subject, with over 40 publications in international journals and conferences, a commendable H-index of 14, and more than 1,200 citations, notably in sustainable building materials and infrastructure durability.

Atheek Ahamed – Civil Engineer & Project Management Enthusiast Atheek Ahamed is a dedicated civil engineering professional with a strong foundation in construction project management, Building Information Modeling (BIM), and architectural design. Currently pursuing a Bachelor of Technology in Civil Engineering at Marwadi University, he has developed expertise in project planning, estimation, and site execution. His hands-on experience includes internships with leading firms such as Macht-Architecture and Planning, Tvasta Project Management Consultancy, and the Construction Management Institute, where he worked on real-time construction and interior projects. Skilled in Revit, Navisworks, Primavera P6, and MS Project, Atheek specializes in implementing BIM workflows to enhance cost management and efficiency in construction projects. Recognized as an SII-Scholar by the Government of India, he is also a student member of the American Society of Civil Engineers. With a keen interest in sustainable construction and emerging technologies, Atheek continues to expand his knowledge in project management and innovative design solutions.

Salim Barbhuiya, PhD, CEng, MICT, FHEA, is a Senior Lecturer in Civil Engineering at the University of East London. With a PhD in Material Technology from Queen's University Belfast, he brings over 15 years of international academic and research experience from institutions including the University of Leeds, Curtin University, and Loughborough University. His expertise lies in sustainable construction materials and resilience in civil engineering. Dr Barbhuiya has secured substantial research funding and has published extensively in high-impact journals. He actively contributes to professional bodies and conferences worldwide, advancing innovation in concrete technology and sustainable infrastructure.

Tiago Jordão Cardoso is a Specialized Technical Officer at the General Secretariat of the Environment in Portugal, where he focuses on the integration of artificial intelligence and chatbots within public administration. Currently completing a master's in public administration, his research centers on technological acceptance, particularly in the public sector's use of AI to enhance sustainability and resilience. With strong involvement in climate change initiatives, Tiago explores the intersection of AI-driven innovation and sustainable practices, co-authoring articles on these topics and preparing additional proposals on AI's role in environmental resilience.

Bibhuti Bhusan Das is currently serving as Professor at National Institute of Technology Karnataka, Surathkal. Before joining NITK, he has served as Centre Head for National Institute of Construction Management and Research, Goa and Indore Campus. He has been working as a Post-Doctoral Research Associate and Adjunct Faculty in the Department of Civil Engineering at Lawrence Technological University, Southfield, Michigan, USA. His area of research focuses on the "Sustainability of Construction and Building Materials" such as microstructure characterization of materials, non-destructive testing of concrete structures, corrosion of reinforcement and durability studies on concrete and sustainability in construction project management.

Pedro Pinheiro Gomes is a Technical Adviser at the Secretary General for the Environment in Portugal, where he currently focuses on project management and communication within the Environment Programme of the EEA Grants. He holds a master's degree in Environmental Engineering, specializing in environmental systems and policy. His professional journey also includes extensive experience in statistics and data analysis within the European Comparison Programme (ECP) for Purchasing Power Parities. In addition to his core roles, Pedro has authored multiple publications, including chapters in scientific books on environmental management, policy, and the integration of Artificial Intelligence in sustainability. He has contributed to high-impact projects, such as the development of mobile applications for coastal and tourism analysis aligned with the Inspire Directive, as well as tools for natural disaster alerts and forecasting.

Sudeepan. J has done his Ph.D. on "STUDY OF TRIBOLOGICAL PROPERTIES OF SOME THERMOPLASTIC COMPOSITES" at Jadavpur University, Kolkata, in 2015. The Co-PI is basically a Plastic Technologist with expertise in Polymer processing, Polymer composites and mold design. He has 11 years of teaching and research experience in polymer related works with around 11 publications to his credit in various scientific journals and conference proceedings. His research & teaching at the institute was on polymer processing, die and mold design and polymer composites from 2009 onwards for UG and PG in his department. Based on the Ph.D. and teaching experiences, his work is focused on the development of polymer composites for the mechanical and tribological characteristics. The current proposal is based on his experience to explore the area of sustainable bio-based high performance polymer composites for automobile and other industrial applications. In the present proposal, PI is responsible for the fabrication of bio-based polymer composites and its investigation on mechanical and tribological characteristics also tribo-informatics evaluation of developed composites and its optimization using machine learning techniques as PI as expertise in machine learning techniques also with around 2 publications to support.

Kaushik Kumar holds a Ph.D. in Engineering from Jadavpur University, India, an MBA in Marketing Management from Indira Gandhi National Open University, India and a Bachelor of Technology from Regional Engineering College (Now National Institute of Technology), Warangal, India. For 11 years, he worked in a manufacturing unit of Global repute. He is currently working as a Professor in the Department of Mechanical Engineering, Birla Institute of Technology, Mesra, Ranchi, India. He has 23 Years of Teaching and Research Experience. His research interests include Composites, Optimization, Non-Conventional Machining, CAD / CAM, Rapid Prototyping and Quality Management Systems towards product development for societal and industrial usage and has received 35+ patents for them most of which are donated to Start-ups and Industrial fraternity through Jharkhand Small Industries Association (JSIA). He has published 55+ Books (They are referred as Text Books and Reference Books by 40+ Universities / Institutes in their academic curriculum), 95+ Book Chapters and 200+ Research Papers in peer-reviewed reputed national and international journals. Kaushik has also served as Editor–in–Chief, Series Editor, Guest Editor, Editor, Editorial Board Member and Reviewers for International and National Journals. He has been felicitated with many awards and honours including Distinguished Alumnus Award for Professional Excellence 2023 under Academic and Research from National Institute of Technology, Warangal, India. He has also received Sponsored Research and Consultancy Projects of more than 1 Crore from Govt. of India and abroad. Kaushik has delivered expert lectures as Keynote Speaker at International & National Conferences, Resource Person at Various Workshops, FDP's and Short-Term Courses. He has guided many students of Doctoral, Masters and Under Graduate programmes of his home and other institutions in India and abroad. He also has served as Reviewer and Examiner of Doctoral and Masters dissertation for institutes in India and abroad.

Sofia Melero-Tur PhD Lecturer and Researcher in Building Construction at Architecture and Design Department, CEU San Pablo University and member of the ARIE Research Group. Her main research interests focus on sustainability, energy efficiency and indoor environmental quality and health in buildings. She has participated in 14 research and innovation projects with public and private funding. First six-year research period by CNEAI (2011-2024). Registered architect.

Chikesh Ranjan is currently serving as a Project Engineer for Project SwaYaan – Capacity Building in Drone/UAS at the Department of Mechanical Engineering, NIT Rourkela, India. He holds a PhD in Mechanical Engineering from the Department of Mechanical Engineering at the Birla Institute of Technology, Mesra, Ranchi, Jharkhand, India, where he also completed his Master of Engineering (M.E.) in Mechanical Engineering, specializing in the Design of Mechanical Equipment. Additionally, he earned a Bachelor of Engineering (B.E.) in Mechanical Engineering from the Marathwara Institute of Technology, Aurangabad, Maharashtra, India, graduating with First Class with Distinction. Chikesh Ranjan has made significant contributions to the field of mechanical engineering through his research, having published 17 journal papers, authored 12 books, and written 16 book chapters. He has been actively involved in organizing 10 workshops and coordinating 13 events, in addition to participating in 31 training programs and attending 34 national and international workshops, seminars, and webinars. His innovative work has been recognized with 13 design patents and 8 foreign patents, underscoring his commitment to advancing mechanical design and technology. With 9 years of teaching and research experience, Chikesh has built a global reputation in his areas of interest, which include composites, non-conventional machining, CAD/CAM, 3D printing, robotics and drone technology. He is a proud member of the Institution of Engineers, India, and has received 7 prestigious awards and honors for his outstanding contributions to the field of Mechanical Engineering

Divya Sabapathi completed his full-time Ph.D. research in the Civil Engineering stream at PSG College of Technology under Anna University, Chennai, India. His research areas include graphene nanoplatelets, supplementary cementitous materials and nano carbon.

Otoniel Itsá Sabillón-Ordóñez is a Honduran architect, graphic artist and social-commentary illustrator. Studied at the National School of Fine Arts and later in architecture at UNAH (Universidad Nacional Autónoma de Honduras). Lecturer at Universidad Tecnológica Centroamericana.

Praveen Kumar Shanmugam is currently working as Associate Professor at Department of Civil Engineering, PSG College of Technology. He completed his

PhD dissertation from the Civil Engineering stream at Anna University, Chennai, India His research area includes Cement and Concrete-FRC-Microstructure analysis of construction materials- Alternative construction materials-Industrial By-products, Nano-silica, ECC, Graphene, Cost effective housing system.

N. Shanmugasundaram is currently working as an Assistant Professor in the Department of Civil Engineering at Sri Krishna College of Technology, Coimbatore. He completed his full-time Ph.D. research in the Civil Engineering stream at PSG College of Technology under Anna University, Chennai, India. His research areas include Engineered cementitious composites (ECC), Supplementary Cementitious Materials, Sustainable Building Materials, Concrete Technology, Material Characterization, Structural Engineering, and Machine Learning Applications in Civil Engineering.

Jeenendra Kumar Shiv is an accomplished academic and researcher in Mechanical Engineering, currently pursuing a PhD at Birla Institute of Technology, Mesra (since 2023). With a strong educational foundation, Jeenendra holds an M.Tech in Mechanical Engineering from the Institute of Technology, Guru Ghasidas University, Bilaspur (2012-2014), and a B.E. in Mechanical Engineering from Chhattisgarh Swami Vivekanand Technical University, Bhilai (2005-2009). With extensive teaching and industry experience, Jeenendra has contributed significantly to academia. He has served as an Assistant Professor in Mechanical Engineering at Krishna Engineering College, Bhilai (2018-2022), Raipur Institute of Technology, Raipur (2016-2018), and Parthivi College of Engineering and Management, Bhilai-3 (2015-2016). Earlier in his career, he worked as a Lecturer at G.D. Rungta College of Engineering and Technology, Bhilai (2011-2012) and gained industrial exposure as a Graduate Engineer Trainee in Power Plant Utility at Bhushan Power and Steel Limited, Sambalpur (2009-2010). His research interests include CAD/CAM, Robotics, Polymer Nanocomposites, Mechanical Behavior of Materials, and Tribology. Jeenendra has demonstrated a keen interest in advancing mechanical engineering through research and innovation throughout his academic and professional journey.

Amit Yadav Aspiring civil engineer, Final year student . Project management, Revit, Autocad, MS Word and MS Excel

Index

M

N

P

R

S

T

V

Z

www.ingramcontent.com/pod-product-compliance
Lightning Source LLC
LaVergne TN
LVHW060005031026
849161LV00025B/1486
* 9 7 9 8 3 3 7 3 2 4 1 4 2 *